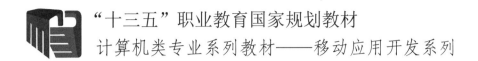

"十三五"职业教育国家规划教材

计算机类专业系列教材——移动应用开发系列

Android 应用开发技术
（第2版）

主　编　查英华
副主编　王　辰
主　审　胡光永

电子工业出版社

Publishing House of Electronics Industry

北京·BEIJING

内 容 简 介

本书系统介绍了基于 Android 10.0 和 Android Studio 4.1 集成环境开发 Android 应用程序的基础知识和实际应用。全书按照项目开发的技能训练逻辑分为 9 章，包括 Android 基础入门、Android 基础界面设计、Activity 与 Fragment、Android 高级界面设计、数据存储、服务与广播、网络编程、多媒体开发和进阶技术，由浅入深、循序渐进地阐述 Android 开发的基础知识、关键技术和进阶技术，对目前流行的 RecyclerView、Material Design 控件、Fragment 之间的数据传递、Android 异步处理技术，网络请求框架 OkHttp 及今后的发展方向 Jetpack 工具库等进行了较详尽的讲述，有较强的实用性，以达到快速提升读者的编程水平和项目开发能力的期望。

本书案例都在 Android 10.0 手机或模拟器成功运行。另外，本书提供配套的教学资源，包括教学大纲、教学课件、习题及答案和案例程序源码等，极大地方便了教学的开展。

本书既可以作为高校计算机相关专业的学生学习 Android 应用开发的教材，也可作为 Android 初学者的自学用书和参考用书。

未经许可，不得以任何方式复制或抄袭本书之部分或全部内容。
版权所有，侵权必究。

图书在版编目（CIP）数据

Android 应用开发技术 / 查英华主编. —2 版. —北京：电子工业出版社，2021.6
ISBN 978-7-121-41260-8

Ⅰ. ①A… Ⅱ. ①查… Ⅲ. ①移动终端—应用程序—程序设计—高等学校—教材 Ⅳ. ①TN929.53

中国版本图书馆 CIP 数据核字（2021）第 098925 号

责任编辑：贺志洪
印　　刷：三河市鑫金马印装有限公司
装　　订：三河市鑫金马印装有限公司
出版发行：电子工业出版社
　　　　　北京市海淀区万寿路 173 信箱　邮编 100036
开　　本：787×1092　1/16　印张：21　字数：537.6 千字
版　　次：2017 年 2 月第 1 版
　　　　　2021 年 6 月第 2 版
印　　次：2024 年 12 月第 9 次印刷
定　　价：56.00 元

凡所购买电子工业出版社图书有缺损问题，请向购买书店调换。若书店售缺，请与本社发行部联系，联系及邮购电话：（010）88254888，88258888。
质量投诉请发邮件至 zlts@phei.com.cn，盗版侵权举报请发邮件至 dbqq@phei.com.cn。
本书咨询联系方式：（010）88254609，hzh@phei.com.cn。

前　　言

随着移动互联网的普及，智能手机已成为人们生活的重要组成部分。Android 手机已占据智能手机 85%以上的市场份额，成为绝对主流的产品。Android 的应用也从智能手机拓展到智能穿戴设备、智能电视、智能家居、自动驾驶及工业互联网等多个领域，高校也纷纷开设 Android 移动应用开发课程以跟上移动互联网技术的快速发展。

Android 系统是基于 Linux 的开源的手机操作系统，已广泛应用于包括手机在内智能设备的开发。Android Studio 是 Android 的集成开发环境，本书系统介绍了基于 Android 10.0 和 Android Studio 4.1 集成环境开发 Android 应用程序的基础知识和实际应用，全书按照项目开发的技能要求，将内容分成三个学习阶段：第一阶段是第 1~3 章，重点学会搭建并熟悉开发环境，掌握基础的界面设计和功能实现；第二阶段是第 4~8 章，重点学会 Android 推出的高级控件和布局，以及学会真实项目开发的典型技术，包括数据存储、服务与广播、网络编程和多媒体开发；第三阶段是第 9 章，重点讲解一些进阶技术，包括手势、传感器及 Android Jetpack 技术发展的最新进展。全书的 9 个章节分别为：Android 基础入门、Android 基础界面设计、Activity 与 Fragment、Android 高级界面设计、数据存储、服务与广播、网络编程、多媒体开发和进阶技术。

通过对 Android 应用开发相关的职业岗位及核心能力的广泛调研，深入研讨移动应用开发课程的教学目标和教学内容，精心设计教学项目和典型案例，增加企业主流技术的知识点，贴近真实的开发场景，帮助学习者尽快适应职场要求。

本书写作特色鲜明：一是教材结构合理，教学内容经过多次推敲，结合岗位能力设计教学案例；二是知识点和案例结合紧密，使学习者能够根据案例实践深入理解知识点；三是通过综合项目的设计与分析，让学习者能综合运用学过的知识点；四是及时跟进 Android 版本的新功能，增加 Material Design、运行时权限、常用的数据库框架技术和网络框架技术、Jetpack 工具库等内容；五是提供丰富的教学资源，给教学实施提供便利。

本书是一本 Android 开发的入门书籍，秉持突出编程基础、学以致用和编程实践的教学理念，将基本理论融入案例进行编写。针对初学者，建议从头开始循序渐进行学习，书中的案例都手动练习，完成每个章节之后的习题练习，以达到学有所用的目的。有基础的开发人员可以选择感兴趣的章节进行学习，遇到问题可以参考提供的案例资源，最大限度地发挥本书的作用。

给授课教师的教学建议如下表所示，实施教学可选择 48~72 课时，应根据课程安排情况灵活处理。

章节名称	建议课时	可选章节课时	可选章节
第 1 章 Android 基础入门	4		
第 2 章 Android 基础界面设计	12		
第 3 章 Activity 与 Fragment	8	4	可选：3.5、3.6 节
第 4 章 Android 高级界面设计	6	2	可选：4.4 节
第 5 章 数据存储	10	2	可选：5.5 节
第 6 章 服务与广播	4		
第 7 章 网络编程	4	2	可选：7.4 节
第 8 章 多媒体开发	4	2	可选：8.4 节
第 9 章 进阶技术		8	可选：9.1～9.3 节
合计	52	20	

  本书由南京工业职业技术大学的查英华组织编写，参加本书编写的还有王辰、曹晓燕等，企业工程师杜警、马勇负责案例的设计。编写过程中参考、借鉴了众多 IT 技术专家、学者发表的博客、相关著作和开源项目，并得到了电子工业出版社的大力支持，在此无法一一列举，谨向各位技术专家、学者和编辑们致以诚挚的谢意！

  本书既可以作为高校计算机专业学生学习 Android 应用开发的教材，也可作为 Android 初学者的自学用书和参考用书。

  本书提供的配套资源包括：所有章节的程序源码、教学课件和习题及参考答案等，读者可以登录电子工业出版社华信教育资源网进行下载，网址为 https://www.hxedu.com.cn。

  由于 Android 技术发展迅速，鉴于编者水平有限，书中难免存在不足和错误之处，敬请广大读者多提宝贵意见和建议，任何意见和建议请发邮件至 zhayh@niit.edu.cn，以便再版时更正和改进。

编者

2021 年 4 月

目 录

第 1 章 Android 基础入门 ·· 1

    1.1 Android 简介 ·· 1
        1.1.1 智能手机及操作系统 ··· 1
        1.1.2 Android 的优势 ·· 2
        1.1.3 Android 的版本 ·· 2
    1.2 Android 系统架构 ··· 3
    1.3 搭建 Android 开发环境 ·· 4
        1.3.1 安装 Android Studio ·· 5
        1.3.2 Android SDK ·· 7
    1.4 开发第一个 Android 项目 ·· 9
        1.4.1 创建项目 ·· 9
        1.4.2 创建模拟器 ··· 10
        1.4.3 运行项目 ··· 12
    1.5 项目文件结构 ·· 13
        1.5.1 java 目录 ··· 14
        1.5.2 res 目录 ··· 15
        1.5.3 AndroidManifest.xml 文件 ··· 15
    1.6 本章小结 ·· 17
    习题 ·· 17

第 2 章 Android 基础界面设计 ·· 19

    2.1 Android 布局文件 ·· 19
        2.1.1 创建 Android 布局文件 ·· 19
        2.1.2 使用 Android 布局文件 ·· 20
    2.2 Android 基本布局 ··· 21
        2.2.1 LinearLayout ··· 21
        2.2.2 FrameLayout ··· 23
        2.2.3 RelativeLayout ··· 24
        2.2.4 GridLayout ·· 27
    2.3 Android 基本控件 ··· 30

    2.3.1 界面控件的基本结构 ·············· 30
    2.3.2 TextView ························ 31
    2.3.3 EditText ························ 32
    2.3.4 Button ·························· 32
    2.3.5 ImageView ······················ 34
    2.3.6 基本控件应用 ··················· 36
    2.3.7 CheckBox ······················· 39
    2.3.8 RadioButton ···················· 40
    2.3.9 Snackbar ························ 41
    2.3.10 TextInputLayout ················ 42
    2.3.11 控件综合应用 ·················· 43
    2.3.12 视图绑定 ······················ 48
  2.4 Notification ······························· 49
    2.4.1 Notification 简介 ················ 50
    2.4.2 Notification 实现 ················ 51
    2.4.3 PendingIntent ··················· 58
  2.5 菜单 ·· 59
    2.5.1 使用 XML 定义菜单 ············ 59
    2.5.2 选项菜单 ························ 61
    2.5.3 上下文菜单 ····················· 63
    2.5.4 弹出菜单 ························ 66
  2.6 常用资源与样式 ··························· 67
    2.6.1 资源目录结构 ··················· 67
    2.6.2 样式和主题 ····················· 68
    2.6.3 Drawable 资源 ·················· 68
  2.7 本章小结 ·································· 70
  习题 ··· 70
第 3 章 Activity 与 Fragment ···················· 72
  3.1 Activity 基础 ······························ 72
    3.1.1 什么是 Activity ················· 72
    3.1.2 创建 Activity ···················· 73
    3.1.3 Activity 生命周期 ··············· 75
  3.2 Android 的事件处理机制 ················ 82
    3.2.1 基于监听的事件处理 ··········· 82
    3.2.2 基于回调的事件处理 ··········· 85
  3.3 Activity 使用 Intent ······················· 87
    3.3.1 显式 Intent ······················ 87
    3.3.2 隐式 Intent ······················ 88
    3.3.3 隐式 Intent 案例 ················ 89
  3.4 Activity 的数据传递 ····················· 91

  3.4.1 Intent 数据传递 ·········· 92
  3.4.2 Activity 的数据回传 ·········· 95
3.5 Activity 启动模式 ·········· 97
3.6 Fragment ·········· 98
  3.6.1 Fragment 简介 ·········· 98
  3.6.2 使用 Fragment ·········· 99
  3.6.3 Fragment 与 Activity 的交互 ·········· 103
  3.6.4 Fragment 新特性 ·········· 106
  3.6.5 Fragment 的生命周期 ·········· 107
  3.6.6 DialogFragment 对话框 ·········· 110
3.7 本章小结 ·········· 115
习题 ·········· 116

第 4 章 Android 高级界面设计 ·········· 117

4.1 Material Design ·········· 117
4.2 高级 UI 布局 ·········· 118
  4.2.1 ConstraintLayout ·········· 118
  4.2.2 CoordinatorLayout ·········· 123
  4.2.3 TabLayout ·········· 125
  4.2.4 DrawerLayout ·········· 132
4.3 高级 UI 组件 ·········· 136
  4.3.1 RecyclerView ·········· 136
  4.3.2 CardView ·········· 141
  4.3.3 FloatingActionButton ·········· 143
  4.3.4 NavigationView ·········· 145
  4.3.5 ViewPager ·········· 146
  4.3.6 Toolbar ·········· 147
4.4 自定义 View ·········· 151
4.5 本章小结 ·········· 157
习题 ·········· 157

第 5 章 数据存储 ·········· 159

5.1 SharedPreferences 存储 ·········· 159
  5.1.1 存储数据 ·········· 159
  5.1.2 读取数据 ·········· 160
5.2 文件存储 ·········· 164
  5.2.1 文件存储简介 ·········· 164
  5.2.2 内部存储 ·········· 164
  5.2.3 外部存储 ·········· 166
5.3 SQLite 数据库存储 ·········· 170
  5.3.1 SQLite 数据库简介 ·········· 170

   5.3.2 创建数据库 ··· 171
   5.3.3 SQLite 数据库操作 ·· 172
  5.4 内容提供者 ··· 182
   5.4.1 内容提供者简介 ·· 182
   5.4.2 创建 ContentProvider ··· 183
   5.4.3 访问其他应用程序的数据 ·· 188
  5.5 数据库框架 Room ··· 190
  5.6 本章小结 ··· 195
  习题 ·· 195

第 6 章 服务与广播··· 197

  6.1 服务 ·· 197
   6.1.1 服务的基本概念 ··· 198
   6.1.2 服务的生命周期 ··· 198
   6.1.3 Activity 和 Service 的交互 ·· 203
   6.1.4 前台服务 ··· 206
  6.2 广播机制 ··· 210
   6.2.1 广播机制简介 ·· 211
   6.2.2 广播接收器 ··· 211
   6.2.3 自定义广播 ··· 213
   6.2.4 最佳实践 ·· 220
  6.3 本章小结 ··· 221
  习题 ·· 221

第 7 章 网络编程··· 223

  7.1 Android 的多线程 ··· 223
   7.1.1 多线程的概念 ·· 223
   7.1.2 Handler 消息传递机制 ··· 226
   7.1.3 ThreadPoolExecutor 线程池技术 ··· 229
  7.2 WebView 控件 ··· 232
  7.3 基于 HTTP 的网络访问 ·· 234
   7.3.1 HTTP 协议简介 ··· 234
   7.3.2 使用 HttpURLConnection ··· 235
   7.3.3 解析 JSON 数据 ··· 238
  7.4 网络访问框架 ·· 244
   7.4.1 OkHttp 框架 ··· 244
   7.4.2 Glide 图片加载框架 ·· 245
  7.5 本章小结 ··· 247
  习题 ·· 247

第 8 章 多媒体开发·· 249

  8.1 多媒体简介 ··· 249

8.2 音频播放 ………………………………………………………………………… 250
8.3 视频播放 ………………………………………………………………………… 255
    8.3.1 VideoView ………………………………………………………………… 255
    8.3.2 SurfaceView ……………………………………………………………… 257
8.4 动画和过渡 ……………………………………………………………………… 259
    8.4.1 逐帧动画 ………………………………………………………………… 260
    8.4.2 补间动画 ………………………………………………………………… 260
    8.4.3 属性动画 ………………………………………………………………… 265
    8.4.4 布局动画 ………………………………………………………………… 270
8.5 本章小结 ………………………………………………………………………… 271
习题 …………………………………………………………………………………… 272

第9章 进阶技术 ……………………………………………………………………… 274

9.1 手势处理 ………………………………………………………………………… 274
    9.1.1 手势简介 ………………………………………………………………… 274
    9.1.2 手势检测 ………………………………………………………………… 275
    9.1.3 手势识别 ………………………………………………………………… 279
9.2 传感器开发 ……………………………………………………………………… 282
    9.2.1 传感器简介 ……………………………………………………………… 282
    9.2.2 使用传感器 ……………………………………………………………… 283
9.3 Android Jetpack ………………………………………………………………… 286
    9.3.1 Jetpack 简介 …………………………………………………………… 286
    9.3.2 Jetpack 架构组件 ……………………………………………………… 287
    9.3.3 综合应用 ………………………………………………………………… 309
9.4 本章小结 ………………………………………………………………………… 320
习题 …………………………………………………………………………………… 320

附录 A　Android 项目开发规范 ……………………………………………………… 321

第 1 章　Android 基础入门

Android 智能手机已占据智能手机的绝大部分市场份额。Android 是基于 Linux 的智能手机的操作系统，引领了智能手机应用 App 的技术发展。从现在开始，本书将带领读者由浅入深地学习 Android 应用开发的基础知识及进阶技术。

通过本章的学习，使读者了解 Android 的发展，对 Android 系统架构有初步了解，掌握 Android 开发环境的搭建，通过创建第一个 Android 应用程序熟悉 App 开发的流程，加深理解。

本章学习目标：

- 理解 Android 系统的体系架构；
- 掌握 Android 应用开发环境的搭建；
- 掌握 Android 应用的开发及部署；
- 了解 Android 项目打包发布的步骤。

1.1　Android 简介

1.1.1　智能手机及操作系统

早期的手机没有操作系统，使用的软件都是由设备生产商在设计手机时定制完成的，除了打电话、发短信及少量的内置附加功能之外，基本不具备扩展能力，只能满足消费者的基本需求。

2007 年，美国苹果公司发布的全"触屏控制"的 iPhone 手机以全新的概念重新定义了手机，成为第一代的智能手机。iPhone 手机不仅集 iPod、智能手机和便携式电脑于一体，也开创了全新的手机交互模式，引领了智能手机的发展潮流。随着移动互联网的普及，智能手机已成为最重要的终端设备，渗透到人们工作、生活的方方面面，成为不可取代的必需品。

智能手机最大的特点是可以自由安装应用软件，目前智能手机的主流操作系统是 Android 和 iOS。iOS 是 Apple 公司开发的手机操作系统，Android 是 Google 公司发布的手机操作系统。根据 Android 智能手机的市场占有率，Android 已成为当前应用最广泛的智能手机操作系统。

1.1.2 Android 的优势

Android 是基于 Linux 内核、开源的操作系统,由 Android 之父安迪·罗宾(Andy Rubin)开发完成,并于 2005 年 7 月被 Google 公司收购,应用于便携设备,如智能手机、平板电脑等。第一部 Android 智能手机发布于 2008 年 10 月,经过十几年的发展,Android 已经发布了十几个版本,目前主流的版本是 Android 10,手机的性能得到了极大的提升。

与 iPhone 的封闭系统相比,Android 具备以下优势。

1. 开源

Android 操作系统的开源意味着平台的开放性,允许任何移动端的厂商加入 Android 联盟。同时开发者也可以利用其开放的源代码进行二次开发,打造出个性化的 Android。

2. 自由度高

Android 操作系统给予用户很高的自由度,用户可以根据自己的喜好设置手机界面,Android 的应用市场甚至还有各式各样的启动器供用户选择,让自己的手机与众不同。

3. 选择多样化

由于 Android 的开放性,手机厂商为了迎合大众会推出层出不穷的新产品。迄今为止,以 Android 为操作系统的机型已经达到上百种。但是这些功能、机型的差异并不会影响到软件的兼容性,给消费者更多的购机选择。

4. 广泛的开发群体

Android 的开源吸引更多的开发者加入 Android 应用开发的大家庭,使得 Android 会被应用到更为广泛的领域和行业。

1.1.3 Android 的版本

Android 1.0 版本于 2008 年 9 月发布,之后每年 Google 公司一般都会进行一个重大的增量升级,2020 年 9 月发布了最新的 Android 11.0。

Google 公司在 2014 I/O 大会推出了版本改动最大的 Android 5.0,Dalvik 虚拟机被 ART 运行时环境替代,手机应用程序的运行速度有了明显的提升。与此同时,Google 公司使用新的设计风格 Material Design 统一了 Android 系统的界面设计风格,并在官网上推出了全新的 Material Design 设计指南,完整讲解了 Material Design 的实现规范。除此之外,Google 公司还在这个版本推出了可穿戴设备、汽车和电视对应的 Android 系统,将 Android 系统推广到这些全新的领域。

Android 6.0 提出了运行时权限、低电耗模式等省电功能。Android 7.0 加入了多窗口模式功能。Android 8.0 增加了画中画、通知通道及获得 Java 8 API 的支持。Android 9.0 对电源、隐私权进行更精细地管理。Android 10.0 增强了对用户隐私的保护,更好地支持可折叠设备和 5G 网络。Android 11.0 则在隐私方面做出了更为严格的规定。

截至 2020 年 11 月,Android 各版本与 API 等级的对应关系及市场占有率如图 1.1 所示,图中数据表明 Android 5.0 及以上版本的市场占有率已达 94%以上,因此本书所有案例都基于 Android 5.0 以上版本，不再兼容 Android 4.x 及以下版本。

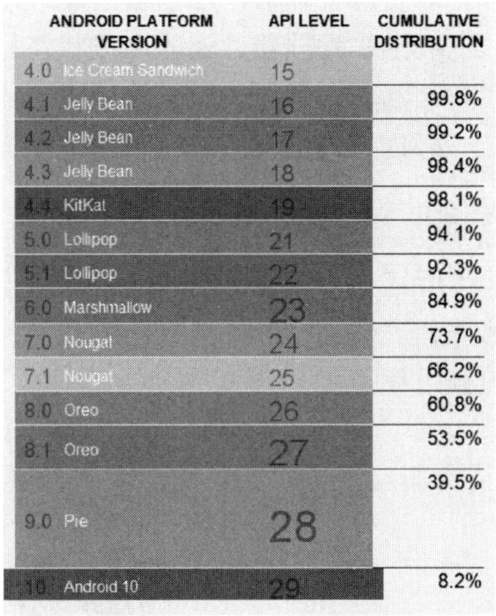

1.2 Android 系统架构

Android 基于 Linux 操作系统，采用了软件堆栈架构，从下往上依次为 Linux 内核层、硬件抽象层、系统 Native 库和 Android 运行时库、应用框架层及应用程序层等五层架构，其中每一层都包含大量的子模块或子系统，如图 1.2 所示。

1. Linux 内核层

Android 系统的基础是 Linux 内核，Linux 内核为 Android 系统提供了安全性、内存管理、进程管理、网络协议栈等核心系统服务，也为 Android 设备提供底

图 1.1 Android 的主要版本市场占有率

层驱动，如蓝牙驱动、显示驱动、相机驱动、WiFi 驱动等，使得 Android 系统实现底层硬件与上层软件之间的抽象。

2. 硬件抽象层（HAL）

硬件抽象层（Hardware Abstraction Layer，HAL）位于 Linux 内核之上，为上层的 Java API 提供设备硬件功能的标准接口。HAL 的每个库模块都为特定类型的硬件实现一组接口，如相机、蓝牙、传感器等模块。当框架 API 请求访问设备硬件时，Android 系统将为该硬件加载相应的库模块。

3. Android 运行时库与系统 Native 库

Android 系统库是用 C/C++语言编写的函数库集合，为 Android 系统提供特性支持，如 SQLite 库提供数据库支持，OpenGL 提供 2D/3D 绘图的支持。Android 应用开发者无法直接调用这套函数库，需要通过应用程序框架提供的 API 进行调用。

这一层还有 Android 运行时库 ART，它主要提供了使用 Java 进行 Android 应用开发的核心库，并且为每个应用提供单独的 Dalvik 虚拟机实例，使得每个 Android 应用程序都能运行在独立的进程中。Dalvik 虚拟机专为移动设备定制，与 Java 虚拟机相比，它在手机内存、CPU 性能、垃圾回收等方面做了优化处理。

Android 还包含一套核心运行时库，可提供 Java API 框架所使用的 Java 编程语言中的大部分功能，包括一些 Java 8 语言功能。

4. 应用框架层

应用框架层以 Java 类的形式为应用程序提供丰富的 API 用于构建各种应用程序，简化系统核心组件和服务的重用，主要包括以下组件和服务：

- 视图系统，用于构建应用的 UI，包括列表、按钮、文本框等；

- 资源管理器，用于访问非代码资源，如本地化的字符串、布局文件等；
- 通知管理器，让应用在状态栏中显示自定义的提醒；
- 活动管理器，用于管理应用的生命周期，提供常见的导航返回栈；
- 内容提供程序，让应用访问其他应用中的数据或共享其自己的数据，如联系人。

5. 应用程序层

所有安装在手机上的应用都属于这一层。应用程序层是核心应用程序的集合，包括电子邮件、短信、日历、浏览器和联系人等，也包括从应用商店下载的应用程序及我们自己开发的应用程序。

图 1.2　Android 系统架构

 ## 1.3　搭建 Android 开发环境

Android 程序开发的语言包括 Java 和 Kotlin，本书将以 Java 语言进行讲解，因此，在开始学习之前，需要有一定的 Java 基础，否则会有难度。如果你是一位 Java 的初学者，在学习 Android 应用开发的同时补充 Java 知识，也是一种非常有效的学习路径。

开发 Android 程序需要以下工具。

（1）JDK：JDK（Java SE Development Kit）是 Java 语言的软件开发工具包，包含 Java 的运行时环境、基础类库和工具集合。Android 5.0 以上版本的 Android 开发需要 JDK 7.0 及以上版本。

（2）Android SDK：Android SDK 是用于构建和编译 Android 应用程序的工具包，包括 Android build-tools、SDK platform-tools 和 SDK tools。Android build-tools 构建工具是编译 Android 应用所需 Android SDK 的一个组件，包括 Android 的开发和调试工具，如模拟器、APK 生成器等。SDK 平台工具包含与 Android 平台进行交互的工具，如 adb、fastboot。SDK 平台工具包含适用于 Android 的开发和调试工具，是与模拟器、智能手机设备进行通信的桥梁。

（3）Android Studio：Android Studio 是 Google 公司在 2013 年推出的官方 IDE 集成开发工具，截至到现在已推出 4.1 版本，本书的所有代码基于该版本的 Android Studio。

Android Studio 是一套基于 Intellij IDEA 的 Android 集成开发和调试环境，与原有的 Eclipse＋ADT 插件模式相比，具有如下优势。

- 继承了 Intellij 强大的智能提示和代码编辑功能。
- 整合了 Gradle 构建工具。
- 强大的 UI 编辑器，自带多设备的实时预览功能。
- 快速且功能丰富的模拟器。
- 完美整合版本控制系统，如 Git、GitHub 等。
- 内在终端，方便命令行操作。
- 更完善的插件系统。

1.3.1　安装 Android Studio

Google 公司提供的 Android Studio 已经集成了开发所需的工具包，只需到 Android 的官方网站（网址为 https://developer.android.google.cn/studio）下载最新版本的安装包进行安装即可。根据操作系统不同下载相应的安装文件，如图 1.3 所示。

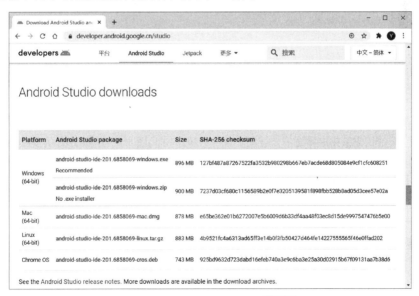

图 1.3　Android Studio 下载页面

Windows 操作系统下载 exe 文件，双击即可启动安装，安装过程非常简单，根据安装过程设置向导的提示安装推荐的所有 SDK 软件包，选择安装组件的界面如图 1.4 所示。单击 Next 按钮进入 Android Studio 安装路径的选择界面，如图 1.5 所示。

图 1.4 选择安装组件　　　　　　　图 1.5 选择 Android Studio 的安装路径

接下来依次单击 Next 按钮即可完成安装,如图 1.6 所示。

安装完成后启动 Android Studio,首次运行会进入配置界面,选择 Customer 进入个性化配置,进行 Java JRE、Android SDK 的安装,如图 1.7、图 1.8 所示,图 1.8 中的 Android SDK Location 用于选择 SDK 安装的目录。

单击 Next 按钮进入文件下载及解压安装,完成后进入图 1.9 所示的界面。单击 Finish 按钮进入如图 1.10 所示的启动页面,此页面中的选项功能说明如下。

图 1.6 安装完成界面　　　　　　　　图 1.7 选择 JRE 的安装目录

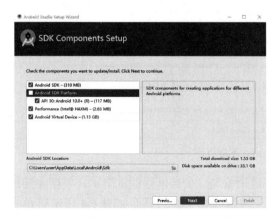

图 1.8 选择 Android SDK 的安装组件　　图 1.9 Android SDK 安装完成

- Create New Project：新建一个 Android Studio 项目。
- Open an Existing Project：打开一个已有的 Android Studio 项目。
- Get from Version Control：从版本服务器中获取一个项目。
- Profile or Debug APK：分析或调试 APK。
- Import Project（Gradle，Eclipse ADT，etc）：导入 Gradle、Eclipse ADT 等类型的项目。
- Import an Android Code Sample：导入 Android 代码示例。

图 1.10　Android Studio 的启动页面

至此，Android 的开发环境就全部搭建完成了，Android Studio 创建了以下主要目录。
- Android Studio 默认安装目录 C:\Program Files\Android\Android Studio。
- Android Studio 自带 JRE 1.8 的默认目录 C:\Program Files\Android\Android Studio\jre。
- Android SDK 默认安装目录 C:\Users\xxx\AppData\Local\Android\Sdk。
- 新建 Android 项目的默认保存目录 C:\Users\xxx\AndroidStudioProjects。
- Android Studio 默认配置目录 C:\Users\xxx\.AndroidStudio4.1。
- Android SDK 生成的模拟器存放目录 C:\Users\xxx\.android。
- Gradle 默认目录 C:\Users\xxx\.gradle。

其中，xxx 为 Windows 安装时设定的用户名。

需要特别注意的是：安装 Windows 操作系统需创建英文的用户名，避免中文名称的乱码问题，造成 Android 项目无法生成。

1.3.2　Android SDK

Android SDK（Software Development Kit）是 Android 软件开发包，它为开发者提供 Windows、Linux 或 MAC 操作系统开发 Android 应用程序的库文件及其开发工具的集合，目录列表如图 1.11 所示，具体含义如下。

- add-ons：存放 Android 扩展库，如 Google API 等。
- build-tools：存放各版本的 Android 编译工具，创建项目时使用这个包。
- docs：存放 Android SDK 的帮助文档，包括 SDK 平台、ADT、开发指南、API 文件等。
- emulator：存放 Android 模拟器所需的资源。

● extras：存放 Android 附加的支持文件，包括 Android 的 support 支持包、Google 提供的工具及 Intel 虚拟化驱动。

● platforms：存放 Android 各版本 SDK 的相关文件，包括系统的 jar 文件、字体、res 资源、模板等。

● platform-tools：存放 Android 平台的相关工具，如 adb.exe、sqlite3.exe 等。

● sources：存放 Android SDK API 各版本的源代码。

● system-images：存放创建 Android 模拟器时的镜像文件。

● temp：存放 SDK 更新安装时自动生成的一些文件。

● tools：存放 Android 各版本通用的开发调试工具和 Android 手机模拟器。

名称	修改日期	类型
add-ons	2017-1-25 20:21	文件夹
build-tools	2020-6-23 14:08	文件夹
cmdline-tools	2020-5-16 12:53	文件夹
docs	2016-12-1 9:34	文件夹
emulator	2020-5-16 14:12	文件夹
extras	2017-3-4 10:29	文件夹
fonts	2017-10-27 8:03	文件夹
licenses	2020-5-16 14:13	文件夹
metrics	2018-7-17 9:53	文件夹
patcher	2017-8-23 14:30	文件夹
platforms	2020-6-23 20:20	文件夹
platform-tools	2020-6-22 16:04	文件夹
skins	2020-6-23 11:03	文件夹
sources	2020-1-16 14:49	文件夹
system-images	2020-1-16 14:43	文件夹
temp	2017-8-22 9:30	文件夹
tools	2020-5-16 13:27	文件夹

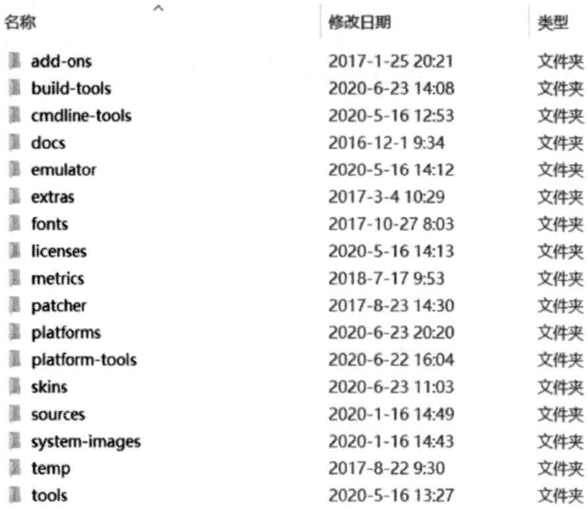

图 1.11　Android SDK 目录

在 Android Studio 主菜单中单击 Tools->SDK Manager，打开 Android SDK 配置界面，如图 1.12 所示，它包括 SDK 版本、模拟器镜像、SDK 工具等的安装、卸载及更新。

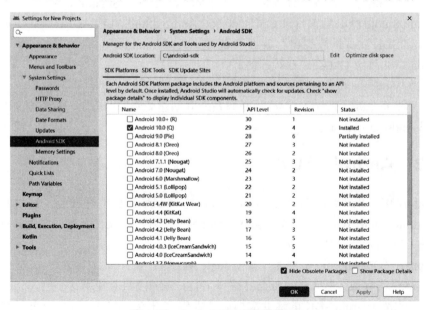

图 1.12　Android SDK 配置界面

1.4 开发第一个 Android 项目

1.4.1 创建项目

【案例 1-1】Hello World 项目是学习编程的经典开端，第一个 Android 项目也从 Hello World 开始。创建一个新的 Android 应用项目，项目名为 D0101_Hello World，该项目的运行结果是在模拟器上显示 "Hello World!" 字符串。

步骤 1：启动 Android Studio

启动 Android Studio，进入图 1.10 所示的启动页面。

步骤 2：选择项目模板

选择 Create New Project，打开选择项目模板的界面，如图 1.13 所示。

Android Studio 不仅可以开发手机和平板项目，还能用于可穿戴设备（Wear OS）、电视（TV）、自动驾驶（Automotive）和物联网（Android Things，此项目已终止）项目开发。Android 手机项目选择 Phone and Tablet，此界面提供了多种 Activity（活动）模板的选项，初学者一般选择最简单的 Empty Activity，创建一个空 Activity。

步骤 3：配置项目信息

单击 Next 按钮进入 Configure Your Project 窗口，设置项目主界面的基本信息，如图 1.14 所示。

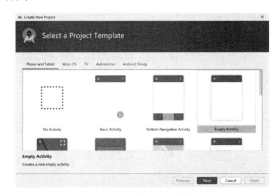
图 1.13 选择项目模板

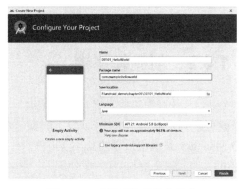
图 1.14 设置项目的基本信息

其中，

- Name：表示项目名称，填写本项目名称 HelloWorld。
- Package name：表示项目的包名，包名用于区分不同的项目，一般使用公司域名区分唯一性，对于个人开发者可以直接使用 com.example 加上项目名称。
- Save location：表示项目代码存放的目录，单击右侧的目录图标可以进行修改。
- Language：表示 Android 项目开发使用的语言，有 Java 和 Kotlin 两种选择，Kotlin 是开发 Android 项目的另一种编程语言。
- Minimum SDK：表示 Android 项目支持的 SDK 的最低版本，根据选择的版本会给出项目适配手机的百分比；如果不清楚如何选择最低版本，也可以单击 Help me choose 超链接了解更详细的信息。本书的所有项目的最低版本设为 API 21，不再兼容 Android 4.x 及以下版本。

● Use legacy android.support.libraries：表示是否兼容 Android 的版本支持库，自 Android 9.0 版本开始推出 AndroidX 对 Android Support Library 进行了升级，将原有的 Android Support Library 中的 API 的包名 android.support.*统一修改为 androidx.*，使得这些 API 不再依赖 Android 操作系统的具体版本，同时命名规则也不再包含版本号，统一为 appcompat 库。

步骤 4：系统构建项目

单击 Finish 按钮，耐心等待一段时间，项目就会创建成功，代码界面如图 1.15 所示。其中①至⑥分别为菜单栏、工具栏、导航栏、项目面板、编辑区和其他面板切换区等。

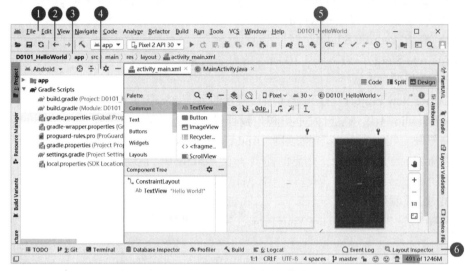

图 1.15　项目创建成功界面

1.4.2　创建模拟器

Android 项目创建成功之后，无须编写一行代码，只需要将项目加载到手机或 Android 模拟器就可运行。接下来创建一个 Android 模拟器。创建和启动模拟器可以选择菜单 Tools-> AVD Manager 菜单项或单击工具条中的 ![] 的第 2 个按钮，即可打开模拟器的欢迎界面，如图 1.16 所示。

单击 Create Virtual Device 进入模拟器的选择设备界面，如图 1.17 所示。

图 1.16　创建模拟器的欢迎界面

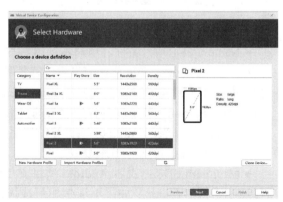

图 1.17　选择设备界面

模拟器选择设备界面包括设备类型和对应类型的型号，可选的类型包括：TV、Phone、Wear OS、Tablet 和 Automotive，我们开发的是 Android 智能手机的应用程序，选中 Phone 后可以选择不同名称、大小和分辨率的手机，选择目前大小为 5 英寸、分辨率为 1080px×1920px 的主流手机即可。单击 Next 按钮进入图 1.18 所示的 Android 系统镜像的界面。

在选择 Android 系统镜像时，如果没有下载 image 文件，单击 download 按钮进行下载，本书模拟器的镜像为 API 28 的 Android 9.0 的手机。

接下来单击 Next 按钮进入 Android 模拟器（AVD）的属性配置界面，如图 1.19 所示。

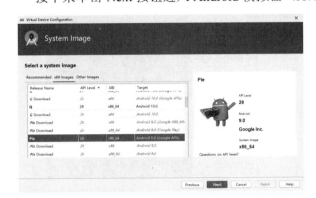
图 1.18　Android 系统镜像选择

图 1.19　AVD 的属性设置

界面中列出了选择的模拟器的手机型号和 Android 系统镜像，单击 Change 按钮可以重新选择，其他的属性设置包括：

● AVD Name：设置模拟器的名称。

● Startup orientation：模拟器竖屏（Portrait）还是横屏（Landscape）的选择。

● Emulated Performance：模拟器性能选择，有 Automatic、Hardware - GLES 2.0 和 Software - GLES 2.0 三个选项，分别是自动、硬件加速和软件加速，选择默认选项即可。

● Show Advanced Settings：显示更多设置项的按钮，单击后显示摄像头、网络、内核个数、内存、存储空间和键盘的选择，如图 1.20 所示。

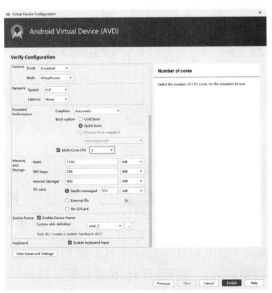
图 1.20　模拟器更多选项的设置界面

单击 Finish 按钮即可完成模拟器的创建，跳转到模拟器列表界面，如图 1.21 所示，单击三角形按钮即可启动模拟器，经过一段时间的初始化，显示如图 1.22 所示的手机模拟器首页。

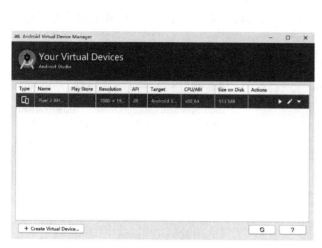

图 1.21　模拟器列表界面

图 1.22　手机模拟器首页

1.4.3　运行项目

模拟器创建成功后，运行 Android 项目有三种方式，包括：单击 Android Studio 工具条上的绿色三角运行按钮（见图 1.23）、选择菜单 Run->Run app 菜单项或按 Shift+F10 快捷键。HelloWorld 项目运行成功的界面如图 1.24 所示。

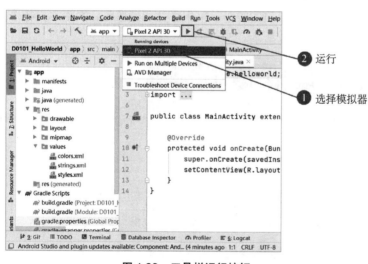

图 1.23　工具栏运行按钮

图 1.24　运行 HelloWorld 项目

通过设置向导，未编写一行代码就能运行项目，那么 Android Studio 是如何帮我们把项目的基本代码自动生成的呢？这就是我们接下来要学习和探讨的内容，详细分析一下 HelloWorld 代码。

1.5 项目文件结构

Android Studio 提供了多种模式展示项目的文件结构，如图 1.25 所示。

在这几种项目结构类型中，最常用的是 Project 和 Android 两种模式，其中，Project 模式下的项目结构都是真实的目录，如图 1.26 所示；而 Android 模式的最大优点就是隐藏了一些自动生成的文件和目录，并且把一些资源文件、源文件清晰地合并在一起，呈现的文件结构清晰简洁、一目了然，如图 1.27 所示。从两张图的对比可以清晰地看出 Android 模式隐藏了哪些文件。

图 1.25 项目结构类型

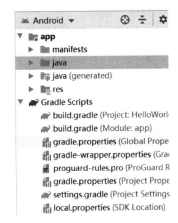

图 1.26 Project 模式的文件结构　　　图 1.27 Android 模式的文件结构

这么多文件、目录看上去是不是有点懵，别着急，接下来我们一一加以讲解。

● .gradle 和.idea 目录：它们是 Android Studio 自动生成的文件列表，不用手动进行编辑。

● app 目录：项目相关的代码、资源基本都存放在此目录中，开发过程编写的代码也都放在此目录中。

● build 目录：包含编译时自动生成的一些文件，可以忽视。

● gradle 目录：包含 gradle wrapper 的配置文件，Android Studio 会根据本地的缓存情况决定是否联网下载 gradle 文件，可以通过 Settings 的 Gradle 选项进行配置。

● .gitignore 文件：存放不受版本控制的文件或目录的列表。

● build.gradle 文件：项目的全局 gradle 编译脚本，一般无须修改。

● gradle.properties 文件：项目的全局 gradle 配置文件，此文件配置的属性会对项目中所有的 gradle 编译脚本产生影响。

● gradlew 和 gradlew.bat 文件：用于在命令行执行 gradle 命令。其中，gradlew 用于 Linux 或 Mac 系统，gradlew.bat 用于 Windows 系统。

● local.properties 文件：用于配置本机 Android SDK 的路径，内容是自动生成的，一般无须修改。

● settings.gradle 文件：用于项目所有模块的引入，通常情况下模块引入都是自动完成的。

Project 模式的文件结构除了 app 目录，大多数文件都是自动生成而无须修改的，所以 Android 模式通过隐藏这些文件让开发者将重心放到 app 目录。

Android 模式下该目录的详细结构如图 1.28 所示，常用的文件及目录如下。

- AndroidManifest.xml 文件：项目的配置文件，用于配置应用程序的公共属性、权限声明及四大组件的注册信息，接下来会详细说明。
- java 目录：项目的源代码及测试代码，展开目录可以看到，创建的 HelloWorld 项目的 MainActivity 文件就存放在对应的包下面。
- res 目录：项目的资源目录，存储图片、布局等资源。这个目录包含多个子目录。
  ◇ drawable：存放一些自定义形状和按钮切换颜色之类的 xml，以及图片资源等。
  ◇ layout：存放界面的布局文件。
  ◇ mipmap：存放图标相关的资源，其他的图片资源存放在 drawable 目录中。
  ◇ values：存放 app 引用的颜色、字符串、样式等的数值，例如，colors.xml、dimens.xml、strings.xml、styles.xml。
- Gradle Scripts：gradle 编译相关的脚本文件，主要包括项目和 app 模块的 build.gradle 配置文件。

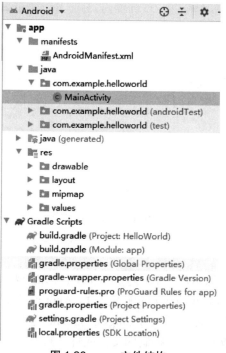

图 1.28　app 文件结构

1.5.1　java 目录

java 目录是源代码目录，存放了项目所有的 Java 源代码文件，HelloWorld 项目创建的 Activity 文件就存放在相应的包下。Android 项目与 Java 项目类似，都是通过 Java 包机制进行程序逻辑及功能的区分。

Activity 用于提供界面与用户交互，Android 项目运行时首先提供一个主 Activity 用于显示用户界面，图 1.29 中的 MainActivity 就是 HelloWorld 项目的主 Activity。双击该文件名，在右侧窗格中会打开这个 MainActivity.java 的源代码，如图 1.28 所示，在 onCreate()方法中调用 setContentView()方法绑定显示界面的布局文件 activity_main.xml，并在界面中显示该布局。

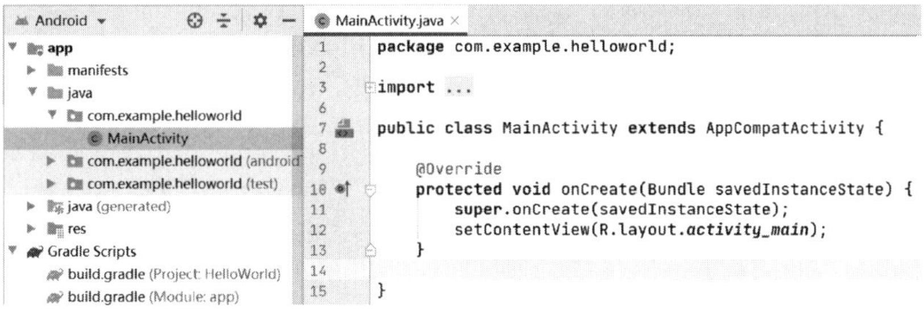

图 1.29　MainActivity.java 文件的源代码

1.5.2　res 目录

res 目录存放项目资源，如动画、图像、布局、图标、字符串、数组及样式资源等，该目录按照资源种类默认分为多个子目录，包括 animation（注：图中未显示）、drawable、layout、mipmap 和 values 等。MainActivity 绑定的布局文件存放在 layout 目录下的 activity_main.xml 中，单击图 1.28 源代码窗格左边的 图标，就能直接跳转到 MainActivity 的布局文件。布局文件采用 XML 方式定义了 Activity 的界面及界面中组件的结构，代码如下：

```xml
1.  <?xml version="1.0" encoding="utf-8"?>
2.  <androidx.constraintlayout.widget.ConstraintLayout
3.      xmlns:android="http://schemas.android.com/apk/res/android"
4.      xmlns:app="http://schemas.android.com/apk/res-auto"
5.      xmlns:tools="http://schemas.android.com/tools"
6.      android:layout_width="match_parent"
7.      android:layout_height="match_parent"
8.      tools:context=".MainActivity">
9.      <TextView
10.         android:layout_width="wrap_content"
11.         android:layout_height="wrap_content"
12.         android:text="Hello World!"
13.         app:layout_constraintBottom_toBottomOf="parent"
14.         app:layout_constraintLeft_toLeftOf="parent"
15.         app:layout_constraintRight_toRightOf="parent"
16.         app:layout_constraintTop_toTopOf="parent" />
17. </androidx.constraintlayout.widget.ConstraintLayout>
```

该示例代码采用 Android Studio 默认采用的 ConstraintLayout 约束性布局，使用 TextView 控件显示字符串信息，关于布局及控件的用法将在第 2、4 章的 UI 布局及控件章节详细讲解。

1.5.3　AndroidManifest.xml 文件

每个 Android 应用程序项目都必须有文件名为 AndroidManifest.xml 的全局配置清单文件，存储在 app 模块的 src/main 根目录中，此文件详细描述了以下内容。

● 应用的包名称：通常与代码的命名控件相匹配，构建项目时会使用此信息确定代码的位置。

● 应用的组件：包括所有 Activity、服务、广播接收器和内容提供程序。每个组件都必须定义 Java 类的名称等基本属性，还可以声明描述组件如何启动的 Intent 过滤器等功能。

- 应用为访问系统或其他应用的受保护部分所需的权限。

Android Studio 构建项目时会自动创建此文件,并添加大部分基本元素,示例代码如下:

```xml
1.  <?xml version="1.0" encoding="utf-8"?>
2.  <manifest xmlns:android="http://schemas.android.com/apk/res/android"
3.      package="com.example.helloworld">
4.
5.      <application
6.          android:allowBackup="true"
7.          android:icon="@mipmap/ic_launcher"
8.          android:label="@string/app_name"
9.          android:roundIcon="@mipmap/ic_launcher_round"
10.         android:supportsRtl="true"
11.         android:theme="@style/AppTheme">
12.         <activity android:name=".MainActivity">
13.             <intent-filter>
14.                 <action android:name="android.intent.action.MAIN"/>
15.                 <category android:name="android.intent.category.LAUNCHER"/>
16.             </intent-filter>
17.         </activity>
18.     </application>
19. </manifest>
```

表 1.1 对 AndroidManifest.xml 文件中的常用标签进行了详细说明。

表 1.1 AndroidManifest.xml 文件的常用标签描述

标 签	描 述
<manifest>	配置文件的根元素,包含包名、版本号、版本名称等属性
<application>	声明每一个应用程序的组件及其属性,描述了应用程序由哪些 Activity、Service、BroadcastReceiver 和 ContentProvider 组成,并指定实现这些组件的类及如何调用
<activity>	声明 Activity 组件,其中的 name 属性是该 Activity 的名称
<intent-filter>	intent 过滤标签,描述组件启动的类型及执行环境等

AndroidManifest.xml 文件的代码含义说明如下。

- 第 1 行声明了 XML 的版本及编码。
- 第 2~3 行声明了根元素<manifest>的命名空间、package 等信息,它们的具体含义介绍如下。

  ◇ xmlns:android:定义 android 命名空间,一般为 http://schemas.android.com/apk/res/android,该命名空间提供了大部分元素的数据,使得 Android 中的各种标准属性能被直接使用。

  ◇ package:指定项目的 Java 主程序包的包名,该属性值与项目的目录结构相匹配,它也是应用进程的默认名称。

- 第 4~11 行定义了<application>元素,<mainfest>根元素仅包含一个<application>元素,该元素声明了 Android 程序的组成部分,包括 Activity、BroadcastReceiver、Service 和 ContentProvider。该元素的属性影响所有的组成部分,常用属性包括如下几项。

  ◇ allowBackup:当 allowBackup 标识为 true 时,用户即可通过 adb backup 和 adb restore 对应用数据进行备份和恢复,这可能会带来一定的安全风险。

  ◇ icon:声明 App 的图标,图片一般都放在 drawable 目录下。

  ◇ label:声明 App 的名称字符串,该字符串常量放在 values 目录的 strings.xml 中。

  ◇ supportsRtl:支持从右往左显示的布局。

  ◇ theme:声明 App 的主题风格,它定义一个默认的主题风格给所有的 Activity,也可以在 styles.xml 中自定义 theme 的样式,有点类似 style。

● 第 12～17 行用于声明使用的 Activity 类，本示例为 MainActivity 类，子元素<intent-filter>描述该 Activity 是程序启动时的第一个界面，<action>元素的 android:name 属性指定程序入口 Activity，<category>元素的 android:name 属性指定当前动作（Action）被执行的环境，CATEGORY_LAUNCHER 决定应用程序是否显示在程序列表中。

本节只对 AndroidManifest.xml 中的基本元素进行了简要说明，其他元素将在后续课程的学习过程中陆续加入。

1.6 本章小结

本章主要讲解了 Android 的基础知识，首先介绍了 Android 的发展历史、优势、版本及系统架构，然后讲解了 Android 开发环境的搭建与配置，以及 Android SDK 工具包的配置，最后通过开发一个 HelloWorld 项目了解 Android 项目的结构，并讲解了项目目录的含义。希望通过本章的学习，读者可以对 Android 应用程序的开发有一个基本了解，为后续的学习开启一个良好的开端。

习 题

一、选择题

1. 下面关于模拟器的说法中，正确的是（　　）。
A. 在模拟器上可预览和测试 Android 应用程序
B. 只可以在模拟器上预览 Android 应用程序
C. 只可以在模拟器上测试 Android 程序
D. 模拟器属于物理设备

2. 下列选项中，属于应用框架层的是（　　）。
A. 活动管理器　　　B. 联系人程序　　　C. 短信程序　　　D. 音频驱动

3. 下列选项中，属于 Android Studio 工具中打开项目的选项是（　　）。
A. Create New Project　　　　　　　B. Open an Existing Project
C. Profile or debug APK　　　　　　 D. Get From Version Control

4. 下列选项中，属于 Android 系统架构的是（　　）。
A. 应用程序层　　　　　　　　　　　B. 应用程序框架层
C. 核心类库　　　　　　　　　　　　D. Linux 内核

5. 下面关于 Android 程序结构的描述中，正确的是（　　）。
A. app/src/main/res 目录用于存放程序的资源文件
B. app/src/main/java 用于存放程序的代码文件
C. app/libs 用于存放第三方 jar 包
D. build.gradle 用于配置在 Android 程序中使用到的子项目

二、简答题

1. 简述 Android 开发环境的搭建。
2. 简述 Android 系统架构的层次划分及各层的特点。

3. 简述 Android 应用程序的基本结构。

4. 简述如何创建一个 Android 模拟器。

三、编程题

1. 创建两种 Android Studio 提供的 Activity 类型的默认应用。

2. 使用 Android Studio 创建一个项目，显示"×××同学生日快乐"，名字是当天过生日的同学，并使用模拟器或真机运行。

第 2 章　Android 基础界面设计

上一章完成了 Android Studio 开发环境的搭建和第一个 Android 应用程序的开发,接下来面临的问题就是:如何让应用程序的界面设计更出色?本章将讲解 Android 应用程序的界面设计。通过本章的学习,使开发者了解布局文件的创建方法,掌握常用的 UI 布局及基础 UI 组件的使用方法,掌握通知和菜单的创建与使用方法,了解界面资源和样式,加深对 Android 界面开发的理解。

本章学习目标:
- 掌握 Android 常用的 UI 布局;
- 掌握 Android 基础 UI 组件的使用方法;
- 理解三种菜单的创建和使用方法;
- 熟悉 Android 常用资源与样式。

2.1　Android 布局文件

2.1.1　创建 Android 布局文件

在 Android 应用程序中,每一个有交互作用的 Activity 都需要对应一个布局文件,布局决定了 Activity 所展现的外观样式,布局结构控制着展现给用户的所有元素。在 Android 项目的目录结构中,布局文件的存放路径为 res/layout/xxx.xml,可以通过以下两种方式创建布局。

① 右击 res/layout 目录,依次选择 New->Layout Resource File,输入布局的文件名,即可打开创建资源文件的对话框,然后在 layout 目录下生成一个新的布局文件,如图 2.1 所示。

图 2.1　新建布局文件的第 1 种方法

② 右击 res/layout 目录，依次选择 New->XML->Layout XML File，输入布局的文件名，即可打开创建 XML 组件的对话框，然后在 layout 目录下生成一个新的布局文件，如图 2.2 所示。

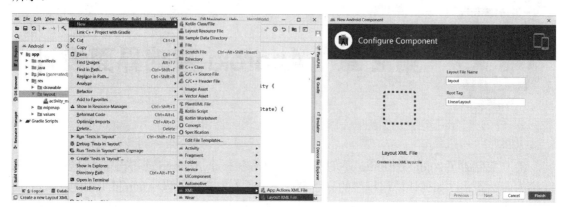

图 2.2　新建布局文件的第 2 种方法

2.1.2　使用 Android 布局文件

布局文件创建后，双击打开该文件，可分别通过图形化界面或代码方式查看布局文件。选择 Design 选项卡打开如图 2.3 所示的图形化界面，就可通过拖曳的方式添加控件。图中标识的①~⑥分别为：XML 布局文件、控件面板、设计及蓝图面板、Code/Split/Design 选项卡、Attributes 选项卡和界面控件的布局结构树。

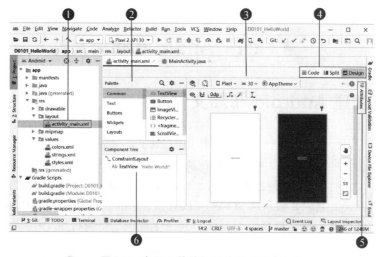

图 2.3　布局文件的图形化开发界面

选择 Code 或 Split 选项卡打开如图 2.4 所示的 XML 文件代码的界面，就可通过代码编写方式添加控件，进行界面设计。

布局文件开发完成后，需要在 java 类文件加载并使用布局中的控件。每个 XML 布局文件都会被编译到一个 View 对象中，可在 Activity 的 onCreate()方法中调用 setContentView()方法加载这个 View 对象。setContentView()将 R.layout.layout_name 作为参数，即可完成该布局文件的加载。例如，已有布局文件 activity_main.xml，则加载该布局文件的方法如下：

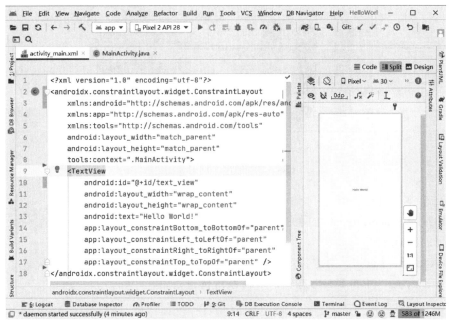

图 2.4　布局文件的代码开发界面

```
1.  public class MainActivity extends AppCompatActivity {
2.      @Override
3.      protected void onCreate(Bundle savedInstanceState) {
4.          super.onCreate(savedInstanceState);
5.          setContentView(R.layout.activity_main);
6.      }
7.  }
```

 ## 2.2　Android 基本布局

在了解布局文件的基础知识之后，接下来需要重点了解的是"如何编写布局文件 layout"。在 Android 开发中 UI 设计十分重要，当用户使用一个手机 App 时，最先感受到的不是这款 App 的功能是否强大，而是界面设计是否赏心悦目、用户体验是否良好。也可以这样说，能吸引用户使用的界面是一个 App 取得成功的关键因素。

因此，Android 的布局至关重要，每种布局既可以单独使用，也可以嵌套使用，在实际应用中进行界面设计时应合理搭配和组合。本节将讲解 Android 应用程序最常用的基础布局：LinearLayout、FrameLayout、RelativeLayout 和 GridLayout。

2.2.1　LinearLayout

LinearLayout 是一种重要的界面布局，开发中经常用到。它采用线性方式进行部署，以行或列的方式添加控件，每个子元素都位于前一个元素之后。该布局容器内的组件一个挨着一个地排列，不仅可以控制各组件横向排列，也可控制各组件纵向排列。如果采用的是垂直排列，那么将是一个 N 行单列的结构，每一行只会有一个元素；而如果是水平排列，那么将是一个单行 N 列的结构，如图 2.5 所示。

horizontal　　　　　　　　　vertical

图 2.5　线性布局示意图

下面结合案例进行讲解。

【案例 2-1】LinearLayout 布局的各种设置方式，如图 2.6～图 2.8 所示。

步骤 1：创建项目

启动 Android Studio，创建名为 D0201_Layout 的项目，选择 Empty Activity，包名改为 com.example.layout，单击 Finish 按钮，等待系统构建完成。

步骤 2：修改布局

右击 activity_main.xml 文件，选择 Refactor->Rename 菜单项或使用 Shift+F6 快捷键，将文件重命名为 activity_linearlayout.xml，然后打开该文件，切换为 Split 模式。通过设置 android:orientation 属性指定线性布局的子视图排列方向，vertical 表示垂直，方向为从上向下，如图 2.6 所示；horizontal 表示水平，方向为从左向右，如图 2.7 所示。

线性布局有个非常重要的属性 gravity，该属性用来指定组件内容的对齐方式，支持 top、bottom、left、right、center、fill 等值，也可以同时指定多种对齐方式，如 bottom|center_horizontal 表示出现在屏幕底部且水平居中，如图 2.8 所示，实现代码如下所示：

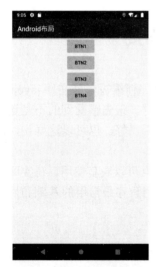  

图 2.6　垂直方向线性布局　　图 2.7　水平方向线性布局　　图 2.8　多种对齐方式的线性布局

```
1.  <LinearLayout xmlns:android="http://schemas.android.com/apk/
                                                       res/android"
2.      android:layout_width="match_parent"
3.      android:layout_height="match_parent"
4.      android:orientation="horizontal"
5.      android:gravity="bottom|center_horizontal">
```

步骤 3：运行项目

启动模拟器，运行项目，查看不同的布局设置。

小技巧：使用 LinearLayout 实现两行两列的结构，可通过先垂直排列两个元素，再向每个元素嵌套一个水平的 LinearLayout 实现。

2.2.2 FrameLayout

FrameLayout 称为帧布局，是常用布局中最简单的布局方式。在这个布局中，整个界面被当成一块空白备用区域，所有子元素的位置都不能够被指定，它们统统放在这块区域的左上角，并且后面的子元素直接覆盖在前面的子元素之上，将前面的子元素部分或者全部遮挡。

因此，帧布局的大小由子控件中最大的子控件决定，如果组件都同样大的话，同一时刻就只能看到最上面的组件。当然我们也可以为组件添加 layout_gravity 属性，从而指定组件的对齐方式，并借助层级视图工具（Hierarchy Viewer tool）查看所有的布局。帧布局在游戏开发中较为常见。

FrameLayout 的主要属性包括以下两个。
- android:foreground：设置帧布局容器的前景图像。
- android:foregroundGravity：设置前景图像显示的位置。

小知识：前景图像是指处于帧布局最上面显示的图片，它不会被覆盖。

【案例 2-1（续）】使用 FrameLayout 布局显示一张背景图片，并在屏幕上方居中显示"你好，Android"，文字覆盖于图片之上。

步骤 1：新建布局文件

打开 D0201_Layout 项目，在 res/layout 目录中创建 activity_framelayout.xml 布局文件，将 Root element 改为 FrameLayout，生成的代码如下：

```
1.  <FrameLayout xmlns:android="http://schemas.android.com/apk/
                                  res/android"
2.      android:layout_width="match_parent"
3.      android:layout_height="match_parent" >
4.  </FrameLayout>
```

步骤 2：添加图片

选择一张手机图片并复制到 drawable 目录中，然后在 FrameLayout 中添加该图片，具体代码如下：

```
1.  <FrameLayout xmlns:android="http://schemas.android.com/apk/
                                  res/android"
2.      android:layout_width="match_parent"
3.      android:layout_height="match_parentt" >
4.      <ImageView
5.          android:id="@+id/image_view"
6.          android:layout_width="wrap_content"
7.          android:layout_height="wrap_content"
8.          android:scaleType="fitXY"
9.          android:src="@drawable/bg" />
10. </FrameLayout>
```

步骤 3：在图片上添加文字

继续在此布局中添加 TextView 控件，显示一段文字，给 TextView 设置一些属性使得文本更好看，具体代码如下：

```xml
1.  <?xml version="1.0" encoding="utf-8"?>
2.  <FrameLayout xmlns:android="http://schemas.android.com/apk/
                                res/android"
3.      android:layout_width="match_parent"
4.      android:layout_height="match_parent">
5.      <ImageView
6.          android:id="@+id/image_view"
7.          android:layout_width="match_parent"
8.          android:layout_height="match_parent"
9.          android:scaleType="fitXY"
10.         android:src="@drawable/bg" />
11.     <TextView
12.         android:id="@+id/tv_content"
13.         android:layout_width="wrap_content"
14.         android:layout_height="wrap_content"
15.         android:layout_gravity="center_horizontal"
16.         android:layout_marginTop="64dp"
17.         android:text="你好, Android"
18.         android:textColor="@android:color/white"
19.         android:textSize="40sp" />
20. </FrameLayout>
```

步骤 4：运行项目

打开 MainActivity 文件，将 onCreate() 的 setContentView() 方法的参数改为 R.layout. activity_ framelayout，运行结果如图 2.9 所示。

```java
1.  public class MainActivity extends AppCompatActivity {
2.      @Override
3.      protected void onCreate(Bundle savedInstanceState) {
4.          super.onCreate(savedInstanceState);
5.          setContentView(R.layout.activity_framelayout);
6.      }
7.  }
```

2.2.3 RelativeLayout

图 2.9 帧布局的界面布局

相对布局 RelativeLayout 是一种非常灵活的布局方式，通过指定界面元素与其他元素的相对位置，确定界面中所有元素的布局位置，它能够最大限度地保证各种屏幕类型设备的适配性。

RelativeLayout 布局中的每个控件不仅可以通过相对于父容器位置的 boolean 类型进行设置，也可以通过另一个子视图的 ID 设置此控件相对于其他视图的位置，常用属性说明如下。

第 1 种方式：相对父容器的位置属性如表 2.1 所示。

表 2.1 使用 boolean 值描述相对父容器位置的属性

属性名称	属性描述	取值
android:layout_centerInParent	在父视图的正中心	true/false
android:layout_centerHorizontal	在父视图的水平中心线上	true/false
android:layout_centerVertical	在父视图的垂直中心线上	true/false
android:layout_alignParentTop	紧贴父视图的顶部	true/false
android:layout_alignParentBottom	紧贴父视图的底部	true/false
android:layout_alignParentLeft	紧贴父视图的左部	true/false
android:layout_alignParentRight	紧贴父视图的右部	true/false

第 2 种方式：相对其他子视图的位置使用 id 引用名称"@id/id-name"，如表 2.2 所示。

表 2.2 使用 id 描述相对其他子视图的属性

属性名称	属性描述	取值
android:layout_alignTop	与指定视图顶部对齐	视图 id，如"@id/***"
android:layout_alignBottom	与指定视图底部对齐	视图 id，如"@id/***"
android:layout_alignLeft	与指定视图左部对齐	视图 id，如"@id/***"
android:layout_alignRight	与指定视图右部对齐	视图 id，如"@id/***"
android:layout_above	在指定视图的上方	视图 id，如"@id/***"
android:layout_below	在指定视图的下方	视图 id，如"@id/***"
android:layout_toLeftOf	在指定视图的左方	视图 id，如"@id/***"
android:layout_toRightOf	在指定视图的右方	视图 id，如"@id/***"

第 3 种方式：相对其他子视图的位置使用具体的像素值，如 30dp，如表 2.3 所示。

表 2.3 使用像素值描述相对其他元素的属性

属性名称	属性描述	取值
android:layout_marginBottom	离某元素底边缘的距离	具体像素值，如 30dp
android:layout_marginLeft	离某元素左边缘的距离	具体像素值，如 30dp
android:layout_marginRight	离某元素右边缘的距离	具体像素值，如 30dp
android:layout_marginTop	离某元素上边缘的距离	具体像素值，如 30dp

小技巧：用好 RelativeLayout 的关键在于找到适当的参照对象。

接下来使用一个具体的案例实现 RelativeLayout 布局。

【案例 2-1（续）】使用 RelativeLayout 布局实现如图 2.10 所示的界面。

步骤 1：新建布局文件

打开 D0201_Layout 项目，在 res/layout 目录中创建 activity_relativelayout.xml 布局文件，修改根元素为 RelativeLayout，生成的代码如下：

```
1.  <RelativeLayout xmlns:android="http://schemas.android.
                                com/apk/res/android"
2.      xmlns:tools="http://schemas.android.com/tools"
3.      android:layout_width="match_parent"
4.      android:layout_height="match_parent" >
5.  </RelativeLayout>
```

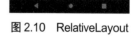

图 2.10 RelativeLayout 界面布局

步骤 2：添加位于屏幕正中间的"中"按钮

根据界面要求，可以先添加位于屏幕正中间的"中"按钮。因为它位于屏幕的正中间，比较容易确定其位置，所以先添加该按钮作为其他控件的参照物，具体代码如下：

```
1.  <RelativeLayout xmlns:android="http://schemas.android.com/apk/res/android"
2.      android:layout_width="match_parent"
3.      android:layout_height="match_parent">
4.      <Button
5.          android:id="@+id/btn_center"
6.          android:layout_width="wrap_content"
7.          android:layout_height="wrap_content"
8.          android:layout_centerHorizontal="true"
```

```
9.          android:layout_centerVertical="true"
10.         android:text="中" />
11. </RelativeLayout>
```

上述代码中，此按钮位于屏幕的正中间通过设置 android:layout_centerHorizontal="true"及 android:layout_centerVertical="true"这两个属性实现。

步骤 3：添加位于"中"按钮上方的按钮

在步骤 2 确定位置参照物——"中"按钮之后，就可以根据上方按钮和此位置的关系，添加位于中间按钮上方的按钮，具体代码如下：

```
1.  <RelativeLayout xmlns:android="http://schemas.android.com/apk/
                                   res/android"
2.      android:layout_width="match_parent"
3.      android:layout_height="match_parent">
4.      <Button
5.          android:id="@+id/btn_center"
6.          android:layout_width="wrap_content"
7.          android:layout_height="wrap_content"
8.          android:layout_centerHorizontal="true"
9.          android:layout_centerVertical="true"
10.         android:text="中" />
11.     <Button
12.         android:id="@+id/btn_above"
13.         android:layout_width="wrap_content"
14.         android:layout_height="wrap_content"
15.         android:layout_above="@+id/btn_center"
16.         android:layout_centerHorizontal="true"
17.         android:text="上" />
18. </RelativeLayout>
```

步骤 4：依次添加界面中的其余按钮

其他按钮也采用"中"按钮为参照对象，根据它们的位置关系，依次添加界面中的剩余按钮，参考表 2.2 所示的属性进行属性设置，方法和步骤 3 一致，此处不再一一详细描述。完整的布局代码如下：

```
1.  <RelativeLayout xmlns:android="http://schemas.android.com/apk/
                                   res/android"
2.      android:layout_width="match_parent"
3.      android:layout_height="match_parent">
4.      <Button
5.          android:id="@+id/btn_center"
6.          android:layout_width="wrap_content"
7.          android:layout_height="wrap_content"
8.          android:layout_centerHorizontal="true"
9.          android:layout_centerVertical="true"
10.         android:text="中" />
11.     <Button
12.         android:id="@+id/btn_above"
13.         android:layout_width="wrap_content"
14.         android:layout_height="wrap_content"
15.         android:layout_above="@+id/btn_center"
16.         android:layout_centerHorizontal="true"
17.         android:text="上" />
18.     <Button
19.         android:id="@+id/btn_below"
20.         android:layout_width="wrap_content"
21.         android:layout_height="wrap_content"
22.         android:layout_below="@+id/btn_center"
```

```
23.        android:layout_centerHorizontal="true"
24.        android:text="下" />
25.    <Button
26.        android:id="@+id/btn_left"
27.        android:layout_width="wrap_content"
28.        android:layout_height="wrap_content"
29.        android:layout_centerVertical="true"
30.        android:layout_toLeftOf="@+id/btn_center"
31.        android:text="左" />
32.    <Button
33.        android:id="@+id/btn_right"
34.        android:layout_width="wrap_content"
35.        android:layout_height="wrap_content"
36.        android:layout_centerVertical="true"
37.        android:layout_toRightOf="@+id/btn_center"
38.        android:text="右" />
39.    <Button
40.        android:id="@+id/btn_above_left"
41.        android:layout_width="wrap_content"
42.        android:layout_height="wrap_content"
43.        android:layout_above="@+id/btn_center"
44.        android:layout_toLeftOf="@+id/btn_above"
45.        android:text="左上" />
46.    <Button
47.        android:id="@+id/btn_above_right"
48.        android:layout_width="wrap_content"
49.        android:layout_height="wrap_content"
50.        android:layout_above="@+id/btn_center"
51.        android:layout_toRightOf="@+id/btn_above"
52.        android:text="右上" />
53.    <Button
54.        android:id="@+id/btn_below_left"
55.        android:layout_width="wrap_content"
56.        android:layout_height="wrap_content"
57.        android:layout_below="@+id/btn_center"
58.        android:layout_toLeftOf="@+id/btn_above"
59.        android:text="左上" />
60.    <Button
61.        android:id="@+id/btn_below_right"
62.        android:layout_width="wrap_content"
63.        android:layout_height="wrap_content"
64.        android:layout_below="@+id/btn_center"
65.        android:layout_toRightOf="@+id/btn_above"
66.        android:text="右上" />
67. </RelativeLayout>
```

步骤 4：运行项目

打开 MainActivity 文件，将 onCreate()的 setContentView()方法的参数改为 R.layout. activity_relativelayout，运行结果如图 2.10 所示。

2.2.4 GridLayout

Android 4.0 版本新增了 GridLayout 网格布局控件。在 Android 4.0 版本之前，如果想要达到网格布局的效果，一般都使用 LinearLayout 布局，但是这种布局会产生如下几点问题：

- 控件不能同时在 X、Y 轴方向上进行对齐。
- 当多层布局嵌套时会涉及性能问题。

也可以使用表格布局 TabelLayout，它把包含的元素以行和列的形式进行排列，每行为一个 TableRow 对象，也可以是一个 View 对象，而在 TableRow 中还可以继续添加其他的控件，每添加一个子控件就成为一列。但是这种布局可能会出现控件不能占据多个行或列的问题，而且渲染速度也不能得到很好的保证。

Android 4.0 的 GridLayout 布局解决了以上问题，它使用虚细线将布局划分为行、列和单元格，也支持一个控件在行、列上都有交错排列。

GridLayout 与 LinearLayout 布局类似，也分为水平和垂直两种方式，默认的是水平布局，一个控件挨着一个控件从左到右依次排列，但是通过指定 android:columnCount 设置列数的属性后，控件会自动换行排列。另外，对于 GridLayout 布局中的子控件，默认按照 wrap_content 的方式显示，只需要在 GridLayout 布局中显式声明即可。

GridLayout 的属性分为两种：其本身的属性和子元素属性，常用属性的说明如表 2.4 所示。

表 2.4 GridLayout 布局的属性描述

分类	属性	描述	取值说明
自身属性	android:alignmentMode	设置 alignMargins 属性，使视图的外边界之间进行校准	alignBounds - 对齐子视图边界 alignMargins - 对齐子视距内容
	android:columnCount	最大列数	值为 4 表示每行有 4 列
	android:rowCount	最大行数	值为 5 表示总共有 5 行
	android:orientation	所含子元素的布局方向	水平：horizontal，竖直：vertical
子元素属性	android:layout_column	显示该子元素所在的列	值为 0 表示在第 1 列
	android:layout_row	显示该子元素所在的行	值为 2 表示在第 3 行
	android:layout_columnSpan	该子元素所占的列数	值为 2 表示占两列
	android:layout_rowSpan	该子元素所占的行数	值为 2 表示占两行
	android:layout_columnWeight	该子元素的列权重	若一行有 2 列，值为 1 表示平分
	android:layout_rowWeight	该子元素的行权重	若共有 3 行，值为 1 表示占 1/3

图 2.11 GridLayout 布局的界面

下面通过案例讲解 GridLayout 布局的实际应用。

【案例 2-1（续）】使用 GridLayout 布局实现如图 2.11 所示的计算器界面。

步骤 1：新建布局文件

打开 D0201_Layout 项目，在 res/layout 目录中新建 activity_gridlayout.xml 布局文件，Root element 改为 GridLayout，columnCount 属性设为 4 列，rowCount 属性设为 6 行，具体代码如下：

```
1.    <GridLayout xmlns:android="http://schemas.
                                 android.com/apk/
                                 res/android"
2.        android:layout_width="match_parent"
3.        android:layout_height="match_parent"
4.        android:columnCount="4"
5.        android:orientation="horizontal"
6.        android:rowCount="6">
7.    </GridLayout>
```

步骤 2：添加 EditText、Button 等控件

在 GridLayout 布局内进行计算器输入框和各个按钮的布局，部分核心代码如下所示：

```xml
1.  <GridLayout xmlns:android="http://schemas.android.com/apk/
                            res/android"
2.      android:layout_width="match_parent"
3.      android:layout_height="match_parent"
4.      android:background="#ece7e7"
5.      android:columnCount="4"
6.      android:orientation="horizontal"
7.      android:rowCount="5">
8.      <EditText
9.          android:id="@+id/tv_result"
10.         android:layout_rowWeight="3"
11.         android:layout_columnSpan="4"
12.         android:layout_columnWeight="4"
13.         android:gravity="end|bottom"
14.         android:text="0"
15.         android:textSize="48sp" />
16.     <Button
17.         android:id="@+id/btn_c"
18.         android:layout_rowWeight="1"
19.         android:layout_columnWeight="1"
20.         android:layout_margin="1dp"
21.         android:background="#ffffff"
22.         android:gravity="center"
23.         android:text="C"
24.         android:textColor="@android:color/holo_blue_light"
25.         android:textSize="28sp" />
26.     <Button
27.         android:id="@+id/btn_divide"
28.         android:layout_rowWeight="1"
29.         android:layout_columnWeight="1"
30.         android:layout_margin="1dp"
31.         android:background="#ffffff"
32.         android:gravity="center"
33.         android:text="/"
34.         android:textColor="@android:color/holo_blue_light"
35.         android:textSize="28sp" />
36.     <Button
37.         android:id="@+id/btn_multipy"
38.         android:layout_rowWeight="1"
39.         android:layout_columnWeight="1"
40.         android:layout_margin="1dp"
41.         android:background="#ffffff"
42.         android:gravity="center"
43.         android:text="x"
44.         android:textColor="@android:color/holo_blue_light"
45.         android:textSize="28sp" />
46.     <Button
47.         android:id="@+id/btn_back"
48.         android:layout_rowWeight="1"
49.         android:layout_columnWeight="1"
50.         android:layout_margin="1dp"
51.         android:background="#ffffff"
52.         android:gravity="center"
53.         android:text="退格"
54.         android:textColor="@android:color/holo_blue_light"
55.         android:textSize="28sp" />
56.     <!-- 部分代码省略 -->
57. </GridLayout>
```

步骤3：运行项目

打开 MainActivity 文件，将 onCreate()中的 setContentView()方法的参数改为 R.layout.

activity_gridlayout，运行结果如图 2.11 所示。

2.3 Android 基本控件

布局创建之后，还需要在布局中添加相应的控件。本节将讲解 Android 应用程序中常见基本控件的使用方法及应用场景，并通过一个实现登录、个人信息维护等的案例综合应用它们。

2.3.1 界面控件的基本结构

Android 的 UI 界面都是由 View、ViewGroup 及其派生类组合而成的。其中，View 是所有 UI 组件的基类，ViewGroup 是 View 及其派生类的容器，也是从 View 派生的。它们之间的继承关系如图 2.12 所示。

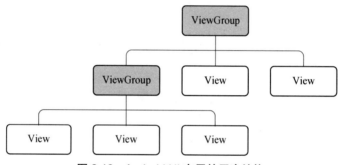

图 2.12　Android UI 布局的层次结构

Android 应用程序常用的界面控件如表 2.5 所示。

表 2.5　常用界面控件

名称	描述
TextView	显示文本信息
Button	普通按钮
EditText	可编辑的文本框组件（输入框）
ImageView	用于显示图片
ImageButton	图片按钮
CheckBox	复选框
RadioGroup	单选框
Spinner	下拉列表组件
ProgressBar	进度条
SeekBar	拖动条
RatingBar	评分组件
ListView	列表
Dialog	对话框
Toast	信息提示组件

2.3.2 TextView

TextView 继承自 View 类，在 android.widget 包中。TextView 控件的功能是向用户展示文本的内容，但不允许用户编辑，其常用属性如表 2.6 所示。

表 2.6 TextView 控件常用属性

属性名称	作用描述
android:layout_width	设置控件的宽度
android:layout_height	设置控件的高度
android:id	设置组件的 id
android:text	设置文本内容
android:textColor	设置文本颜色
android:textSize	设置文本大小
android:background	设置控件的背景色
android:gravity	设置文本相对控件的位置
android:layout_gravity	设置控件相对于其所在容器的位置

TextView 控件在使用时，首先要将其添加到布局文件，即 res/layout/activity_main.xml 文件中。添加 TextView 控件后，若需要修改 TextView 的显示内容、字体大小等，有以下两种方式。

1. layout 布局设置方式

```
1.  <TextView
2.      android:id="@+id/ textview"
3.      android:layout_width="wrap_content"
4.      android:layout_height="wrap_content"
5.      android:background="#0000ff"
6.      android:text="@string/hello_world"
7.      android:textColor="#ffffff"
8.      android:textSize="18sp" />
```

通过在布局文件中设置控件的属性实现，示例代码的第 2 行的 android:id 属性声明 TextView 的 id，这个 id 是该控件的唯一识别，用于在 Java 代码中引用这个 TextView 对象。属性值"@+id/textview"中@表示 id 资源的字符串，加号（+）表示需要建立新资源名称，项目编译后生成 id 的整型值添加到 R.java 文件中，斜杠后面的字符串 textview 表示这个 TextView 的名称。

第 5 行代码用于设置 TextView 的背景色，第 6～8 行代码用于设置 TextView 文本的内容、颜色和字号。

2. Java 编码方式

获取 TextView 的内容字符串是通过布局文件设置控件 id 的，调用 findViewById(int id)实例化控件，然后可以获取文本内容、设置属性等。

```
1.  import android.widget.TextView;
2.  public class MainActivity extends Activity {
3.      @Override
4.      protected void onCreate(Bundle savedInstanceState) {
5.          super.onCreate(savedInstanceState);
```

```
6.      setContentView(R.layout.activity_main);
7.      TextView tv = findViewById(R.id.textview);
8.      tv.setText("hello world");
9.      tv.setTextSize(20);
10.     tv.setTextColor(0xFFFFFFFF);
11.     tv.setBackgroundColor(0xFF0000FF);
12. }
```

2.3.3 EditText

输入框 EditText 是用户和 App 交互的重要控件之一，有了它就等于有了一扇和 App 交互的"门"。通过它，用户可以把数据传给 App，获取想要的信息。

EditText 继承自 android.widget.TextView 类，具有 TextView 的属性特点，表 2.7 所示的是 EditText 控件的常用属性，requestFocus()方法用于获取输入焦点。

表 2.7 EditText 控件常用属性

属性名称	作用描述
android:inputType	设置文本的类型
android:digits	设置允许输入哪些字符
android:hint	设置编辑框内容为空时，显示的提示信息
android:password	设置为密码类型，以小点"."显示文本
android:singleLine	设置文本单行显示
android:editable	设置是否可编辑
android:phoneNumber	设置为电话号码的输入方式
android:ems	设置控件的宽度为 N 个字符的宽度
android:imeOptions	监控软键盘的事件，如 actionNext 表示单击软键盘上的下一项会跳到该组件

2.3.4 Button

Button 是按钮控件，用户可以在该控件上单击触发相应的事件处理函数。Button 也继承自 android.widget.TextView，它的常用子类包括 CheckBox、RadioButton、ToggleButton 等。

下面介绍 Button 的基本使用方法。

1. 添加 Button 控件到 layout 布局，也可通过代码动态添加

在布局文件中设置按钮的属性，如位置、宽高、按钮上显示的文字、颜色和字号等。最重要的属性是 id 值，为按钮添加单击事件监听、动态设置按钮属性时都需要通过 id 实例化对象。示例代码如下：

```
1.  <Button
2.      android:id="@+id/button"
3.      android:layout_margin="10dp"
4.      android:background="#ffffff"
5.      android:gravity="center"
6.      android:text="C"
7.      android:textColor="@android:color/holo_blue_light"
8.      android:textSize="28sp" />
```

通过 Java 代码动态添加的示例代码如下：

```
1.  private void addButton() {
2.      // 获取按钮所在的布局
3.      LinearLayout linearLayout = findViewById(R.id.linear_layout);
4.      // 创建 Button 对象
5.      Button button = new Button(this);
6.      // 设置 Button 对象的文本、大小等属性
7.      button.setText("按钮");
8.      // 获取布局的宽高参数
9.      LinearLayout.LayoutParams params = new LinearLayout.LayoutParams(
10.             LayoutParams.WRAP_CONTENT, LayoutParams.WRAP_CONTENT);
11.     button.setLayoutParams(params);
12.     // 将 Button 对象添加到布局中
13.     linearLayout.addView(button);
14. }
```

2. 处理按钮的单击事件

按钮单击事件有如下的处理方法。

● 第 1 种方法。通过 onClick 属性设置处理单击事件的方法名，并在 Activity 中实现这个方法。在 layout 布局文件中设置 Button 的属性：android:onClick="click"，然后在该布局文件对应的 Acitivity 中实现该方法。

```
1.  public void click(View view) {
2.      // 添加逻辑代码
3.  }
```

● 第 2 种方法。使用 setOnClickListener 添加监听器对象，通过内部匿名类实现 OnClickListener 接口，并重写 onClick()方法，实现单击按钮时需要完成的业务逻辑。

```
1.  Button btnSend = findViewById(R.id.btn_send);
2.  btnSend.setOnClickListener(new View.OnClickListener() {
3.      public void onClick(View v) {
4.          // 添加逻辑代码
5.      }
6.  }
```

● 第 3 种方法。Activity 类实现 OnClickListener 接口，重写 onClick()方法，然后 View 对象调用 setClickListener()添加监听器对象，实现单击按钮时需要完成的业务逻辑。代码结构如下：

```
1.  public class LoginActivity extends AppCompatActivity implements View.OnClickListener{
2.      @Override
3.      protected void onCreate(Bundle savedInstanceState) {
4.          super.onCreate(savedInstanceState);
5.          setContentView(R.layout.activity_login);
6.          // 按钮的事件监听
7.          Button btnSend = findViewById(R.id.btn_send);
8.          btnsend.setOnClickListener(this);
9.      }
10.     @Override
11.     public void onClick(View v) {
12.         // 添加逻辑代码
13.     }
14. }
```

2.3.5 ImageView

ImageView 控件用于展示图片，可以展示的图片分为两类：普通静态图片、gif 格式的动态图片。ImageView 控件的常用属性如表 2.8 所示。

表 2.8 ImageView 控件常用属性

属性名称	属性描述
android:adjustViewBounds	是否保持宽高比，需要与 maxWidth、maxHeight 一起使用，单独使用无效
android:cropToPadding	是否截取指定区域用空白代替，单独设置无效果，需要与 scrollY 一起使用
android:maxHeight	设置 View 最大高度，单独使用无效，需要与 setAdjustViewBounds 一起使用
android:maxWidth	设置 View 最大宽度，单独使用无效，需要与 setAdjustViewBounds 一起使用
android:src	设置 ImageView 中展示图片的资源路径，Android 推荐使用 png 格式图片
android:scaleType	设置图片的填充方式
android:tint	将图片渲染成指定的颜色

android:scaleType 有 8 种取值，最常用的是 fitXY 属性。
- fitXY：图片拉伸或收缩填满 ImageView，不保持比例。
- fitStart：图片按比例扩大、缩小至 ImageView 宽度，在 ImageView 上方显示。
- fitEnd：图片按比例扩大、缩小至 ImageView 宽度，在 ImageView 下方显示。
- fitCenter：图片按比例扩大、缩小至 ImageView 宽度，在 ImageView 居中显示。
- center：以图片和 ImageView 的中心点居中显示图片原尺寸，裁剪超出 ImageView 的部分。
- centerCrop：以图片和 ImageView 的中心点居中显示按比例扩大或缩小的图片，图片的宽高有一边等于 ImageView 的宽高，长出的部分被裁剪。

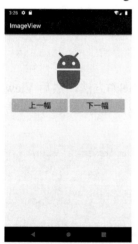

图 2.13 Imageview 实现图片循环浏览

- centerInside：以图片和 ImageView 的中心点居中显示按比例缩小图片，使图片的宽高小于等于 ImageView 的宽高，直到将图片的内容完整居中显示。
- matrix：不改变原图大小，从 ImageView 左上角开始绘制，超过 ImageView 的部分被裁剪。

【案例 2-2】利用 ImageView 控件实现图片循环浏览的功能，当单击"下一幅"按钮时，依次向后浏览不同的图片。当单击"上一幅"按钮时，则依次向前浏览不同的图片，如图 2.13 所示。

步骤 1：创建项目

启动 Android Studio，创建名为 D0202_ImageView 的项目，选择 Empty Activity，将包名改为 com.example.imageview，单击 Finish 按钮，等待系统构建完成。

步骤 2：完成界面布局

将 4 张本地图片复制至 res/drawables 目录中，图片的大小统一为 128px×128px。界面布局使用 LinearLayout，部分代码如下所示：

```
1.    <LinearLayout xmlns:android="http://schemas.android.com/apk/
                            res/android"
2.        android:layout_width="match_parent"
3.        android:layout_height="match_parent"
4.        android:gravity="center_horizontal"
```

```
5.         android:orientation="vertical">
6.     <ImageView
7.         android:id="@+id/iv_pic"
8.         android:layout_width="wrap_content"
9.         android:layout_height="wrap_content"
10.        android:layout_gravity="center_horizontal"
11.        android:layout_marginTop="48dp"
12.        android:src="@drawable/image1" />
13.    <LinearLayout
14.        android:layout_width="match_parent"
15.        android:layout_height="wrap_content"
16.        android:layout_marginTop="20dp"
17.        android:orientation="horizontal">
18.        <Button
19.            android:id="@+id/btn_previous"
20.            android:layout_width="0dp"
21.            android:layout_height="wrap_content"
22.            android:layout_marginStart="20dp"
23.            android:layout_weight="1"
24.            android:text="@string/pre"
25.            android:textSize="24sp" />
26.        <Button
27.            android:id="@+id/btn_next"
28.            android:layout_width="0dp"
29.            android:layout_height="wrap_content"
30.            android:layout_gravity="right"
31.            android:layout_marginEnd="20dp"
32.            android:layout_weight="1"
33.            android:text="@string/next"
34.            android:textSize="24sp" />
35.    </LinearLayout>
36. </LinearLayout>
```

为了实现界面效果，上述布局采用了 LinearLayout 布局嵌套的方法，即在一个垂直方向的线性布局中嵌套一个水平方向的线性布局，这种方法可以使界面布局更加灵活多样，且易于实现，但多层嵌套会影响应用程序运行的性能，2016 年，Google 公司发布了 ConstraintLayout 布局以解决大型复杂界面的嵌套问题。

步骤 2：上一幅、下一幅按钮事件的编码

打开 MainActivity 类文件，先声明 ImageView 对象、Button 对象和图片 id 整型数组，代码如下：

```
1. public class MainActivity extends AppCompatActivity
2.         implements View.OnClickListener {
3.     private ImageView ivPhoto;
4.     private int[] imgIds = new int[]{ R.drawable.image1,
                      R.drawable.image2,
5.           R.drawable.image3, R.drawable.image4};
6.     private int currentIndex = 0;
7.     @Override
8.     protected void onCreate(Bundle savedInstanceState) {...}
9.     @Override
10.    public void onClick(View view) {...}
11. }
```

然后在 onCreate()方法中编写初始化代码，具体包括获取三个控件对象、为按钮对象设置监听器，代码如下：

```
1. protected void onCreate(Bundle savedInstanceState) {
2.     super.onCreate(savedInstanceState);
```

```
3.        setContentView(R.layout.activity_main);
4.        // 获取 ImageView、Button 对象
5.        ivPhoto = findViewById(R.id.iv_pic);
6.        Button btnPrevious = findViewById(R.id.btn_previous);
7.        Button btnNext = findViewById(R.id.btn_next);
8.        // 设置按钮的监听器
9.        btnPrevious.setOnClickListener(this);
10.       btnNext.setOnClickListener(this);
11. }
```

最后重写 OnClickListener 的 onClick()方法，实现当用户单击按钮时依次向前或向后展示图片，按钮单击事件的监听逻辑的代码如下：

```
1.  public void onClick(View view) {
2.       switch (view.getId()) {
3.           case R.id.btn_previous:
4.               currentIndex--;
5.               if (currentIndex < 0) {
6.                   currentIndex = imgIds.length - 1;
7.               }
8.               break;
9.           case R.id.btn_next:
10.              currentIndex++;
11.              if (currentIndex == imgIds.length) {
12.                  currentIndex = 0;
13.              }
14.              break;
15.      }
16.      ivPhoto.setImageResource(imgIds[currentIndex]);
17. }
```

2.3.6 基本控件应用

图 2.14 登录界面

【案例 2-3】使用以上控件完成登录界面的布局及功能，单击"登录"按钮弹出消息提示框显示输入的信息，界面如图 2.14 所示。

步骤 1：创建项目

启动 Android Studio，创建名为 D0203_Login 的项目，选择 No Activity，将包名改为 com.example.login，单击 Finish 按钮，等待系统构建完成。

步骤 2：创建矢量图片

Android Studio 提供一个名为 Vector Asset Studio 工具，帮助添加 Material 图标或可缩放矢量图片，使用矢量图片代替位图不仅可以减小 APK 的大小，而且不会因为手机分辨率的改变而降低图片质量。Vector Asset Studio 会将矢量图片作为描述图片的 XML 文件添加到项目中。

在 res/drawable 目录上单击右键，选择 New->Vector Asset，进入创建 Vector 的界面，如图 2.15 所示。

在此界面中可设置文件名称、选择图标、设置尺寸和颜色等，单击 Clip Art 框会弹出如图 2.16 所示的 Icon 选择框，单击 Color 框会弹出如图 2.17 所示的颜色选择框。设置完成后，单击 Next 按钮就会在 drawable 目录中生成相应名称的 xml 文件，然后在 xml 代码中以@drawable 的形式引用该对象。

图 2.15　创建 Vector 图标的界面

图 2.16　Icon 选择框

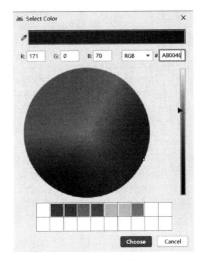

图 2.17　颜色选择框

步骤 3：创建 Activity

Activity 直观理解就是手机屏幕的界面，Activity 的主要作用是将界面呈现出来。Activity 是 Android 系统的四大组件之一，可以用于显示 View 控件。Activity 是一个与用户交互的系统模块，几乎所有的 Activity 都是和用户进行交互的。交互的目的一是显示，二是人机互动。

使用向导创建一个 Empty Activity，生成文件名为 LoginActivity 和 activity_login 的布局文件。

步骤 4：设置界面布局

打开 activity_login.xml 的图形化界面进行界面布局，采用 LinearLayout 线性布局，将 android:orientation 属性设为 vertical，然后从左侧面板中依次拖放 ImageView、EditText、和 Button 控件放到布局中。

设置 ImageView 的 src 属性为 res/mipmap 目录下的 ic_launcher_round 图标，设置 layout_gravity 属性为 center，使图片居中显示，上下间距都设为 96dp，代码如下：

```
1.  <ImageView
2.      android:layout_width="96dp"
3.      android:layout_height="96dp"
4.      android:layout_gravity="center"
5.      android:layout_marginTop="96dp"
6.      android:layout_marginBottom="96dp"
7.      android:src="@mipmap/ic_launcher_round" />
```

用户名和密码使用 EditText 文本框，前端图标通过设置 android:drawableStart 属性实现，设置密码输入框的 inputType 属性为 textPassword，实现密码不可见，设置两个输入框的背景色 android:background 属性值为#EBEBEB，设置 padding、margin 等属性使得 EditText 更美观，具体代码如下：

```
1.  <EditText
2.      android:id="@+id/et_username"
3.      android:layout_width="match_parent"
4.      android:layout_height="wrap_content"
5.      android:layout_marginLeft="48dp"
6.      android:layout_marginRight="48dp"
7.      android:background="#EBEBEB"
8.      android:drawableStart="@drawable/ic_username"
9.      android:drawablePadding="8dp"
10.     android:hint=" 请输入用户名"
11.     android:inputType="text"
12.     android:padding="8dp" />
13. <EditText
14.     android:id="@+id/et_password"
15.     android:layout_width="match_parent"
16.     android:layout_height="wrap_content"
17.     android:layout_marginLeft="48dp"
18.     android:layout_marginTop="8p"
19.     android:layout_marginRight="48dp"
20.     android:background="#EBEBEB"
21.     android:drawableStart="@drawable/ic_password"
22.     android:drawablePadding="8dp"
23.     android:hint=" 请输入密码"
24.     android:inputType="textPassword"
25.     android:maxEms="20"
26.     android:padding="8dp" />
```

两个按钮采用 LinearLayout 的横向布局，设置它们的 android:layout_weight 都为 1，完成横向布局的平分，它们的 android:layout_width 都必须设为 0dp，否则权重无效，代码如下：

```
1.  <LinearLayout
2.      android:layout_width="match_parent"
3.      android:layout_height="wrap_content"
4.      android:layout_marginLeft="48dp"
5.      android:layout_marginTop="48dp"
6.      android:layout_marginRight="48dp"
7.      android:gravity="center"
8.      android:orientation="horizontal">
9.      <Button
10.         android:id="@+id/btn_login"
11.         android:layout_width="0dp"
12.         android:layout_height="wrap_content"
13.         android:layout_weight="1"
14.         android:text="登录"
15.         android:textSize="20sp" />
16.     <Button
17.         android:id="@+id/btn_register"
18.         android:layout_width="0dp"
19.         android:layout_height="wrap_content"
20.         android:layout_weight="1"
21.         android:text="注册"
22.         android:textSize="20sp" />
23. </LinearLayout>
```

步骤 5：编写控件初始化的代码

打开 MainActivity 类文件，重写父类的 onCreate()方法，完成两个输入框对象的初始化，设置登录按钮的事件监听器。onCreate()方法为必须重写的方法，主要工作包括：完成布局界面的显示和控件的初始化和完成单击事件的监听器。

（1）完成布局界面的显示和控件的初始化

```
1.  protected void onCreate(Bundle savedInstanceState) {
2.      super.onCreate(savedInstanceState);
3.      setContentView(R.layout.activity_login);
4.      // 初始化控件对象
5.      etUsername = findViewById(R.id.et_username);
6.      etPassword = findViewById(R.id.et_password);
7.      Button btnLogin = findViewById(R.id.btn_login);
8.  }
```

（2）完成单击事件的监听器

```
1.  // 登录按钮的事件监听
2.  Button btnLogin = findViewById(R.id.btn_login);
3.  btnLogin.setOnClickListener(new View.OnClickListener() {
4.      @Override
5.      public void onClick(View v) {
6.      }
7.  });
```

步骤6：添加"登录"按钮的单击事件处理代码

当用户单击"登录"按钮时，弹出提示框显示用户输入的用户名与密码，其单击事件的处理逻辑为：

```
1.  btnLogin.setOnClickListener(new View.OnClickListener() {
2.      @Override
3.      public void onClick(View view) {
4.          // 获取输入的用户名和密码
5.          String username = etUsername.getText().toString();
6.          String password = etPassword.getText().toString();
7.          // 检查用户名或密码是否为空
8.          if (TextUtils.isEmpty(username) || TextUtils.isEmpty(password)){
9.              Toast.makeText(LoginActivity.this, "用户名或密码不能为空",
10.                     Toast.LENGTH_SHORT).show();
11.         } else {
12.             // 弹框显示输入的用户名和密码字符串
13.             Toast.makeText(LoginActivity.this, "用户名: " + username +
14.                     ", 密码: " + password, Toast.LENGTH_SHORT).show();
15.         }
16.     }
17. });
```

步骤7：运行项目

启动模拟器运行项目，输入用户名和密码，查看弹框显示的用户名和密码是否正确。

在本案例中，首先介绍了创建 App 的基本步骤，然后通过对登录界面的实战演练，要求掌握 TextView、EditText 和 Button 三个基本控件的使用方法。

2.3.7 CheckBox

CheckBox 和 Button 一样，也是一种常见的控件。它是 CompoundButton 的子类，是一个带有选中/未选中状态的按钮，可用于多项选择的场景，也可以用于只有一个选项的情况，例如，注册时是否同意使用协议的选项。

CheckBox 的优点在于无须填写具体信息，只要单击选择框即可，缺点在于只有"选择"和"不选择"两种情况，我们可以利用它的这个特性获取选择的信息。

CheckBox 的关键属性及方法如下。

- android:text：用于设置 CheckBox 控件提示文字。
- android:checked：用于设置此标签的初始状态为选中。
- isChecked()：用于判断按钮是否处于被选中状态。
- setChecked(Boolean flag)：通过传递一个布尔参数设置按钮的状态。

CheckBox 的布局示例代码如下：

```
1.  <CheckBox
2.      android:id="@+id/cb_android"
3.      android:layout_width="wrap_content"
4.      android:layout_height="wrap_content"
5.      android:checked="true"
6.      android:text="Android"
7.      android:textSize="18sp" />
```

改变 CheckBox 的选择状态有以下三种方式。

- 在 XML 布局文件中设置 checked 属性。
- 调用 setChecked()方法动态改变。
- 用户触摸单击事件。

Checkbox 状态的改变会触发 OnCheckedChange 事件，可以使用 OnCheckedChangeListener 监听器监听这个事件。示例代码如下：

```
1.  CheckBox cbAndroid = findViewById(R.id.cb_android);
2.  cbAndroid.setOnCheckedChangeListener(new CompoundButton.
                                   OnCheckedChangeListener(){
3.      @Override
4.      public void onCheckedChanged(CompoundButton buttonView,
                                   boolean isChecked){
5.
6.      }
7.  });
```

2.3.8 RadioButton

RadioButton 控件是一个单选按钮，也是 CompoundButton 的子类，它主要应用于单项选择的场景，与 RadioGroup 控件一起使用实现单选效果。

RadioGroup 是单选组合框，用于将 RadioButton 组合成选择互斥的选项。在没有 RadioGroup 的情况下，RadioButton 可以全部都选中；当多个 RadioButton 被 RadioGroup 包含时，RadioButton 就只能选择一个，实现单选的效果。

RadioButton 和 RadioGroup 在使用过程中需要注意以下几点。

（1）RadioButton 表示单个圆形单选框，可以单独使用；而 RadioGroup 是可以容纳多个 RadioButton 的容器，使 RadioButton 实现单选功能。

（2）RadioGroup 中的 RadioButton 同时只能有一个被选中。

（3）不同 RadioGroup 中的 RadioButton 互不相干，即如果 RadioGroup A 中有一个选中，RadioGroup B 中依然可以有一个被选中。

（4）通常一个 RadioGroup 中至少有两个及以上的 RadioButton。

（5）一般一个 RadioGroup 中的 RadioButton 默认会有一个被选中，通常建议将它放在 RadioGroup 的起始位置。

RadioGroup 和 RadioButton 的布局示例代码如下：

```
1.  <RadioGroup
2.      android:id="@+id/rg_sex"
3.      android:layout_width="match_parent"
4.      android:layout_height="wrap_content"
5.      android:orientation="horizontal">
6.      <RadioButton
7.          android:id="@+id/rb_male"
8.          android:layout_width="wrap_content"
9.          android:layout_height="wrap_content"
10.         android:layout_marginStart="16dp"
11.         android:checked="true"
12.         android:text="男"
13.         android:textSize="18sp" />
14.     <RadioButton
15.         android:id="@+id/rb_female"
16.         android:layout_width="wrap_content"
17.         android:layout_height="wrap_content"
18.         android:layout_marginStart="16dp"
19.         android:text="女"
20.         android:textSize="18sp" />
21. </RadioGroup>
```

RadioButton 控件的常用事件也是 OnCheckedChange，当选项发生变化时触发该事件。该事件需实现 RadioGroup 的 OnCheckedChangedListener 接口，重写回调方法 onCheckedChanged()，设置监听对象。实际开发中一般使用 RadioGroup 的 getCheckedRadioButtonId()方法来获取 RadioGroup 中具体是哪一个 RadioButton 被选中。

RadioButton 和 CheckBox 都属于 CompoundButton 的子类，它们之间的区别如表 2.9 所示。

表 2.9 RadioButton 和 CheckBox 的区别

区别点	RadioButton	CheckBox
选中后再次单击	无法变为未选中的状态	可以变为未选中的状态
选择项	同时只能选中一个	同时能选中多个
图标形状	默认用圆形表示	默认用矩形表示

2.3.9　Snackbar

Snackbar 是 Android 5.0 推出的 Material Design 控件，用于替代 Toast 在界面下方弹出提示信息。它与 Toast 的主要区别是：Snackbar 可以与用户交互，允许用户滑动退出，也可以通过设定 Action 处理用户的交互或单击事件。另外，Toast 是悬浮在包括键盘在内的所有布局之上的，而 Snackbar 是在 View 上直接调用 addView()方法显示的，因此在调用 Snackbar.show()之前，要注意先调用 Keyboard.hide()隐藏键盘，不然键盘就会遮住 Snackbar 的提示信息。

Snackbar 的特性包括：
- Snackbar 在超时或者用户在屏幕的任何地方触摸后会自动消失。
- Snackbar 可以在屏幕上滑动关闭。
- Snackbar 显示时不会阻碍用户在屏幕上输入。

- 屏幕上最多只能显示一个 Snackbar。
- 可以在 Snackbar 中添加一个按钮，处理用户单击事件。
- Snackbar 一般使用 CoordinatorLayout 作为父容器，CoordinatorLayout 保证 Snackbar 可以右滑退出，CoordinatorLayout 在后续章节中详细讲解。

Snackbar 的语法规则如下所示：

```
1. Snackbar.make(view, message, duration)
2.     .setAction(action message, click listener).show();
```

Snackbar 的常用方法如表 2.10 所示。

表 2.10 Snackbar 的常用方法说明

方法名称	方法描述
make()	构造 SnackBar，有三个参数：父容器、提示信息和持续时间。需要注意的是，该方法的第一个参数 view 不能是 ScrollView，因为 SnackBar 的实现逻辑是向这个 View 添加 View，而 ScrollView 只能有一个 Child，会报异常
setAction(CharSequence, View.OnClickListener)	给 SnackBar 设定一个 action，单击后会回调 OnclickListener 的 OnClick()方法，处理相应的逻辑
show()	显示 SnackBar
setActionTextColor()	设置 Action 的字体颜色

Snackbar 的示例代码如下：

```
1. Snackbar.make(view, "是否撤销删除？", Snackbar.LENGTH_LONG)
2.     .setAction("是的", new View.OnClickListener() {
3.         @Override
4.         public void onClick(View v) {
5.             Toast.makeText(MainActivity.this, "已取消删除！",
6.                 Toast.LENGTH_SHORT).show();
7.         }
8.     }).show();
```

2.3.10 TextInputLayout

Google 推出的 Material Design 对文本输入的界面效果进行了全面升级，大大改善了文本输入的用户体验。TextInputLayout 是继承自 LinearLayout 的新布局，使用时只能包含一个 EditText 或者其子类的控件，也就是 EditText 的容器，通过设置 hint 和 Error 显示浮动标签，hint 会随着 EditText 获取或失去焦点而浮动，可以伴随动画。

在使用过程中，TextInputLayout 可以看成是 EditText 加装了一层外壳，具体使用步骤如下。

（1）添加 TextInputLayout 控件所需要的依赖，以 Android Studio 4.1 为例：

```
1. dependencies {
2.     implementation 'com.google.android.material:material:1.2.1'
3. }
```

（2）在布局文件中先添加 TextInputLayout，然后在其中添加 EditText：

```
1. <LinearLayout xmlns:android="http://schemas.android.com/apk/
                                res/android"
2.     android:layout_width="match_parent"
3.     android:layout_height="match_parent"
4.     android:orientation="vertical">
```

```
5.     <com.google.android.material.textfield.TextInputLayout
6.         android:layout_width="match_parent"
7.         android:layout_height="wrap_content">
8.         <EditText android:id="@+id/et_name"
9.             android:layout_width="match_parent"
10.            android:layout_height="wrap_content"
11.            android:saveEnabled="false"
12.            android:maxLength="48"
13.            android:hint="提示文字"/>
14.    </com.google.android.material.textfield.TextInputLayout>
15. </LinearLayout>
```

TextViewInputLayout 的常用属性描述如表 2.11 所示。

表 2.11　TextViewInputLayout 的常用属性描述

属性名称	属性描述
app:hintAnimationEnabled	是否显示 hint 的动画，默认为 true
app:hintEnabled	是否使用 hint 属性，默认为 true
app:counterEnabled	是否显示计数器，默认为 false
app:counterMaxLength	设置计数器的最大值，与 counterEnabled 同时使用
app:counterTextAppearance	计数器的字体样式
app:counterOverflowTextAppearance	输入字符大于限定个数时的字体样式
app:passwordToggleEnabled	是否显示密码开关图片，需要 EditText 设置 inputType
app:passwordToggleDrawable	设置密码开关图片，与 passwordToggleEnabled 同时使用
app:errorEnabled	是否显示错误信息
app:errorTextAppearance	错误信息的字体样式
app:hintTextAppearance	设置 hint 的文字样式（指运行动画效果之后的样式）

我们将在接下来的综合案例中使用它。

2.3.11　控件综合应用

【案例 2-4】综合应用 ImageView、CheckBox、RadioButton 和 TextInputLayout 等控件设计如图 2.18 所示的个人信息维护界面。

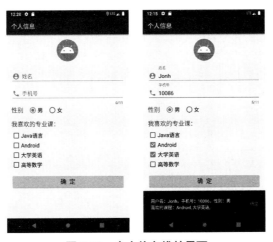

图 2.18　个人信息维护界面

步骤 1：创建项目

启动 Android Studio，创建名为 D0204_Info 的项目，选择 Empty Activity，将包名改为 com.example.info，单击 Finish 按钮，等待系统构建完成。

步骤 2：添加头像布局

打开 activity_main.xml 布局文件，修改默认布局为垂直方向 LinearLayout，然后添加头像，并使其居中，代码如下所示：

```xml
1.  <LinearLayout xmlns:android="http://schemas.android.com/apk/res/android"
2.      android:id="@+id/ll_main"
3.      android:layout_width="match_parent"
4.      android:layout_height="match_parent"
5.      android:orientation="vertical">
6.      <ImageView
7.          android:layout_width="72dp"
8.          android:layout_height="72dp"
9.          android:layout_gravity="center"
10.         android:layout_marginTop="24dp"
11.         android:layout_marginBottom="24dp"
12.         android:src="@mipmap/ic_launcher_round" />
13. </LinearLayout>
```

步骤 3：添加姓名和手机号的布局

在头像布局之后，添加姓名和手机号的布局，直接使用 TextInputLayout 布局内嵌 EditText，代码如下：

```xml
1.  <com.google.android.material.textfield.TextInputLayout
2.      android:layout_width="match_parent"
3.      android:layout_height="wrap_content"
4.      android:layout_marginStart="16dp"
5.      android:layout_marginEnd="16dp">
6.      <EditText
7.          android:id="@+id/et_name"
8.          android:layout_width="match_parent"
9.          android:layout_height="wrap_content"
10.         android:drawableStart="@drawable/ic_username"
11.         android:drawablePadding="8dp"
12.         android:hint="姓名"
13.         android:imeOptions="actionNext"
14.         android:inputType="text" />
15. </com.google.android.material.textfield.TextInputLayout>
16. <com.google.android.material.textfield.TextInputLayout
17.     android:layout_width="match_parent"
18.     android:layout_height="wrap_content"
19.     android:layout_marginStart="16dp"
20.     android:layout_marginEnd="16dp"
21.     app:counterEnabled="true"
22.     app:counterMaxLength="11">
23.     <EditText
24.         android:id="@+id/et_phone"
25.         android:layout_width="match_parent"
26.         android:layout_height="wrap_content"
27.         android:drawableStart="@drawable/ic_phone"
28.         android:drawablePadding="8dp"
29.         android:hint="手机号"
30.         android:inputType="phone" />
31. </com.google.android.material.textfield.TextInputLayout>
```

第 12、29 行分别设置 Android:hint 提示字符串的属性；第 10、27 行的 android:drawableStart

属性用于设置输入框左边的图标,图标是使用 Vector Asset 创建的矢量图;第 21~22 行的 app:counterEnabled 和 app:counterMaxLength 属性用于设置字符计数标志;第 30 行的 inputType 属性设为 phone,当输入手机号时会弹出数字软键盘,提升用户的体验。

步骤 4:添加用户性别

使用嵌套一个横向排列的 LinearLayout 布局,添加性别的标签及性别的单选按钮,代码如下:

```xml
1.  <LinearLayout
2.      android:layout_width="match_parent"
3.      android:layout_height="wrap_content"
4.      android:layout_marginStart="16dp">
5.      <TextView
6.          android:layout_width="wrap_content"
7.          android:layout_height="wrap_content"
8.          android:layout_gravity="center_vertical"
9.          android:gravity="center"
10.         android:text="性别"
11.         android:textSize="20sp" />
12.     <RadioGroup
13.         android:id="@+id/rg_sex"
14.         android:layout_width="match_parent"
15.         android:layout_height="wrap_content"
16.         android:orientation="horizontal">
17.         <RadioButton
18.             android:id="@+id/rb_male"
19.             android:layout_width="wrap_content"
20.             android:layout_height="wrap_content"
21.             android:layout_marginStart="16dp"
22.             android:checked="true"
23.             android:text="男"
24.             android:textSize="18sp" />
25.         <RadioButton
26.             android:id="@+id/rb_female"
27.             android:layout_width="wrap_content"
28.             android:layout_height="wrap_content"
29.             android:layout_marginStart="16dp"
30.             android:text="女"
31.             android:textSize="18sp" />
32.     </RadioGroup>
33. </LinearLayout>
```

使用 RadioGroup 组合两个 RadioButton,使其互斥,RadioGroup 默认的是纵向布局,所以需要设置它的 orientation 属性为 horizontal,使得两个 RadioButton 为横向排列。

步骤 5:添加用户喜欢的专业课列表

使用嵌套一个纵向排列的 LinearLayout 布局,添加专业课的文字说明及多选按钮,代码如下:

```xml
1.  <LinearLayout
2.      android:layout_width="match_parent"
3.      android:layout_height="wrap_content"
4.      android:layout_marginStart="16dp"
5.      android:layout_marginTop="16dp"
6.      android:orientation="vertical">
7.      <TextView
8.          android:layout_width="wrap_content"
9.          android:layout_height="wrap_content"
10.         android:layout_marginBottom="8dp"
```

```
11.        android:text="我喜欢的专业课："
12.        android:textSize="20sp" />
13.    <CheckBox
14.        android:id="@+id/cb_java"
15.        android:layout_width="wrap_content"
16.        android:layout_height="wrap_content"
17.        android:text="Java 语言"
18.        android:textSize="18sp" />
19.    <CheckBox
20.        android:id="@+id/cb_android"
21.        android:layout_width="wrap_content"
22.        android:layout_height="wrap_content"
23.        android:text="Android"
24.        android:textSize="18sp" />
25.    <CheckBox
26.        android:id="@+id/cb_english"
27.        android:layout_width="wrap_content"
28.        android:layout_height="wrap_content"
29.        android:text="大学英语"
30.        android:textSize="18sp" />
31.    <CheckBox
32.        android:id="@+id/cb_math"
33.        android:layout_width="wrap_content"
34.        android:layout_height="wrap_content"
35.        android:text="高等数学"
36.        android:textSize="18sp" />
37. </LinearLayout>
```

此布局代码相对简单，不再一一说明。

步骤 6：编写初始化控件的功能代码

打开 MainActivity 类，首先定义各个控件对象并初始化，代码如下：

```
1.  private LinearLayout mainLayout;
2.  private EditText etUsername, etPhone;
3.  private RadioGroup rgSex;
4.  // CheckBox 选中的文本字符串
5.  private String selected = "";
6.  @Override
7.  protected void onCreate(Bundle savedInstanceState) {
8.      super.onCreate(savedInstanceState);
9.      setContentView(R.layout.activity_main);
10.     // 初始化布局对象
11.     mainLayout = findViewById(R.id.ll_main);
12.     // 初始化输入框、单选控件对象
13.     etUsername = findViewById(R.id.et_name);
14.     etPhone = findViewById(R.id.et_phone);
15.     rgSex = findViewById(R.id.rg_sex);
16.     // 初始化复选按钮控件
17.     CheckBox cbAndroid = findViewById(R.id.cb_android);
18.     CheckBox cbJava = findViewById(R.id.cb_java);
19.     CheckBox cbEnglish = findViewById(R.id.cb_english);
20.     CheckBox cbMath = findViewById(R.id.cb_math);
21.     // 获取按钮对象，设置它的单击事件监听器
22.     Button btnConfirm = findViewById(R.id.btn_confirm);
23. }
```

步骤 7：添加事件监听器的功能代码

在 MainActivity 类的 onCreate()方法中添加 CheckBox 的 OnChekedChange 事件和

Button 的 OnClick 事件的监听器。本案例采用 MainActivity 类实现 View.OnClickListener 和 CompoundButton.OnCheckedChangeListener 接口方式设置事件监听器，因此需要重写 onClick() 和 onCheckedChanged() 方法，代码如下：

```java
1.  public class MainActivity extends AppCompatActivity
2.              implements View.OnClickListener, CompoundButton.
                OnCheckedChangeListener {
3.      // 定义成员变量...
4.
5.      @Override
6.      protected void onCreate(Bundle savedInstanceState) {
7.          super.onCreate(savedInstanceState);
8.          setContentView(R.layout.activity_main);
9.          // 初始化布局对象...
10.         // 设置事件监听器
11.         cbJava.setOnCheckedChangeListener(this);
12.         cbAndroid.setOnCheckedChangeListener(this);
13.         cbEnglish.setOnCheckedChangeListener(this);
14.         cbMath.setOnCheckedChangeListener(this);
15.         btnConfirm.setOnClickListener(this);
16.     }
17.     @Override
18.     public void onCheckedChanged(CompoundButton buttonView,
         boolean isChecked) {
19.     }
20.     @Override
21.     public void onClick(View v) {
22.     }
23. }
```

步骤 8：编写 onCheckedChanged() 的功能代码

当用户选中或取消选中都会触发 OnCheckedChange 事件，通过 onCheckedChanged() 方法进行处理，使用 Snackbar 实时显示当前所选课程的名称，其具体逻辑如下：

```java
1.  public void onCheckedChanged(CompoundButton buttonView, boolean
     isChecked){
2.      CheckBox checkbox = (CheckBox) buttonView;
3.      if (isChecked) {
4.          selected += checkbox.getText().toString() + ", ";
5.      } else {
6.          selected = selected.replace(checkbox.getText().
             toString() + ", ", "");
7.      }
8.      Snackbar.make(mainLayout, selected, Snackbar.LENGTH_LONG).show();
9.  }
```

onCheckedChanged() 方法中，第 2 行将 buttonView 对象强制转换为 CheckBox 类型，然后根据 isChecked 的值为 true 或 false，判断选项是否被选中，如果选中，第 4 行代码将 CheckBox 对象的文本添加到 selected 字符串的后面，若取消选中，第 6 行代码则调用 String 的 replace() 方法删除 CheckBox 对象的文本，实现根据选项的选中与否动态更新选课的文本内容。

步骤 9：编写 onClick() 的功能代码

单击"确定"按钮，获取输入的姓名和手机号的数据。第 7 行代码通过 RadioGroup 的 getCheckedRadioButtonId() 方法获取 RadioButton 的 id，然后获取 RadioButton 按钮的文本值；第 12～13 行将所有的数据组合成一个字符串，然后使用 Snackbar 显示这个字符串。代码如下：

```
1.  public void onClick(View v) {
2.      // 获取输入的值
3.      String username = etUsername.getText().toString().trim();
4.      String phone = etPhone.getText().toString().trim();
5.      // 获取 RadioButton 选项的值
6.      String sex = "男";
7.      int id = rgSex.getCheckedRadioButtonId();
8.      If (id == R.id.rb_female) {
9.          sex = "女";
10.     }
11.     // 将数据组合成字符串
12.     String info = "用户名：" + username + "，手机号：" + phone
13.             + "，性别：" + sex + "\n 喜欢的课程：" + selected;
14.     // 使用 Snackbar 显示信息
15.     Snackbar.make(mainLayout, info, Snackbar.LENGTH_LONG)
16.             .setAction("确定", new View.OnClickListener() {
17.                 @Override
18.                 public void onClick(View v) {
19.                     Toast.makeText(MainActivity.this,
20.                             "信息已确认", Toast.LENGTH_SHORT).show();
21.                 }
22.             }).show();
23. }
```

步骤 10：运行项目

启动模拟器，运行项目，检查选项切换使得多选的文本内容的变化，单击"确定"按钮查看输出的数据是否正确。

2.3.12 视图绑定

从以上的代码中可以得知，获取布局控件对象需要使用 findViewById()方法，参数为布局文件中的 android:id 属性值，但从 Android Studio 3.6 开始，Android 提供了视图绑定（View Binding）功能，可以更轻松地编写与 View 交互的代码。在模块中启用视图绑定之后，系统会为该模块中的每个 XML 布局文件生成一个绑定类,绑定类的实例包含的布局控件就可以直接引用。视图绑定的步骤如下。

（1）启用视图绑定

在 app/build.gradle 文件中添加 viewBinding 元素，Android Studio 4.0 以下版本的代码如下：

```
1.  android {
2.      ...
3.      viewBinding {
4.          Enabled true
5.      }
6.  }
```

Android Studio 4.0 及以上版本的代码如下：

```
1.  android {
2.      ...
3.      buildFeatures {
4.          viewBinding true
5.      }
6.  }
```

（2）设置布局控件的 Android:id 属性值

启用视图绑定后，系统会为该模块中的每个 XML 布局文件生成绑定类，绑定类名称的命名是将 XML 的名称转为首字母大写的驼峰式，并在末尾添加"Binding"，如 activity_main.xml 的绑定类的名称为 ActivityMainBinding。

（3）Activity 类使用视图绑定

在 Activity 类的 onCreate()方法中调用生成的绑定类的 inflate()方法创建实例，然后调用 getRoot()方法获取对根视图的引用，将它传递给 setContentView()完成绑定。示例代码如下：

```
1.  private ActivityMainBinding binding;
2.  @Override
3.  protected void onCreate(Bundle savedInstanceState) {
4.      super.onCreate(savedInstanceState);
5.      binding = ActivityMainBinding.inflate(getLayoutInflater());
6.      View view = binding.getRoot();
7.      setContentView(view);
8.  }
```

（4）引用布局文件中的控件

使用定义的 binding 对象引用控件对象，布局定义及类引用的代码如下：

```
1.  <Button
2.      android:id="@+id/btn_login"
3.      android:layout_width="wrap_content"
4.      android:layout_height="wrap_content" />
```

```
1.  binding.btnLogin.setOnClickListener(new View.OnClickListener() {
2.      @Override
3.      public void onClick(View v) {
4.          Toast.makeText(MainActivity.this, "登录", Toast.LENGTH_
                      SHORT).show();
5.      }
6.  });
```

View Binding 的处理不仅更简单，而且与 findViewById 相比，还具备以下一些显著的优点。

- Null 安全：视图绑定是创建对 View 视图的直接引用，因此不存在因视图 id 无效而引发 Null 指针异常的风险。
- 类型安全：每个绑定类中的字段均具有与它们在 XML 布局文件中引用的 View 相匹配的类型，因此不存在发生类转换异常的风险。

2.4 Notification

人们在使用手机的过程中，经常会收到电池电量低、短信通知等各种类型的消息提醒，它们通常以小图标的方式显示在手机状态栏的通知区域，也有一些是显示在应用程序图标上的圆点提示，如微信、QQ 等。Android 使用 Notification 实现消息提醒，提供通知栏和通知抽屉查看通知及通知详情。Android 的 Notification 机制应用非常广泛，如系统事件提醒、App

更新通知、新闻类 App 的推送、社交类 App 的及时消息等。通知一般在广播或服务中创建，实现后台服务等的提醒，在 Activity 中创建通知相对较少，此处借助 Activity 创建通知，学习广播和服务之后可以自行扩展。

2.4.1 Notification 简介

Notification 俗称通知，是指在应用程序之外显示的消息，其作用是以一种醒目的方式提醒用户。Android 的 Notification 的布局由系统模板决定，结构如图 2.19 所示，开发者只需要定义模板中各部分的内容，通知详情就会在点按后展开视图显示设置的内容。

图 2.19　包含详情的通知结构

图中展示了通知最常见的结构，具体含义为：
① 小图标，必须提供，调用 setSmallIcon()设置。
② 应用名称，由系统提供。
③ 时间戳，由系统提供，可以调用 setWhen()替换或者调用 setShowWhen（false）隐藏。
④ 大图标，可选内容，调用 setLargeIcon()进行设置。
⑤ 消息标题，可选内容，调用 setContentTitle()设置。
⑥ 消息文本，可选内容，调用 setContentText()设置。

从 Android 4.1 到 Android 8.0，Notification 也经历了一系列的行为变化，展现形式也呈现多样化，目前大体可分为状态栏图标通知、悬挂提醒式通知、锁屏通知及圆点通知，如表 2.12 所示。

表 2.12　不同 Android 版本的通知功能变化

Android 版本	通知的功能变化
Android 4.1	引入展开式通知模板，向上或向下滑动即可展开
Android 5.0	引入锁屏和提醒式通知，可以调用 setPriority()方法设置通知的优先级
Android 7.0	引入直接在通知内进行回复，可以重新设置通知模板的样式
Android 8.0	引入通知通道和圆点通知，可以按通道关闭通知、设置通知背景色及自动超时等

从 Android 8.0 开始，必须为所有通知分配一个或多个通道，否则通知将不会显示。通过将通知归类为不同的通道，用户就可以针对通道进行通知的停用或设置，如微信的通知通道及通知设置如图 2.20、图 2.21 所示。Android 利用通知的重要程度决定提醒用户的程度，重要程度越高，提醒力度越大，Android 8.0 及以上版本通过通道的 importance 属性设置重要性，Android 8.0 以下版本每条通知通过调用 setPriority()方法进行设置，重要程度等级设置如表 2.13 所示。

图 2.20 微信的通知通道设置　　　　　图 2.21 微信的通知设置

表 2.13　不同版本 Android 通知的重要性

用户可见的重要性级别	重要性（Android 8.0 及更高版本）	优先级（Android 7.1 及更低版本）
紧急，发出提示音并以浮动通知显示	IMPORTANCE_HIGH	PRIORITY_HIGH 或 PRIORITY_MAX
高，发出提示音	IMPORTANCE_DEFAULT	PRIORITY_DEFAULT
中，不发出提示音	IMPORTANCE_LOW	PRIORITY_LOW
低，不发出提示音且不在状态栏显示	IMPORTANCE_MIN	PRIORITY_MIN

2.4.2　Notification 实现

Android 系统提供了 Notification 类实现通知功能，官方推荐使用 Notifaction 的内部构造器类 Builder 创建 Notification 对象，Notification 类封装了通知的标题、内容、时间、小图标等信息。

创建 Notification 的基本步骤如下：

① 使用 getSystemService()获取 NotificationManager 管理类。
② 使用 Notification 类创建通知对象，设置标题、内容、图片等样式。
③ 调用 NotificationManager 的 notify()调出通知栏，显示通知。
④ Android 8.0 及以上版本，需要创建通知通道对通知进行管理。

Notification 类常用的方法及说明如表 2.14 所示。

表 2.14　Notification 类的常用方法

方法定义	功能说明
setSmallIcon(long)	设置通知小图标，必选
setContentTitle(String)	设置通知标题，可选
setContentText(String)	设置通知内容，可选
setWhen(long)	设置通知时间，默认为系统发出通知的时间，一般不用设置
setSound(Uri)	设置自定义音乐

续表

方法定义	功能说明
setLargeIcon(Bitmap)	设置通知栏下拉后的大图标
setContentIntent(PendingIntent)	设置单击通知后跳转的 Activity，即延迟 Intent 实现
setAutoCancel(boolean)	设置单击通知后是否消失，true 为消失
setStyle(Notification.Style)	设置通知的样式，如大文本样式、大视图样式
setPriority(int)	设置通知优先级，用于 Android 8.0 以下版本
setDefault(int)	设置通知的 LED 灯、音乐、震动等默认设置

通知不仅有如图 2.19 所示的基本样式，官方还提供了多种样式的展开式通知，如大图片、大段文本、收件箱样式、对话样式和媒体控制样式等，甚至还能自定义通知视图及直接回复等。下面通过案例讲解几种常用的通知的创建过程。

【案例 2-5】创建通知，包含基本样式通知、展开式通知和自定义布局通知，界面如图 2.22 所示。

步骤 1：创建项目

启动 Android Studio，创建名为 D0205_notification 的项目，选择 Empty Activity，将包名改为 com.example.notification，单击 Finish 按钮，等待系统构建完成。

步骤 2：设计主界面

打开 activity_main.xml 文件，使用 Spinner 控件列出如图 2.22 所示的通知样式。

图 2.22　主界面布局

步骤 3：初始化界面控件对象

打开 MainActivity 类文件，首先定义一些数据常量，代码如下：

```
1.  // 通知 Id
2.  private final static int NOTIFIATION_ID = 100;
3.  // 通知列表
4.  private final static String ID_BASIC = "basic";
5.  private final static String ID_SUBSCRIBE = "subscribe";
6.  // Spinner 列表数据
7.  private static final String BASIC_STYLE = "基本样式";
8.  private static final String FULL_STYLE = "悬浮样式";
9.  private static final String CUSTOM_STYLE = "自定义View";
10. private static final String BIG_TEXT_STYLE = "长文本样式";
11. private static final String BIG_PICTURE_STYLE = "大图样式";
12. private static final String INBOX_STYLE = "收件箱样式";
13. // Spinner 的数据源数组
14. private static final String[] NOTIFICATION_STYLES = {
15.      "--请选择--", BASIC_STYLE, FULL_STYLE, CUSTOM_STYLE,
16.      BIG_TEXT_STYLE, BIG_PICTURE_STYLE, INBOX_STYLE
17. };
18. // TextView 的描述数据
19. private static final String[] NOTIFICATION_STYLES_DESC = {
20.      "",
21.      "基本样式的通知",
22.      "高优先级的悬浮样式的通知",
23.      "自定义View的通知",
```

```
24.         "大文本样式的通知",
25.         "大图样式的通知",
26.         "收件箱样式的通知"
27. };
```

然后，在 onCreate()方法中初始化 Spinner、TextView 控件，为 Spinner 控件加载列表数据。创建 Spinner 的数据，使用 ArrayAdapter 适配器加载数据，设置监听器，代码如下：

```
1.  @Override
2.  protected void onCreate(Bundle savedInstanceState) {
3.      super.onCreate(savedInstanceState);
4.      setContentView(R.layout.activity_main);
5.      // 初始化控件
6.      tvDesc = findViewById(R.id.tv_description);
7.      Spinner spinner = findViewById(R.id.sp_style);
8.      // 设置 Spinner 适配器
9.      ArrayAdapter<String> adapter = new ArrayAdapter<>(this,
10.             android.R.layout.simple_spinner_item, NOTIFICATION_STYLES);
11.     // 设置 Spinner 的下拉样式
12.     adapter.setDropDownViewResource(android.R.layout.simple_spinner_
                                        dropdown_item);
13.     spinner.setAdapter(adapter);
14.     // 设置 Spinner 选项监听器
15.     spinner.setOnItemSelectedListener(this);
16. }
```

步骤 4：创建通知通道

通知通道是 Android 8.0 引入的新特性，目的是将 Notification 进行更为细化的分类管理。首先获取 NotificationManager 对象，然后创建 NotificationChannel 对象，设置通道的 id、name 和 importance 三个属性，通过 NotificationManager 对象在系统中注册应用的通知通道，才能在 Android 8.0 及更高版本上发送通知。创建通道的方法的代码如下：

```
1.  @TargetApi(Build.VERSION_CODES.O)
2.  private void createNotificationChannel(String channelId,
3.                                         String channelName, int
                                           importance) {
4.      NotificationChannel channel = new NotificationChannel(channelId,
5.          channelName, importance);
6.      NotificationManager manager = (NotificationManager)
7.          getSystemService(NOTIFICATION_SERVICE);
8.      manager.createNotificationChannel(channel);
9.  }
```

然后，打开 MainActivity 类文件，在 onCreate()中添加创建通知两个通道，分别是"订阅消息"和"一般消息"，设置重要性分别为 IMPORTANCE_HIGH 和 IMPORTANCE_DEFAULT，代码如下：

```
1.  @Override
2.  protected void onCreate(Bundle savedInstanceState) {
3.      ...
4.      // Android 8.0 创建两种通知通道
5.      if (Build.VERSION.SDK_INT >= Build.VERSION_CODES.O) {
6.          String channelId = ID_BASIC;
7.          String channelName = "一般消息";
8.          int importance = NotificationManager.IMPORTANCE_DEFAULT;
9.          createNotificationChannel(channelId, channelName, importance);
10.
11.         channelId = ID_SUBSCRIBE;
```

```
12.        channelName = "订阅消息";
13.        importance = NotificationManager.IMPORTANCE_HIGH;
14.        createNotificationChannel(channelId, channelName, importance);
15.    }
16. }
```

步骤 5：实现 Spinner 的 OnItemSelected 事件

在 Spinner 类的 onItemSelected()事件处理方法中，根据选项调用显示通知的方法，代码如下：

```
1.  @Override
2.  public void onItemSelected(AdapterView<?> parent, View view,
                               int position, long id) {
3.      tvDesc.setText(NOTIFICATION_STYLES_DESC[position]);
4.      switch (NOTIFICATION_STYLES[position]) {
5.          case BASIC_STYLE:
6.              showBasicNotification();
7.              break;
8.          default:
9.              // continue below
10.     }
11. }
```

步骤 6：创建 Notification 并发送

首先创建一个小图标、标题、内容和大图标等基本样式的通知，第 8～17 行调用 NotificationCompat 的 Builder 对象构建通知，Android 8.0 版本之后使用 Builder 构造通知对象时都需要增加通知通道 id，最后调用 build()方法获取 Notification 对象。代码如下：

```
1.  public void showBasicNotification() {
2.      // 定义单击通知启动 Activity 的 Intent
3.      Intent intent = new Intent(this, MainActivity.class);
4.      intent.setFlags(Intent.FLAG_ACTIVITY_NEW_TASK | Intent.FLAG_
                        ACTIVITY_CLEAR_TASK);
5.      PendingIntent pendingIntent = PendingIntent.getActivity(this, 0,
                             intent,
6.              PendingIntent.FLAG_CANCEL_CURRENT);
7.      // 创建通知
8.      Notification notification = new NotificationCompat.Builder
                              (this, ID_BASIC)
9.              .setSmallIcon(R.drawable.ic_message)
10.             .setContentTitle("会议时间")
11.             .setContentText("研讨会于今天下午1:30开始")
12.             .setWhen(System.currentTimeMillis())
13.             .setContentIntent(pendingIntent)
14.             .setAutoCancel(true)
15.             .setLargeIcon(BitmapFactory.decodeResource(getResources(),
                        R.drawable.pic))
16.             .setDefaults(Notification.DEFAULT_VIBRATE | Notification.
                        DEFAULT_SOUND)
17.             .build();
18.     // 发送通知
19.     NotificationManager notificationManager = (NotificationManager)
20.             getSystemService(NOTIFICATION_SERVICE);
21.     notificationManager.notify(NOTIFIATION_ID, notification);
22. }
```

当单击通知时，通知应做出某种响应，通常是打开相应于该通知的 Activity，使用 setContentIntent()方法实现，该方法的参数是 PendingIntent 对象，第 3～6 行代码创建了

PendingInteng 对象。PendingIntent 是对 Intent 的封装，意为即将发生的 Intent，即为当某些特定的条件发生后，执行指定的行为，它与 Intent 之间的区别将在下小节详细描述。Notifiaction 创建后调用 NotificationManager 对象的 notify()方法显示通知。

步骤 7：运行项目

完成以上基本样式通知的创建后，运行程序，单击 Spinner 的"基本样式"选项，得到的运行结果如图 2.23、图 2.24 所示，单击图 2.24 中的通知会跳回 Activity 界面，同时通知会消失。

图 2.23 Spinner 选择后的界面

图 2.24 基本样式通知界面

接下来描述不同的通知样式的实现。

（1）高优先级的悬浮通知

需要立即引起用户注意的通知，可以通过创建高优先级的通知来实现，通知发出后会悬浮在状态栏，直到用户单击才会消失。不同 Android 版本的优先级设置会有所不同，参考表 2.13 的说明，Android 8.0 及以上版本通过创建通知通道时设置 importance 为 IMPORTANCE_HIGH 实现，还需要通过调用 setFullScreenIntent（PendingIntent）实现悬浮样式；Android 8.0 以下版本则通过 setPriority（NotificationCompat.PRIORITY_HIGH）方法实现。代码如下：

```
1.  Notification notification = new NotificationCompat.Builder(this, ID_HIGH)
2.          .setSmallIcon(R.drawable.ic_importance)
3.          .setLargeIcon(BitmapFactory.decodeResource(getResources(),
                    R.drawable.pic))
4.          .setAutoCancel(true)
5.          .setContentTitle("重要消息")
6.          .setContentText("紧急通知：计算机等级考试本周末开考！")
7.          .setFullScreenIntent(pendingIntent, true)
8.          .build();
```

运行项目，选择 Spinner 的"悬浮样式"选项的运行效果如图 2.25 所示，本应用程序的通知通道如图 2.26 所示。

图 2.25　悬浮样式通知　　　　　　图 2.26　项目设置的通知通道

（2）展开式通知

展开式通知通过下滑手势将通知展开，直接获取通知的更多信息，Android8.0 及以上版本提供了多种可展开的通知样式，通过调用 setStyle()方法进行设置，展开为大图片、大文本和收件箱样式的代码如下：

```
1.  // 大文本
2.  Notification notification = new NotificationCompat.Builder(this, ID_BASIC)
3.          ...
4.          .setStyle(new NotificationCompat.BigTextStyle().bigText(msg))
5.          .build();
6.  // 大图片
7.  Notification notification = new NotificationCompat.Builder(this, ID_BASIC)
8.          ...
9.          .setStyle(new NotificationCompat.BigPictureStyle()
10.                 .bigPicture(bitmap).bigLargeIcon(null))
11.         .build();
12. // 收件箱
13. Notification notification = new NotificationCompat.Builder(this, ID_BASIC)
14.         ...
15.         .setStyle(new NotificationCompat.InboxStyle()
16.              .addLine("可以应用 NotificationCompat.InboxStyle 添加多个
                    简短摘要行。")
17.              .addLine("可以添加多条内容文本,并且每条文本均截断为一行。")
18.              .addLine("不显示为 NotificationCompat.BigTextStyle
                    提供的连续文本行。")
19.              .addLine("可以调用 addLine()添加新行")
20.              .addLine("可以使用 HTML 标记添加样式,比如加粗主题,以区分消息
                    主题和内容"))
21.         .build();
```

以上样式未展开与展开之后的对比如图 2.27 所示。

图 2.27 通知展开前后的效果对比

（3）自定义布局的通知

如果系统提供的通知模板不能满足要求，Android 还提供了自定义布局的通知，将自定义布局扩充为 RemoteViews 的实例，在构建通知时调用 setCustomContentView() 设置自定义布局，示例代码如下：

```
1.  public void showCustomViewNotification() {
2.      // 通知的自定义布局
3.      RemoteViews remoteViews=new RemoteViews(getPackageName(),
                                    R.layout.custom_layout);
4.      // 通过控件 id 设置属性
5.      remoteViews.setImageViewResource(R.id.iv_pic, R.mipmap.ic_
                                    launcher_round);
6.      remoteViews.setTextViewText(R.id.tv_title, "热点新闻");
7.      remoteViews.setTextViewText(R.id.tv_content, "全国进入急冻模式,
                                    开始大幅度降温");
8.      // 创建通知
9.      Notification notification = new NotificationCompat.Builder
                                    (this, ID_BASIC)
10.            .setCustomContentView(remoteViews)
11.            .build();
12. }
```

（4）添加操作按钮

除了展示通知信息之外，还可以添加至多三个操作按钮，以便用户快速响应，如暂停提醒、回复短信等，但这些操作按钮不应该重复用户在点按通知时执行的操作。

要添加操作按钮，需要构建一个 PendingIntent 和 Notifaction.Action 对象，在构建通知时调用 addAction() 方法进行设置，PendingIntent 一般用于完成各种任务，如启动后台任务的广播，以便在不干扰当前正在运行的前提下实现快速响应。示例代码如下：

```
1.  public void showBasicNotification() {
2.      // 添加回复输入框
3.      RemoteInput remoteInput = new RemoteInput.Builder("reply").build();
4.      // 添加一个 PendingIntent 的广播
5.      final PendingIntent replyIntent = PendingIntent.getBroadcast(this, 2,
6.          new Intent("com.example.notification"),PendingIntent.
                FLAG_UPDATE_CURRENT);
7.      // 构建回复 Action
8.      NotificationCompat.Action action = new NotificationCompat.
                                    Action.Builder(
9.          R.drawable.ic_send, "回复", replyIntent)
10.         .addRemoteInput(remoteInput).build();
```

```
11.         // 创建通知
12.         Notification notification = new NotificationCompat.Builder
                                        (this, ID_BASIC)
13.                 ...
14.                 .addAction(action)
15.                 .build();
16.         ...
17. }
```

从 Android 5.0 开始，可以通过调用 setVisibility()设置在锁定屏幕上的通知的可见等级，该方法的取值如下。

- VISIBILITY_PUBLIC：显示通知的完整内容。
- VISIBILITY_SECRET：不在锁定屏幕上显示通知的任何部分。
- VISIBILITY_PRIVATE：显示通知的基本信息，如图标、标题等，隐藏通知的完整内容。

通知也可以更新和移除，更新通知只需再次调用 NotificationManagerCompat.notify()，将通知的 id 传递给该方法即可，如果发出的通知已被关闭，则系统会创建新通知。除此之外，还可以调用 setOnlyAlertOnce()方法只设置提醒一次，通知只会在首次出现时通过声音、振动等提醒用户，之后更新则不会再提醒。

移除通知可以调用 NotificationManager 的 cancel()或 cancelAll()方法，cancel()用于移除特定 id 的通知，cancelAll()用于移除所有发出的通知。创建通知时通过 setAutoCancel（true）也可以在用户点按通知后移除。

2.4.3 PendingIntent

Notification 最大的用途是提醒用户，显示的信息有限，一般仅给出概要信息，需要通过点按了解详细通知内容。为了达到这个目的，需要给 Notification 绑定一个 Intent，当用户点按 Notification 时，通过此 Intent 启动 Activity 了解更多信息。而之前学过的 Intent 是即刻发生的意图，无法直接用于 Notification，系统提供了 PendingIntent 用于延迟的 Intent。PendingIntent 可以看作是对 Intent 的封装，它会在将来某个不确定的时刻发生，而不像 Intent 即刻发生。PendingIntent 的典型使用场景包括闹钟、通知和桌面部件。需要注意的是，在使用 PendingIntent 时，禁止使用空 Intent 和隐式 Intent，因为空 Intent 会导致恶意用户劫持修改 Intent 的内容。

PendingIntent 实例对象可以调用 PendingIntent 的 getActivity()、getBroadcast()和 getService()等方法获取，它们分别对应 Activity、Broadcast 和 Service 三个组件的应用，这些方法的参数都是相同的，共有 4 个参数：Context、requestCode、Intent 和 flags，分别对应上下文对象、请求码、请求意图和关键标志位。前三个参数标志一个行为的唯一性。Flags 有以下取值。

- FLAG_CANCEL_CURRENT：如果当前系统已经存在一个相同的 PendingIntent 对象，那么就将已有的 PendingIntent 取消，重新生成一个新的 PendingIntent 对象。
- FLAG_NO_CREATE：如果当前系统中不存在相同的 PendingIntent 对象，系统将不会创建该 PendingIntent 对象而是直接返回 null，如果之前设置过便可以获取。
- FLAG_ONE_SHOT：该 PendingIntent 只作用一次。它通过 send()方法触发后，PendingIntent 将自动调用 cancel()进行销毁，如果再次调用 send()方法，系统将会返回一个 SendIntentException 异常。
- FLAG_UPDATE_CURRENT：如果系统中有一个和描述的 PendingIntent 相同的

PendingInent 对象，那么系统将使用该 PendingIntent 对象，但使用新的 Intent 来更新之前 PendingIntent 中的数据。

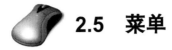

2.5 菜单

菜单是许多应用程序不可或缺的一部分，在 Android 系统中更是如此。从 Android 3.0 开始，Android 设备不必提供一个专用的"菜单"按钮，取而代之的是提供一个应用栏，用来呈现常见的用户操作。Android 有三种基本的菜单类型。

● 选项菜单 OptionsMenu：Activity 的主菜单项集合，一般放置对应用有全局影响的操作，如"搜索""设置"等。

● 上下文菜单 ContextMenu：用户长按某元素时出现的浮动菜单，菜单提供的操作仅影响所选内容或上下文。

● 弹出菜单 PopupMenu：以垂直列表形式显示一系列菜单项，并且该列表会固定在调用该菜单的视图中，它适用于提供与特定内容相关的操作，如 Activity 的内容区域相关的扩展操作。

2.5.1 使用 XML 定义菜单

针对这三种类型的菜单，Android 都提供了标准的 XML 格式定义菜单及其所有项，XML 方式具备以下优点：

● XML 方式可以可视化菜单结构。
● 菜单内容与应用的逻辑代码分离。

在 res 目录上右击选择 Android Resource File，打开 New Resource File 对话框，创建 Resource type 为 menu 的 XML 文件，该菜单文件的构成元素包括：\<menu\>、\<item\>、\<group\>。

● \<menu\>：菜单根节点，能够包含一个或多个\<item\>和\<group\>元素，它是定义菜单项的容器。

● \<item\>：菜单项节点，用于创建 MenuItem 对象，可包含嵌套\<menu\>元素，以便创建子菜单。

● \<group\>：用于对菜单项进行分类，它是\<item\>元素的不可见容器。

XML 菜单文件的示例代码如下：

```
1.  <menu xmlns:android="http://schemas.android.com/apk/res/android"
2.      xmlns:app="http://schemas.android.com/apk/res-auto">
3.      <item
4.          android:id="@+id/item_save"
5.          android:icon="@drawable/ic_settings"
6.          android:title="保存"
7.          app:showAsAction="ifRoom" />
8.      <item
9.          android:id="@+id/item_settings"
10.         android:icon="@drawable/ic_ settings "
11.         android:title="设置" />
12. </menu>
```

\<item\>元素的常用属性包括 android:id、android:icon、android:title、android:showAsAction。

- android:id：设置菜单项的 id，用于识别该菜单项。
- android:icon：设置菜单项的图标，Android 3.0 之后折叠菜单默认不显示图标。
- android:title：设置菜单项的标题字符串。
- android:showAsAction：指定菜单项在标题栏的显示的时机和方式，菜单项只有在 Activity 包含应用栏时才能显示为操作项。取值说明如表 2.15 所示。

表 2.15 showAsAction 的取值说明

取值	说明
ifRoom	只有在标题栏中有空间时才将此项放置其中。如果没有足够的空间容纳标记为 ifRoom 的所有项，标题栏显示不了的菜单项将显示在溢出菜单中
withText	菜单项的文本和图标一起显示，可以将此值与其他值用竖线（|）分隔共同起作用
never	菜单项永远不显示在标题栏，而隐藏在溢出菜单中
always	始终将此菜单项显示在标题栏。除非必须始终显示在标题栏，否则不要使用该值
collapseActionView	将菜单项折叠到一个按钮中，选择按钮展开，一般与 ifRoom 一起使用

菜单项的图标可以通过 Image Asset 工具进行创建，实现步骤为：右击 drawable 目录，选择 New->Image Asset，打开如图 2.28 所示的对话框，下拉选择 Action Bar and Tab Icons，编辑图标名称，然后单击 Clipart Art 单选按钮，单击 Next 按钮和 Finish 按钮完成图标的创建。

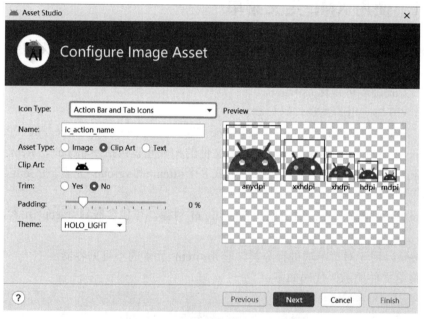

图 2.28 创建菜单项图标的对话框

接下来通过案例讲解三种类型菜单的开发过程。

【案例 2-6】三种类型菜单如何在 Activity 和控件对象上加载、显示及单击事件处理。

步骤 1：创建项目

启动 Android Studio，创建名为 D0206_Menu 的项目，选择 Empty Activity，将包名改为 com.example.menu，单击 Finish 按钮，等待系统构建完成。

步骤 2：创建菜单布局文件

在 res 目录上右击，按照上一小节的讲解创建选项菜单文件 option_menu.xml，代码如下：

```
1.  <menu xmlns:android="http://schemas.android.com/apk/res/android"
2.      xmlns:app="http://schemas.android.com/apk/res-auto">
3.      <item android:id="@+id/item_download"
4.          android:icon="@drawable/ic_download"
5.          android:title="保存"
6.          app:showAsAction="ifRoom" />
7.      <item android:id="@+id/item_settings"
8.          android:icon="@drawable/ic_settings"
9.          app:showAsAction="withText"
10.         android:title="设置" />
11. </menu>
```

上下文菜单 context_menu.xml 的示例代码如下：

```
1.  <menu xmlns:android="http://schemas.android.com/apk/res/android" >
2.      <item android:icon="@drawable/ic_copy"
3.          android:id="@+id/item_copy"
4.          android:title="拷贝" />
5.      <item android:icon="@drawable/ic_paste"
6.          android:id="@+id/item_paste"
7.          android:title="粘贴" />
8.      <item  android:id="@+id/item_clear"  android:title="清空" />
9.  </menu>
```

弹出菜单 popup_menu.xml 的示例代码如下：

```
1.  <menu xmlns:android="http://schemas.android.com/apk/res/android">
2.      <item android:id="@+id/item_copy" android:title="复制"/>
3.      <item android:id="@+id/item_delete" android:title="粘贴"/>
4.  </menu>
```

接下来分三小节讲解三种类型菜单的编码逻辑。

2.5.2 选项菜单

每个 Activity 默认都自带选项菜单，只需要添加菜单项，响应菜单项的单击事件。Android 分别使用 android.view.Menu 接口表示菜单，android.view.MenuItem 接口表示菜单项，android.view.SubMenu 接口表示子菜单。

Activity 通过重写 OnCreateOptionsMenu()方法加载创建的 XML 菜单资源，在此方法中使用 MenuInflater 类的 inflate()方法将 XML 资源加载到 Menu 对象。除此之外，还可以使用 add()添加菜单项，该方法有 4 个参数，分别为：菜单项的组别 id、菜单项 id、排列顺序及菜单项标题，菜单项的图标设置可以调用 setIcon()来实现，使用方式参考第 7 行代码。示例代码如下：

```
1.  @Override
2.  public boolean onCreateOptionsMenu(Menu menu) {
3.      // 加载 xml 菜单资源
4.      MenuInflater inflater = getMenuInflater();
5.      inflater.inflate(R.menu.option_menu, menu);
6.      // 动态加载菜单项
7.      menu.add(Menu.NONE, Menu.FIRST, 3, "帮助").setIcon(R.drawable.
                                                            ic_help);
8.      return true;
9.  }
```

当用户从选项菜单中选择菜单项时，系统将调用 onOptionsItemSelected()，此方法将传递所选的 MenuItem 对象，通过调用该 MenuItem 对象的 getItemId() 方法获取菜单项的 id。例如：

```
1.  @Override
2.  public boolean onOptionsItemSelected(MenuItem item) {
3.      // 处理菜单项的选择
4.      switch (item.getItemId()) {
5.          case R.id.item_download:
6.              Toast.makeText(MainActivity.this, "下载", Toast.LENGTH_
                        LONG).show();
7.              return true;
8.          case R.id.item_settings:
9.              Toast.makeText(MainActivity.this, "设置", Toast.LENGTH_
                        LONG).show();
10.             return true;
11.         case Menu.FIRST:
12.             Toast.makeText(MainActivity.this, "帮助", Toast.LENGTH_
                        LONG).show();
13.             return true;
14.         default:
15.             return super.onOptionsItemSelected(item);
16.     }
17. }
```

选项菜单的运行效果如图 2.29 所示。

图 2.29　选项菜单的运行效果图

默认情况下，即便在菜单 XML 文件中定义了 android:icon 属性，在溢出菜单中图标也不会显示，本案例使用反射的方法获取 Menu 对象的 setOptionalIconsVisible()，调用该方法显示图标 icon，示例代码如下：

```
1.  // 使用反射方法显示图标
2.  private void setIconsVisible(Menu menu, boolean flag) {
3.      // 判断 menu 是否为空
4.      if (menu != null) {
5.          try {
6.              // 如果不为空,使用反射获取 menu对象的setOptionalIconsVisible()
                    方法
```

```
7.            Method method = menu.getClass().getDeclaredMethod(
8.                "setOptionalIconsVisible", Boolean.TYPE);
9.            // 强制访问该方法
10.           method.setAccessible(true);
11.           // 调用该方法显示 icon
12.           method.invoke(menu, flag);
13.       } catch (Exception e) {
14.           e.printStackTrace();
15.       }
16.   }
17. }
```

然后将此方法添加到 onCreateOptionsMenu()方法的 return 语句之前即可，参看以下代码：

```
1. public boolean onCreateOptionsMenu(Menu menu) {
2.     // 加载 xml 菜单资源
3.     MenuInflater inflater = getMenuInflater();
4.     inflater.inflate(R.menu.option_menu, menu);
5.     // 动态加载菜单项
6.     menu.add(Menu.NONE, Menu.FIRST, 3, "帮助").setIcon(R.drawable.ic_help);
7.     // 设置菜单添加图标有效
8.     setIconsVisible(menu, true);
9.     return true;
10. }
```

2.5.3 上下文菜单

当用户长按某一控件时出现的浮动菜单称为上下文菜单。上下文菜单不同于选项菜单，选项菜单服务于 Activity，而上下文菜单则注册在某个 View 对象上。它通常应用于列表中的每一项元素，如：长按列表项弹出删除对话框。

Android 提供了两种上下文操作菜单的方法。
- 使用浮动上下文菜单，如图 2.30 所示，当长按某个 View 时，显示为菜单项的浮动列表。
- 使用上下文操作模式，如图 2.31 所示，它在屏幕顶部栏显示上下文操作栏选项。

图 2.30　浮动上下文菜单

图 2.31　上下文操作栏模式

1. 创建浮动上下文菜单操作步骤

（1）注册菜单

通过调用 registerForContextMenu（View view）注册与上下文菜单关联的 View。如果将 ListView 或 GridView 作为参数传入，那么每个列表项将会有相同的浮动上下文菜单。

```
1.   // 注册浮动上下文菜单
2.   etUsername = findViewById(R.id.et_name);
3.   registerForContextMenu(etUsername);
```

（2）加载菜单资源

在 Activity 或 Fragment 中实现 onCreateContextMenu()，动态加载 Menu 资源。当注册后的 View 收到长按事件时，系统将调用此方法。

```
1.   @Override
2.   public void onCreateContextMenu(ContextMenu menu, View v,
3.                                   ContextMenu.ContextMenuInfo menuInfo) {
4.       super.onCreateContextMenu(menu, v, menuInfo);
5.       if (v.getId() == R.id.et_name) {
6.           // 加载菜单 XML 资源
7.           MenuInflater inflater = getMenuInflater();
8.           inflater.inflate(R.menu.context_menu, menu);
9.           // 动态创建菜单项
10.          // menu.add(Menu.NONE, Menu.FIRST, 0, "拷贝");
11.          // menu.add(Menu.NONE, Menu.FIRST + 1, 0, "粘贴");
12.          // menu.add(Menu.NONE, Menu.FIRST + 2, 0, "清空");
13.      }
14.  }
```

（3）处理菜单项单击事件

在 Activity 或 Fragment 中实现 onContextItemSelected()，实现菜单项的单击逻辑。

```
1.   @Override
2.   public boolean onContextItemSelected(@NonNull MenuItem item) {
3.       switch (item.getItemId()) {
4.         case R.id.item_copy:
5.           Toast.makeText(MainActivity.this, item.getTitle(),
                           Toast.LENGTH_LONG).show();
6.           return true;
7.         case R.id.item_paste:
8.           Toast.makeText(MainActivity.this, item.getTitle(),
                           Toast.LENGTH_LONG).show();
9.           return true;
10.        case R.id.item_clear:
11.          etUsername.setText("");
12.          return true;
13.        default:
14.          return super.onOptionsItemSelected(item);
15.      }
16.  }
```

上下文操作模式是 ActionMode 对象的系统实现，当用户长按控件或选中复选框等组件时调用此模式，会在屏幕顶部出现上下文操作栏，显示用户对所选控件执行的操作。

2. 创建上下文操作栏模式的上下文菜单操作步骤

（1）实现 ActionMode.Callback 接口

在该接口的回调方法中，为上下文操作栏指定操作、响应菜单项的单击事件等。

```java
1.  // 上下文操作模式对象
2.  private ActionMode actionMode;
3.  // 实现 ActionMode 的 CallBack 回调接口
4.  ActionMode.Callback actionModeCallback = new ActionMode.Callback() {
5.      // 在启动上下文操作模式 startActionMode(Callback)时调用
6.      // 在此配置上下文菜单的资源
7.      @Override
8.      public boolean onCreateActionMode(ActionMode mode, Menu menu) {
9.          mode.getMenuInflater().inflate(R.menu.context_menu, menu);
10.         return true;
11.     }
12.     // 在创建方法后进行调用
13.     @Override
14.     public boolean onPrepareActionMode(ActionMode mode, Menu menu) {
15.         return false;
16.     }
17.     // 菜单项被单击,类似 onContextItemSelected()方法
18.     @Override
19.     public boolean onActionItemClicked(ActionMode mode, MenuItem item) {
20.         switch (item.getItemId()) {
21.             case R.id.item_copy:
22.                 Toast.makeText(MainActivity.this, "拷贝", Toast.
                        LENGTH_SHORT).show();
23.                 break;
24.             case R.id.item_paste:
25.                 Toast.makeText(MainActivity.this, "粘贴", Toast.
                        LENGTH_SHORT).show();
26.                 break;
27.             case R.id.item_clear:
28.                 etUsername.setText("");
29.                 break;
30.             default:
31.                 break;
32.         }
33.         return true;
34.     }
35.     // 上下文操作模式结束时被调用
36.     @Override
37.     public void onDestroyActionMode(ActionMode mode) {
38.         actionMode = null;
39.     }
40. };
```

该接口的回调方法与选项菜单的回调方法基本相同,只是需要传递与事件相关联的 ActionMode 对象。onActionItemClicked()方法用于处理菜单项的单击事件,与 onContextItemSelected()类似。当系统销毁操作模式时,需要在 onDestroyActionMode()方法中将 actionMode 变量设置为 null。

(2)启动上下文操作模式

在用户名的 EditView 控件的长按事件中调用 startActionMode()启动上下文操作模式,代码如下:

```java
1.  // 在 View 的长按事件中启动上下文操作模式
2.  etUsername.setOnLongClickListener(new View.OnLongClickListener() {
3.      @Override
4.      public boolean onLongClick(View view) {
5.          if (actionMode != null) {
6.              return false;
```

```
7.      }
8.      actionMode = startActionMode(actionModeCallback);
9.      view.setSelected(true);
10.     return false;
11.   }
12. });
```

调用 startActionMode()方法返回 ActionMode 对象，通过该对象响应上下文操作栏的事件。

2.5.4 弹出菜单

图 2.32 弹出菜单运行效果

弹出菜单 PopMenu 以垂直列表形式显示一系列操作选项，一般由控件元素触发，弹出菜单将显示在对应控件的上方或下方，如图 2.32 所示。弹出菜单适用于：

● 为与特定内容密切相关的操作提供溢出式菜单；
● 提供类似 Spinner 控件，但不保留永久选择的下拉菜单。

使用弹出菜单的步骤如下。

（1）创建菜单资源文件

在 res/menu 目录中创建 PopupMenu 的 XML 资源文件。

（2）创建并加载弹出菜单

在 EditText 控件的单击事件 onClick()方法中实例化 PopupMenu 对象，传递当前上下文对象及绑定的 View 对象，然后调用 MenuInflater.inflate()方法加载菜单 XML 文件，调用 PopupMenu 的 setOnMenuItemClickListener()设置单击菜单项的事件监听器，最后调用 show()方法显示菜单。

```
1.  etPhone.setOnClickListener(new View.OnClickListener() {
2.    @Override
3.    public void onClick(View v) {
4.      // 创建弹出菜单对象(最低版本 11)，第二个参数是绑定的那个 View
5.      PopupMenu popup = new PopupMenu(MainActivity.this, v);
6.      // 获取菜单填充器
7.      MenuInflater inflater = popup.getMenuInflater();
8.      // 填充菜单
9.      inflater.inflate(R.menu.popup_menu, popup.getMenu());
10.     //绑定菜单项的单击事件
11.     popup.setOnMenuItemClickListener(MainActivity.this);
12.     // 显示弹出菜单
13.     popup.show();
14.   }
15. });
```

（3）处理菜单项的 OnMenuItemClick 事件

MainActivity 类实现 PopupMenu.OnMenuItemClickListener 接口，重写 onMenuItemClick()回调方法，当用户选择菜单项时，系统调用 onMenuItemClick()进行处理。

```
1.  public boolean onMenuItemClick(MenuItem item) {
2.    switch (item.getItemId()) {
3.      case R.id.item_copy:
4.        Toast.makeText(this, "复制…", Toast.LENGTH_SHORT).show();
5.        break;
6.      case R.id.item_delete:
```

```
7.                Toast.makeText(this, "删除…", Toast.LENGTH_SHORT).show();
8.                break;
9.            default:
10.                break;
11.        }
12.        return false;
13. }
```

2.6 常用资源与样式

Android 资源是指代码使用的附加文件和静态内容，包括布局、字符串、颜色、数组、动画、位图、图标、音视频和其他应用程序使用的组件等。

res 是存放项目资源的目录，该目录下的资源可以进行可视化编辑，这些资源通过 AAPT（Android Asset Packaging Tool）工具自动生成 gen 目录下的 R.java 资源索引文件，然后可以在 Java 代码和 XML 资源文件中通过索引引用。需要注意的是，该目录下的任何资源出现问题，都会影响 R.java 文件的生成，导致所有的资源都无法引用。

2.6.1 资源目录结构

Android 资源除了 assets 目录是与 res 同级的，其他资源均被放在 res 目录下，该目录下的资源目录并不是随意命名的，需要遵循系统规范，否则编译生成 R.java 过程中会报类似"invalid resource directory name**"的错误提示，并且导致 R.java 自动生成失败。

常用的默认目录和对应资源类型在 SDK 帮助中有详细列出，简单摘抄如表 2.16 所示。

表 2.16 res/目录资源目录和对应资源类型

目录 Directory	资源类型 Resource Types
animator/	定义属性动画的 XML 文件
anim/	定义视图动画的 XML 文件（属性动画也可保存在此目录中，但应首选 animator 目录）
raw/	存放以原始形式保存的文件，以 R.raw.filename 为参数调用 Resources.openRawResource()使用资源。它们无须编译，会直接打包到生成的 APK 文件中
drawable/	存放能转换为绘制资源（Drawable Resource）的位图文件（.png、.9.png、.jpg、.gif）或编译为可绘制对象资源子类型的 XML 文件（状态列表、形状、动画可绘制对象等）
color/	定义颜色状态列表的 XML 文件
layout/	定义用户界面布局的 XML 文件
menu/	定义应用菜单（如选项菜单、上下文菜单或子菜单）的 XML 文件
values/	存放包含字符串、整型数和颜色等简单值的 XML 文件
mipmap/	存放适用于不同启动器图标密度的可绘制对象文件
xml/	存放任意的 XML 文件，在运行时可以通过调用 Resources.getXML()读取
font/	带有扩展名的字体文件（.ttf、.otf 或.ttc）或包含<font-family>元素的 XML 文件

需要特别注意的是，资源文件的名称必须是[a-z0-9_.]范围的字符，不能包含大写字母和中文字符，并且每个目录中存放的文件类型不仅有规定，而且对文件内容也有严格要求。

2.6.2 样式和主题

样式是一个属性集合，如设置字体颜色、字号、背景颜色等属性用于指定单个 View 的样式。样式可以用于布局中的单个 View、Activity 或 Application。主题也称为主题背景，它不仅仅用于单个视图，而是应用于整个 App、Activity 或 View 的样式集合。当将样式作为主题背景时，App 中的每个视图都会使用对应控件的样式属性；主题还可以应用于状态栏、窗口背景等非视图元素。

样式和主题在 res/values/styles.xml 文件中声明，使用@[package:]style/style_name 方式引用资源，而非 XML 文件名，因此，可以把样式资源和其他简单类型资源一起放在<resources>元素下，它的语法结构如下：

```xml
1. <?xml version="1.0" encoding="utf-8"?>
2. <resources>
3.     <style name="style_name" parent="@[package:]style/style_to_inherit">
4.         <item name="[package:]style_property_name">
5.             style_value
6.         </item>
7.     </style>
8. </resources>
```

其中，包含的元素说明如下。

● <style>标签：定义样式，包含<item>元素，其中，name 属性是样式的名称，String 类型，必填项，用作将样式应用于 View、Activity 或 Application 的资源 id；parent 属性是此样式继承的样式的引用。

● <item>标签：定义样式的单个属性，必须是<style>元素的子元素。它有 name 属性，即样式属性的名称，必填项，可以带上包前缀，如：android:textColor。

Android 项目生成的 style.xml 的代码如下：

```xml
1.<style name="AppTheme" parent="Theme.MaterialComponents.DayNight.DarkActionBar">
2.    <item name="colorPrimary">@color/colorPrimary</item>
3.    <item name="colorPrimaryDark">@color/colorPrimaryDark</item>
4.    <item name="colorAccent">@color/colorAccent</item>
5. </style>
```

最佳实践：

● 国际化与本地化：将字符串作为资源使用，避免在布局文件中直接使用字符串常量。开发多版本的应用程序时，需要适应中文、英文、阿拉伯等语言，定义相应的资源文件即可实现国际化开发，中文资源的目录名称为 values-zh。

● 统一样式与风格：界面的文本字体、颜色、背景、权重等属性的样式均可采用 style 设置。

2.6.3 Drawable 资源

Drawable 资源称为可绘制对象资源，指可在屏幕上绘制的图形，包括位图文件、九宫格文件、图层列表、状态列表及各种可绘制对象等，其中，位图文件主要是图片资源，指各类图像文件，支持 png、jpg 或 gif 格式，官方建议"png 最佳、jpg 可接受、gif 不推荐"的原则。

九宫格文件（NinePatch）是一种 png 图片，在其中定义可伸缩区域，当视图中的内容超出正常图像边界时进行缩放，命名为 filename.9.png。使用它时通常指定大小为"wrap_content"，当视图通过扩展适应内容时，九宫格图像也会通过扩展匹配视图的大小。

从 Android SDK 文档关于 Drawable 资源的描述中可知，Drawable 是抽象类，它通过子类操作具体类型的资源，比如：BitmapDrawable 操作位图，ColorDrawable 操作颜色，ClipDrawable 操作剪切板等。

使用 Drawable 资源的方法是把图片放入项目的 res/drawable 目录中，编程环境会自动在 R 类里为此资源创建引用。程序可以通过调用相关的方法获取图片资源，如：访问文件名为 ic_launcher 的图片，可以采用[packagename].R.drawable.ic_launcher)，例如：

```
1. ImageView imageView = findViewById(R.id.img);
2. imageView.setImageResource(R.drawable.ic_launcher);
```

此外，还经常会用到另一种资源——状态列表 StateListDrawable，它是 XML 文件定义的可绘制列表，分配一组 Drawable 资源，以<selector>元素起始，其内部的每个 Drawable 资源内嵌在<item>元素中，每个<item>使用各种属性描述图形的状态。例如：

```
1. <?xml version="1.0" encoding="utf-8"?>
2. <selector xmlns:android="http://schemas.android.com/apk/res/android">
3.     <item android:state_focused="true" android:drawable
                                ="@drawable/star"></item>
4.     <item android:state_pressed="true" android:drawable
                                = "@drawable/moon"></item>
5. </selector>
```

当 StatListDrawable 资源作为控件的背景或者前景资源时，它会根据对象状态自动切换对应的资源。例如，Button 可以根据按下、松开等状态而使用 StateListDrawable 资源提供不同的背景图片对应不同的状态。当组件的状态变化时，会自动遍历 StateListDrawable 对应的 XML 文件查找第一个匹配 item。selector 的<item>元素的常用属性如表 2.17 所示。

表 2.17　item 元素的常用属性及描述

item 状态	含义描述
android:state_focused	对象获得输入焦点时值为 true，在默认的非焦点状态下为 false
android:state_pressed	按下或触摸对象时值为 true，未按下状态为 false
android:state_selected	当定向控件浏览被选择时值为 true，未选择对象时为 false
android:state_checkable	对象可选中时值为 true，不可选中为 false
android:state_checked	对象已选中时值为 true，对象未选中时为 false
android:state_enabled	对象启用时值为 true，对象停用时为 false
android:state_activated	对象激活作为持续选择时值为 true，对象未激活时为 false

本节介绍了 Android 系统中的常用资源，主要包含 style、string、color 等 Values 资源的使用方法、Drawable 资源的使用方法等，前几个章节都有这些资源的应用，本节没有设计专门的案例，可以根据这节的讲解进行重构，进一步加深对常用资源的应用场景及使用方法。

2.7 本章小结

本章主要讲解了 Android 应用程序的基础界面设计，首先介绍了 Android 布局文件的创建和使用，然后讲解了 Android 开发中 4 种常用的基本布局，接着讲解了 Android 应用程序开发中常用的基本控件和菜单，并通过案例的开发加深对它们的理解和应用。希望通过本章的学习，读者对 Android 应用程序的基本界面设计与开发有所了解和掌握，为后续高级界面设计的学习打下良好的基础。

习 题

一、选择题

1. 在 Android 项目结构中，布局文件通常存放在（　　）。
   A. res/layout/　　　　B. res/drawable　　　C. res/LinearLayout　　D. 以上都对

2. 有一种布局，它将整个界面当成一块空白备用区域，所有子元素的位置都不能被指定，统统放于这块区域的左上角，并且后面的子元素直接覆盖在前面的子元素之上，这种布局是（　　）。
   A. LinearLayout　　　B. RelativeLayout　　C. FrameLayout　　　D. GridLayout

3. Android 开发中会用到许多组件，这些 UI 组件的基类都是（　　）。
   A. BaseUI　　　　　　B. View　　　　　　C. AppCompat　　　　D. UIGroup

4. 以下哪个属性用来设置控件相对于其所在容器的位置（　　）。
   A. android:layout_width　　　　　　　　B. android:layout_gravity
   C. android:gravity　　　　　　　　　　D. android:layout_height

5. 以下哪一种菜单不是 Android 开发中常用的菜单（　　）。
   A. 选项菜单　　　　　B. 头菜单　　　　　C. 上下文菜单　　　　D. 弹出菜单

二、简答题

1. 简述 LinearLayout 布局的重要属性。
2. 简述 GridLayout 布局的应用场景。
3. 简述 RelativeLayout 布局的特点及定位属性。
4. 简述 RadioButton 和 CheckBox 的区别。
5. 简述 Snackbar 的主要特性及应用场景。
6. 简述三种菜单的特点及区别。
7. 简述通知的实现步骤。
8. 简述常用的资源。

三、编程题

1. 开发一个 BMI 体质指数计算器，输入体重（单位：kg）和身高（单位：m），BMI 指数的描述如表 2.18 所示，参考界面如图 2.33 所示，计算公式为：BMI = 体重/身高^2。

2. 给 BMI 计算器设计选项菜单，用于清空输入的身高和体重数据。

表 2.18 BMI 与胖瘦的对应表

描述	BMI 值	
	男性	女性
体重过轻	BMI<20	BMI<19
体重正常	20<= BMI<25	19<= BMI<24
体重超重	25<= BMI<27	24<= BMI<26
轻度肥胖	27<= BMI<30	26<= BMI<29
中度肥胖	30<= BMI<35	29<= BMI<34
重度肥胖	BMI>=35	BMI>=34

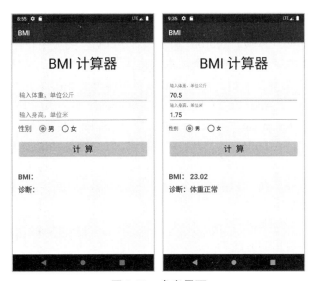

图 2.33 参考界面

第 3 章 Activity 与 Fragment

通过上一章的学习，我们已经基本掌握了常用布局和基础控件的使用，但一个优秀的应用程序还需要有良好的交互体验。Android 系统通过 Activity 完成用户与程序的交互，Activity 也是 Android 四大组件中使用最多的一个。本章将详细讲解 Activity 的基本概念和应用。

作为一名 Android 的开发人员，开发时必须考虑屏幕大小的适配性，最突出的就是手机与平板屏幕的差异。为此，Android 3.0 版本引入 Fragment（碎片或片段），让界面也能在平板上有很好的展示。基于 Fragment 与 Activity 的关联性和编码的相似性，将它们放在同一章进行讲解。

本章学习目标：

- 掌握 Activity 的基本应用
- 理解 Activity 的生命周期
- 掌握 Intent 传递数据的方法
- 理解 Android 的启动模式
- 掌握 Fragment 的基本应用
- 理解 Fragment 的生命周期

 ## 3.1 Activity 基础

3.1.1 什么是 Activity

Activity 的中文含义是"活动"，它提供可视化的用户界面，通过界面的事件监听与用户交互完成某项任务。Activity 是 Android 四大组件中最基础的组件，也是使用频率最高的组件。一个 Activity 对象代表一个窗口，该窗口通常充满屏幕，也可以是非全屏的悬浮窗体或对话框。

对于 Android 应用程序来说，用户交互至关重要，因此 Activity 必不可少，一个应用程序通常由多个 Activity 组成，其中的一个 Activity 被指定为主界面，在应用程序启动时第一个呈现给用户。因此，在新建 Android 项目时选择 Empty Activity，系统就会生成一个名为 MainActivity 的类，该类继承自 androidx.appcompat.app.AppCompatActivity 类，AppCompatActivity 类继承自 v4 包的 FragmentActivity 类，目的是兼容旧版本的设备，它是 Activity 类的子类。Activity 类及其子类的继承关系如图 3.1 所示。

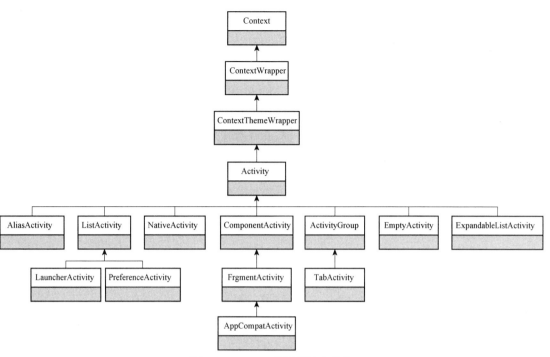

图 3.1 Activity 类继承层次结构

一般情况下，用户创建自定义 Activity 只需继承 Activity 类，也可以根据不同场景选择继承 Activity 的子类，如：只包含列表的界面可以继承 ListActivity，实现操作 Fragment 的界面可继承 FragmentActivity，AppCompatActivity 则能实现带有标题栏的 Activity。为了有更好的向下兼容性，建议继承自 AppCompatActivity 类。Activity 类提供的常用方法如表 3.1 所示。

表 3.1 Activity 类的常用方法

方法名称	功能描述
onCreate()	Activity 被系统创建时调用，完成 Acitivity 的初始化
SetContentView(int layoutResID)	设置布局组件
findViewByid(int id)	根据组件的 ID 取得组件对象
startActivity(Intent intent)	启动一个新的 Activity
startActivityForResult()	启动一个新的 Activity 并返回

3.1.2 创建 Activity

在创建项目时选择 Empty Activity，系统会创建一个主 Activity 类，除此之外，系统提供了以下两种方法用于手动创建一个自定义 Activity。

- 使用系统提供的创建 Activity 类的向导。
- 手动创建 Activity 类、界面布局及注册。

【案例 3-1】使用向导创建自定义 Activity。

步骤 1：创建项目

启动 Android Studio，创建名为 D0301_ActivityCreate 项目，选择 No Activity，将包名改为 com.example.activity，等待系统构建完成，项目结构如图 3.2 所示。

步骤 2：创建 Activity

在左侧导航栏中右键单击 com.example.activity 包，选择 New->Activity->Empty Activity，弹出创建 Activity 的对话框，如图 3.3 所示。

图 3.2　项目初始结构

图 3.3　创建 Activity 的对话框

图 3.3 中的 Activity Name 填写 FirstActivity，勾选 Generate a Layout File 就会同时创建 FirstActivity 对应的布局文件 activity_first.xml，Launcher Activity 选项表示将此 Activity 设置为当前项目的主 Activity，此处勾选。其余的选项参照图中所示进行设置，单击 Finish 按钮即可完成 Activity 的创建。

Android Studio 自动创建 FirstActivity 类文件、activity_first.xml 布局文件及在 AndroidManifest.xml 中注册 FirstActivity 组件，代码如下所示：

```
1.  <?xml version="1.0" encoding="utf-8"?>
2.  <manifest xmlns:android="http://schemas.android.com/apk/res/android"
3.      package="com.example.activity">
4.
5.      <application
6.          android:allowBackup="true"
7.          android:icon="@mipmap/ic_launcher"
8.          android:label="@string/app_name"
9.          android:roundIcon="@mipmap/ic_launcher_round"
10.         android:supportsRtl="true"
11.         android:theme="@style/AppTheme">
12.         <activity android:name=".FirstActivity">
13.             <intent-filter>
14.                 <action android:name="android.intent.action.MAIN" />
15.                 <category android:name="android.intent.category.
                                            LAUNCHER" />
16.             </intent-filter>
17.         </activity>
18.     </application>
19. </manifest>
```

FirstActivity 类的结构非常简单，包括 onCreate()方法，该方法首先调用父类的 onCreate() 方法，使用 setContentView()方法绑定布局文件，代码如下所示：

```
1.  public class FirstActivity extends AppCompatActivity {
2.      @Override
3.      protected void onCreate(Bundle savedInstanceState) {
4.          super.onCreate(savedInstanceState);
5.          setContentView(R.layout.activity_first);
6.      }
7.  }
```

图 3.4 所示的是 activity_first.xml 布局文件的编辑界面。

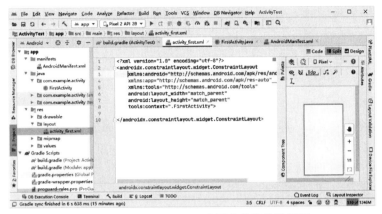

图 3.4 activity_first.xml 布局文件的编辑界面

由此得知，如果采用手动方式创建 Activity，需要做三件事：
- 创建继承自 Activity 或 AppCompatActivity 的自定义类，重写 onCreate()方法。
- 在 res/layout 目录中创建自定义 Activity 类的布局文件。
- 在 AndroidManifest.xml 配置清单文件中进行注册。

由于采用的是一个空布局，所以运行结果除了标题没有任何内容，如图 3.5 所示。

3.1.3 Activity 生命周期

Android 的 Activity 是层叠在一起的，不仅一个应用程序中的 Activity 会在用户浏览、退出和返回时切换状态，不同应用程序的 Activity 也会相互层叠，多个 Activity 构成一个集合。

图 3.5 FirstActivity 运行结果

Android 系统使用栈结构管理 Activity 集合，Activity 按照打开的顺序压入一个返回堆栈中，处于栈顶位置的 Activity 显示在手机当前界面上，在图 3.6 所示的多个 Activity 集合中，在当前 Activity 1 启动 Activity 2 时，Activity 2 就被压入栈顶并获得显示焦点，Activity 1 仍会保留在堆栈中，但处于停止状态，系统会保留它的当前状态。当按手机的返回键退出时，Activity 2 就会出栈被销毁，前一个入栈的 Activity 1 就会重新处于栈顶的位置。

每个 Activity 都具备从启动到销毁的生命周期，从显示到停止的变化会经历 4 种不同的状态。

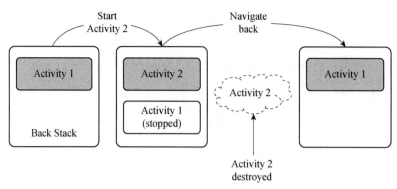

图 3.6 Activity 栈结构

1. 运行状态 Running

当 Activity 位于栈顶时，用户可见且能获得焦点，这时 Activity 就进入运行状态。

2. 暂停状态 Paused

当 Activity 被部分覆盖，即不在栈顶但部分可见时，此时的 Activity 就进入了暂停状态。最常用的例子就是对话框形式的界面，它只占用屏幕中间的部分区域。处于暂停状态的 Activity 仍然保存着所有的状态和信息，当系统内存极低时才会考虑回收此状态的 Activity。

3. 停止状态 Stopped

当 Activity 被其他 Activity 完全覆盖，它不再处于栈顶且完全不可见，此时的 Activity 就进入了停止状态。尽管系统仍会保存该 Activity 的状态和信息，但这并不可靠，当其他系统需要内存时，处于停止状态的 Activity 就可能被系统销毁。

4. 销毁状态 Killed

当 Activity 从活动栈移除后就会变成销毁状态，系统会优先回收这种状态的 Activity，释放占用的内存。

Android 为 Activity 的状态转换提供了 7 个回调方法，如表 3.2 所示。

表 3.2 Activity 的回调方法

方法名称	功能描述
onCreate()	Activity 实例被 Android 系统创建时调用，通常在该方法中完成初始化操作，如加载布局、初始化数据、初始化控件、绑定控件的响应事件等
onStart()	Activity 由不可见变为可见时调用，但还不能与用户交互
onResume()	Activity 初始化完成后，与用户交互时调用，此时 Activity 位于栈顶，并且处于运行状态
onPause()	Activity 由可见变为不可见时调用，新 Activity 的 onResume()必须在这个方法执行完才能调用，所以该方法只能执行一些耗时少的操作，如：保存一些关键数据
onStop()	Activity 完全不可见时调用，它与 onPause()的区别在于：onPause()只在启动一个对话框式的 Activity 时才会调用
onDestroy()	Activity 被销毁之前调用，完成回收和资源释放等操作
onRestart()	Activity 由不可见返回可见状态时调用，如 Home 键切换界面，重新回到界面的过程

从图 3.7 所示的 Activity 生命周期图可以很好地理解 Activity 在不同状态切换时所调用的方法，Activity 完整的生命周期是依次调用 onCreate()->onStart()->onResume()->onPause()->

onStop()->onDestroy()。

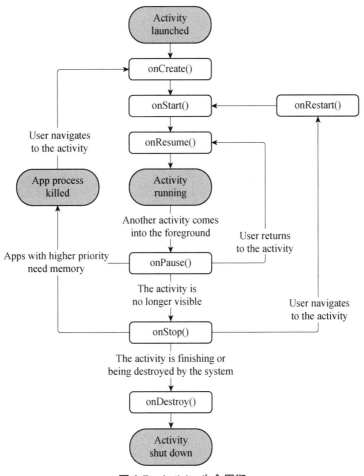

图 3.7　Activity 生命周期

接下来通过一个具体的案例展示 Activity 的生命周期及 4 种状态的转换。

【案例 3-2】通过三个 Activity 之间的跳转理解生命周期方法的转换。

步骤 1：创建项目

启动 Android Studio，创建名为 D0302_ActivityLifeCycle 的项目，选择 No Activity，将包名改为 com.example.activity.lifecycle，单击 Finish 按钮，等待系统构建完成。

步骤 2：创建 Activity

右击 com.example.activity.lifecycle 包，选择 New->Activity->Empty Activity，分别创建 Activity1、Activity2 和 Activity3，在 AndroidManifest.xml 中将 Activity1 设为主 Activity，Activity3 设为对话框模式的 Activity，代码如下：

```xml
1.  <?xml version="1.0" encoding="utf-8"?>
2.  <manifest xmlns:android="http://schemas.android.com/apk/res/android"
3.      package="com.example.activity.lifecycle">
4.
5.      <application
6.          android:allowBackup="true"
7.          android:icon="@mipmap/ic_launcher"
8.          android:label="@string/app_name"
9.          android:roundIcon="@mipmap/ic_launcher_round"
```

```
10.        android:supportsRtl="true"
11.        android:theme="@style/AppTheme">
12.        <activity
13.            android:name=".Activity3"
14.            android:theme="@style/Theme.AppCompat.Light.Dialog.
                              MinWidth" />
15.        <activity android:name=".Activity2" />
16.        <activity android:name=".Activity1">
17.            <intent-filter>
18.                <action android:name="android.intent.action.MAIN" />
19.                <category android:name="android.intent.category.
                              LAUNCHER" />
20.            </intent-filter>
21.        </activity>
22.    </application>
23. </manifest>
```

步骤3：完成Activity的界面布局

界面布局使用ConstraintLayout约束性布局，设置一个TextView控件和一个Button控件，具体样式如图3.8~图3.10所示。

图3.8　Activity1的界面

图3.9　Activity2的界面

图3.10　Activity3的界面

三个界面的布局基本相同，都采用ConstraintLayout约束性布局，使用TextView显示一行文字，使用Button触发Activity跳转事件，三个界面除了id和显示文字不同之外，其他都相同，以activity_1.xml为例的布局代码如下：

```
1.  <?xml version="1.0" encoding="utf-8"?>
2.  <androidx.constraintlayout.widget.ConstraintLayout
3.      xmlns:android="http://schemas.android.com/apk/res/android"
4.      xmlns:app="http://schemas.android.com/apk/res-auto"
5.      xmlns:tools="http://schemas.android.com/tools"
6.      android:layout_width="match_parent"
7.      android:layout_height="match_parent"
8.      tools:context=".Activity1">
9.      <TextView
10.         android:id="@+id/tv_content"
11.         android:layout_width="match_parent"
12.         android:layout_height="wrap_content"
13.         android:gravity="center"
14.         android:text="Activity 1"
```

```
15.         android:textSize="24sp"
16.         app:layout_constraintBottom_toBottomOf="parent"
17.         app:layout_constraintLeft_toLeftOf="parent"
18.         app:layout_constraintRight_toRightOf="parent"
19.         app:layout_constraintTop_toTopOf="parent"
20.         app:layout_constraintVertical_bias="0.4" />
21.     <Button
22.         android:id="@+id/btn_jump"
23.         android:layout_width="wrap_content"
24.         android:layout_height="wrap_content"
25.         android:text="Start Activity 2"
26.         android:textAllCaps="false"
27.         android:textSize="24sp"
28.         app:layout_constraintBottom_toBottomOf="parent"
29.         app:layout_constraintLeft_toLeftOf="parent"
30.         app:layout_constraintRight_toRightOf="parent"
31.         app:layout_constraintTop_toBottomOf="@+id/tv_content"
32.         app:layout_constraintVertical_bias="0.2" />
33. </androidx.constraintlayout.widget.ConstraintLayout>
```

步骤 4：编写事件监听代码

编写三个 Activity 类的 Button 单击事件的注册和监听代码，在 onCreate()方法中注册 Button 的监听器，当按下 Button 时触发 onClick()方法进行处理，使用 startActivity()方法实现 Activity 的跳转。重写生命周期的各个方法，使用日志打印一行文字提示方法已被调用，以 Activity1 为例的代码如下：

```
1.  public class Activity1 extends AppCompatActivity {
2.      private final String TAG = "Activity 1";
3.
4.      @Override
5.      protected void onCreate(Bundle savedInstanceState) {
6.          super.onCreate(savedInstanceState);
7.          setContentView(R.layout.activity_1);
8.          // 输出 onCreate()方法被调用的日志
9.          Log.i(TAG, "onCreate() is called");
10.         // 设置 Start Activity2 按钮的单击事件监听器
11.         Button btnJump = findViewById(R.id.btn_jump);
12.         btnJump.setOnClickListener(new View.OnClickListener() {
13.             @Override
14.             public void onClick(View v) {
15.                 // 跳转到 Activity 2
16.                 startActivity(new Intent(Activity1.this,
                                       Activity2.class));
17.             }
18.         });
19.     }
20.     // 输出 onStart()方法被调用的日志
21.     @Override
22.     protected void onStart() {
23.         super.onStart();
24.         Log.i(TAG, "onStart() is called");
25.     }
26.     // 输出 onResume()方法被调用的日志
27.     @Override
28.     protected void onResume() {
29.         super.onResume();
30.         Log.i(TAG, "onResume() is called");
31.     }
32.     // 输出 onPause()方法被调用的日志
```

```
33.        @Override
34.        protected void onPause() {
35.            super.onPause();
36.            Log.i(TAG, "onPause() is called");
37.        }
38.    // 输出 onStop()方法被调用的日志
39.        @Override
40.        protected void onStop() {
41.            super.onStop();
42.            Log.i(TAG, "onStop() is called");
43.        }
44.    // 输出 onDestory()方法被调用的日志
45.        @Override
46.        protected void onDestroy() {
47.            super.onDestroy();
48.            Log.i(TAG, "onDestroy() is called");
49.        }
50. }
```

步骤 5：观察运行结果

运行程序，观察 LogCat 窗口的日志输出。当启动项目运行时，打印日志如图 3.11 所示。

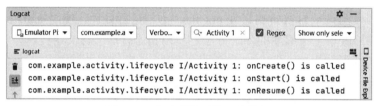

图 3.11 启动项目时的日志输出

可以看到，当 Activity1 第一次创建时会依次执行 onCreate()、onStart()和 onResume()方法，这是 Activity1 界面从创建到前台可见并可交互的过程，单击图 3.8 中的 Activity1 的按钮跳转到 Activity2 界面，输出的日志如图 3.12 所示。

图 3.12 打开 Activity2 时的日志输出

从图 3.12 可以看出，由于 Activity2 已经把 Activity1 完全遮挡，则首先执行 Activity1 的 onPause()方法，然后 Activity2 依次执行 onCreate()、onStart()和 onResume()方法，从创建到前台可见，此时 Activity1 再执行 onStop()方法。单击图 3.9 中的 Activity2 的按钮跳转到 Activity3 界面，输出的日志如图 3.13 所示。

从图 3.13 可以看出，由于 Activity3 没有将 Activity2 完全遮蔽，Activity2 只有 onPause() 方法被执行，onStop()方法并没有被执行，此时的 Activity2 只是进入了暂停状态，并没有进入停止状态。此时按下返回键返回 Activity2 只会执行 onResume()方法，输出日志如图 3.14 所示。

图 3.13　打开 Activity3 时的日志输出

图 3.14　返回 Activity2 时的日志输出

此时再按返回键返回 Activity1 界面，再按返回键退出程序，输出的日志如图 3.15、图 3.16 所示。此时的 Activity2 的 onPause()、onStop()和 onDestroy()方法依次被调用，Activity2 被销毁。Activity1 也会依次调用这三个方法，最终销毁 Activity1，从而结束程序。

图 3.15　返回 Activity1 时的日志输出

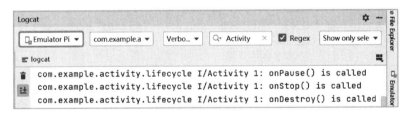

图 3.16　退出程序时的日志输出

通过这个案例的练习和讲解，对于图 3.6 所示的 Activity 栈结构、Activity 的状态转换及完整的生命周期应该有了更深刻的理解。

在手机使用过程中，经常会根据不同场景进行横竖屏切换。当手机横竖屏切换时，Activity 会销毁重建。如果希望 Activity 不被销毁，可以在 AndroidManifest.xml 清单文件中设置 Activity 的 Android:configChangers 属性为"orientation|keyboardHidden|screenSize"。

如果希望 Activity 界面一直处于竖屏或横屏状态，可以在 AndroidManifest.xml 清单文件中通过设置 Activity 的参数实现，具体代码如下所示：

竖屏：android:screenOrientation="portrait"

横屏：android:screenOrientation="landscape"

3.2 Android 的事件处理机制

在上节 Activity 的生命周期案例中，用户与应用程序的交互都是通过事件处理完成的。事件是程序设计过程中用户与应用程序进行交互的一种机制，如单击、长按、滑动等，Android 提供了以下两种人机交互的处理机制。

● 基于回调的处理机制：Android 的 View 都提供事件处理的回调方法，通过重写 View 的回调方法实现事件响应，没有被 View 处理的事件会被 Activity 的回调方法调用。

● 基于监听的处理机制：通过为 Android 的界面组件绑定特定的事件监听器，在事件监听器的方法中实现事件处理。

3.2.1 基于监听的事件处理

熟悉 Java 的 Swing 窗体编程的开发者都会很熟悉基于监听的事件处理机制，它涉及三类对象：事件源（Event Source）、事件（Event）和事件监听器（Event Listener），事件源是发生事件的界面组件，如：按钮、窗口或菜单等；事件封装了界面组件发生的特定事情，事件的相关信息一般通过 Event 对象获得；事件监听器负责监听事件源所发生的事件，并对事件做出响应，处理模型如图 3.17 所示。

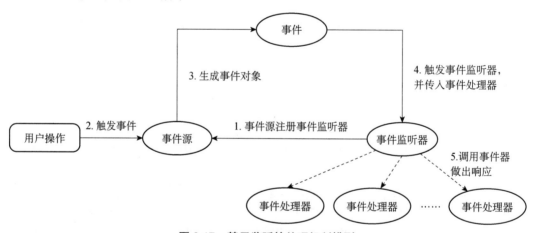

图 3.17 基于监听的处理机制模型

根据图 3.17 可以看出，Android 的事件处理的开发流程如下。

（1）获取作为事件源的界面组件，也就是被监听的对象。
（2）为界面组件设置监听器，监听用户的操作。
（3）实现事件监听器类，该监听器类是一个特殊的 Java 类，必须实现一个 XxxListener 接口。
（4）调用事件源的 setXxxListener 方法将事件监听器对象注册给事件源组件。

Android 系统为不同的界面组件提供了不同的监听器接口，常用的监听器接口如表 3.3 所示。

表 3.3 常用的监听器接口

监听器接口	含义说明	需实现的事件处理方法
OnClickListener	发生单击事件时调用	onClick
OnLongClickListener	发生长按事件时调用	onLongClick
OnKeyListener	按下或释放某个按键时调用	onKey
OnTouchListener	发生触摸屏运动事件时调用	onTouch
OnFocusChangeListener	视图获得或失去焦点时调用	onFocusChange
OnItemSelectedListener	列表项选中时调用	onItemSelected

事件处理就是实现事件监听器的类，它是实现特定接口的 Java 类实例，通过上个生命周期案例讲解事件监听的 5 种实现方式。

1. 匿名内部类

组件设置监听器的参数直接使用匿名内部类创建监听器对象，这是最常用的事件监听器，但只能临时使用一次，复用性低，示例代码如下：

```
1.  public class Activity1 extends AppCompatActivity {
2.      @Override
3.      protected void onCreate(Bundle savedInstanceState) {
4.          super.onCreate(savedInstanceState);
5.          setContentView(R.layout.activity_1);
6.          // 初始化按钮组件
7.          Button btnJump = findViewById(R.id.btn_jump);
8.          // 匿名内部类形式
9.          btnJump.setOnClickListener(new View.OnClickListener() {
10.             @Override
11.             public void onClick(View v) {
12.                 // 响应处理代码
13.             }
14.         });
15.     }
16. }
```

以上代码的第 7 行定义的 btnJump 按钮对象是事件源，第 9 行的 setOnClickListener(new View.OnClickListener)用于注册事件监听器，它的参数是单击事件监听器，也就是 View.OnclickListener(){...}匿名类对象。

2. 直接使用 Activity 类的形式

Activity 类直接实现 OnClickListener 监听器接口，重写相应的事件处理方法，这种方式也很常用，示例代码如下：

```
1.  public class Activity2 extends AppCompatActivity implements
        View.OnClickListener {
2.      @Override
3.      protected void onCreate(Bundle savedInstanceState) {
4.          super.onCreate(savedInstanceState);
5.          setContentView(R.layout.activity_2);
6.
7.          Button btnJump = findViewById(R.id.btn_login);
8.          btnJump.setOnClickListener(this);
9.      }
10.     @Override
```

```
11.    public void onClick(View view) {
12.        // 响应处理代码
13.    }
14. }
```

第 1 行定义的 MainActivity 类必须实现 OnClickListener 接口，第 10~13 行代码重写这个接口的 onClick()方法，在方法内实现事件的响应处理。

3. 内部类形式

将事件监听器类定义为当前 Activity 类的内部类，这种方式不仅复用性高，还可以直接访问外部类的所有组件，示例代码如下：

```
1. public class Activity3 extends AppCompatActivity {
2.     protected void onCreate(Bundle savedInstanceState) {
3.         super.onCreate(savedInstanceState);
4.         setContentView(R.layout.activity_3);
5.         Button btnJump = findViewById(R.id.btn_jump);
6.         btnJump.setOnClickListener(new ButtonClickListener());
7.     }
8.     // 定义单击事件监听器的内部类
9.     class ButtonClickListener implements View.OnClickListener {
10.        @Override
11.        public void onClick(View view) {
12.            // 响应处理代码
13.        }
14. }
```

第 9~14 行自定义 ButtonClickListener 类实现 OnClickListener 接口，重写 onClick()方法，第 6 行 btnLogin 对象设置单击事件监听器时，即可使用该内部类的实例对象。

4. 外部类形式

将上一种形式的内部类定义成 public 的 Java 外部类，使用方法与上一种形式相同。由于外部类不能直接访问 Activity 中的组件，需要通过构造方法将组件传入后使用，所以这种形式用得较少。

```
1. public class ButtonClickListener implements View.OnClickListener {
2.     @Override
3.     public void onClick(View view) {
4.         // 响应处理代码
5.     }
6. }
```

5. 直接绑定布局组件形式

在界面布局 XML 文件中为组件设置 android:onClick 属性，绑定事件处理方法，示例代码如下：

```
1. <Button
2.     ……
3.     android:onClick="clickHandle" />
```

然后在 Activity 类中添加相应的方法定义。需要注意的是，属性定义的方法名务必与类定义的方法名一致，且不能添加@Override 注解，示例代码如下所示：

```
1. public class MainActivity extends AppCompatActivity {
2.     // 方法名与布局中 android:onClick 属性的值一致
```

```
3.      public void clickHandle(View view) {
4.          // 响应处理代码
5.      }
6.  }
```

3.2.2 基于回调的事件处理

当用户按下、滑动或双击界面上某个控件时，会触发该控件特定的方法处理该事件，这个处理过程就是基于回调机制的事件处理。为了实现回调机制的事件处理，Android 为所有的 View 组件提供了回调方法，View 类包含的回调方法示例如下。

- boolean onKeyDown（int keyCode, KeyEvent event）：按下按键时触发的方法。
- boolean onKeyLongPress（int keyCode, KeyEvent event）：长按按键时触发的方法。
- boolean onKeyUp（int keyCode, KeyEvent event）：松开按键时触发的方法。

以上事件处理方法都返回 Boolean 类型的值，该返回值用来标识这个方法是否已经完全处理完该事件了，如果返回 false 则表示该事件未处理完，事件会向上传播，被上一层组件的回调方法继续处理，如组件所在的 Activity；返回 true 则表示该事件已被处理完，不会被其他组件的回调方法再处理，这就是基于回调的事件传播性质。

键盘事件的回调方法的第一个参数 keyCode 表示按下的按键的键盘码，键盘上的每个按键都有一个键盘码值，用于判断用户按下的是哪个按键。第二个参数是封装按键事件的 KeyEvent 对象，它包含事件对象的详细信息，如事件状态、事件类型及事件触发的时间等，它也定义了按键的键盘码常量，如 KeyEvent.KEYCODE_BACK 表示回退键。

除了以上针对手机键盘按键的事件处理，手机屏幕触摸事件由 onTouchEvent（MotionEvent event）方法处理，如在屏幕中的按下、抬起和滑动等事件。该方法在 View 类定义，所有的 View 都重写了该方法，它的参数 event 是屏幕触摸事件封装的对象，用于表示触摸位置、类型及时间等信息，它的返回机制与键盘响应事件相同，此部分知识将在第 9 章的手势处理一节详细讲解。

下面通过一个案例讲解两种事件处理机制。

【案例 3-3】实现事件监听、事件回调两种事件处理机制，界面及运行结果如图 3.18 所示。

图 3.18　界面布局及运行结果

步骤1：创建项目

启动 Android Studio，创建名为 D0303_EventHandle 的项目，选择 Empty Activity，将包名改为 com.example.event，单击 Finish 按钮，等待构建完成。

步骤2：编写界面布局

根据图 3.18 所示的界面布局，在 activity_main.xml 布局文件中添加 TextView 和 Button 控件。

步骤3：编写基于监听机制的事件处理代码

打开 MainActivity 类文件，在 onCreate()方法中初始化 TextView 和 Button 对象，编写 Button 的事件监听代码，示例代码如下：

```java
1.  public class MainActivity extends AppCompatActivity {
2.      private TextView tvResult;
3.      @Override
4.      protected void onCreate(Bundle savedInstanceState) {
5.          super.onCreate(savedInstanceState);
6.          setContentView(R.layout.activity_main);
7.          // 初始化控件对象
8.          tvResult = findViewById(R.id.tv_result);
9.          Button btnConfirm = findViewById(R.id.btn_confirm);
10.         // 设置监听器
11.         btnConfirm.setOnClickListener(new View.OnClickListener() {
12.             @Override
13.             public void onClick(View v) {
14.                 tvResult.setText("按下了按钮-基于监听机制的事件处理");
15.             }
16.         });
17.     }
18. }
```

步骤4. 编写基于回调机制的事件处理代码

在 MainActivity 类中重写 onKeyDown()和 onTouchEvent()两个回调方法，根据 onKeyDown()方法的第一个参数 keyCode 键盘码获悉按下了哪个键，通过 switch…case 代码段完成不同按键的处理。onTouchEvent 的 MotiveEvent 事件对象可以获取当前触摸点的 x、y 坐标值。示例代码如下：

```java
1.  @Override
2.  public boolean onKeyDown(int keyCode, KeyEvent event) {
3.      switch (keyCode) {
4.          case KeyEvent.KEYCODE_0:
5.              tvResult.setText("按下了【0】");
6.              break;
7.          case KeyEvent.KEYCODE_VOLUME_UP:
8.              tvResult.setText("按下了【音量+】");
9.              break;
10.         case KeyEvent.KEYCODE_A:
11.             tvResult.setText("按下了【a】");
12.             break;
13.         default:
14.             tvResult.setText("按下的键盘码【" + keyCode + "】");
15.             break;
16.     }
17.     return super.onKeyDown(keyCode, event);
18. }
19.
20. @Override
21. public boolean onTouchEvent(MotionEvent event) {
```

```
22.         tvResult.setText("当前坐标:【" + String.format("%.2f", event.
                                                        getX()) + ", " +
23.                 String.format("%.2f", event.getY()) + "】");
24.         return super.onTouchEvent(event);
25. }
```

步骤 5：运行项目

运行项目，分别单击按钮、手指在界面滑动及按下"0"、"a"和"音量+"键，查看 TextView 显示的内容。

3.3 Activity 使用 Intent

上节的案例已经告诉我们 App 通常由多个界面组成，界面跳转本质上是 Activity 之间的跳转。Android 系统通过 Intent 实现 Activity 之间的跳转及数据的传递。

Intent 的中文含义是"意图"，对要执行的操作进行描述，是组件交互的重要方式，它不仅用于指定当前组件执行的动作，还可以在组件之间传递数据，一般用于启动 Activity、Service 和发送广播（注：Service 和广播将在后续章节讲解）。

Intent 类封装了如表 3.4 所示的属性，系统会根据这些属性的组合确定对应的行为。

表 3.4 Intent 包含的属性的描述

属性名称	设置方法	说明与用途	备注
Component	setComponent	组件，用于指定 Intent 的来源与目的	
Action	setAction	动作，用于指定 Intent 的操作行为	隐式 Intent 必需
Data	setData	数据，用于指定动作要操作的数据路径，用 Uri 表示	隐式 Intent 必需
Type	SetType	数据类型，用于指定 Data 类型的定义	
Category	setCategory	类别，用于指定 Intent 的操作类别	隐式 Intent 必需
Extras	putExtra	扩展信息，用于指定装载的参数信息	附加信息
Flags	setFlags	标志位，用于指定 Intent 的运行模式	附加信息

Android 系统将 Intent 分为显式 Intent 和隐式 Intent，接下来分别进行讲解。

3.3.1 显式 Intent

显式 Intent 通常用在同一个应用程序中，通过构造函数 Intent（Context context, Class<?> cls）指定目标组件，它的第一个参数 Context 表示启动 Activity 的上下文，第二个参数 Class 则是指定需要启动的目标组件，然后将创建的 Intent 对象作为 startActivity()方法的参数启动目标 Activity，案例 3-2 中 Activity1 跳转到 Activity2 的具体代码如下：

```
1.  btnJump.setOnClickListener(new View.OnClickListener() {
2.      @Override
3.      public void onClick(View v) {
4.          // 创建 Intent 对象
5.          Intent intent = new Intent(Activity1.this, Activity2.class);
6.          // 启动 Activity 2
7.          startActivity(intent);
8.      }
9.  });
```

第 5 行代码构造了一个 Intent 对象，传入 Activity1.this 作为上下文，传入 Activity2.class 作为目标组件，那我们的"意图"就是在 Activity1 这个 Activity 的基础上打开 Activity2，然后通过第 7 行的 startActivity()方法执行"意图"，这就是显式 Intent。

除了通过指定类名启动组件的方法之外，显式 Intent 还可以通过 setClassName()方法设置目标组件的包名、全限定名进行指定，具体代码如下：

```
1.  btnJump.setOnClickListener(new View.OnClickListener() {
2.      @Override
3.      public void onClick(View v) {
4.          Intent intent = new Intent();
5.          intent.setClassName("com.example.activity.lifecycle",
6.              "com.example.activity.lifecycle.Activity2");
7.          startActivity(intent);
8.      }
9.  });
```

3.3.2 隐式 Intent

相对于显式 Intent，隐式 Intent 并不明确指定目标组件，而是过滤 action、category 等信息，交给系统分析，从而找到匹配的目标组件。在 AndroidManifest.xml 清单文件中配置 Activity 的 action、category 的具体代码如下：

```
1.  <activity android:name=".Activity2" >
2.      <intent-filter>
3.          <action android:name="com.example.activity.lifecycle.
                              ACTION_START" />
4.          <category android:name="android.intent.category.DEFAULT" />
5.      </intent-filter>
6.  </activity>
```

上述代码的<action>标签指定 Activity2 可以响应"com.example.activity.lifecycle.ACTION_START"动作，而<category>标签包含类别信息，当这两个信息同时匹配时，Activity2 才会启动。使用隐式 Intent 启动 Activity2 的代码如下：

```
1.  btnJump.setOnClickListener(new View.OnClickListener() {
2.      @Override
3.      public void onClick(View v) {
4.          Intent intent = new Intent();
5.          intent.setAction("com.example.activity.lifecycle.ACTION_START");
6.          startActivity(intent);
7.      }
8.  });
```

使用隐式 Intent 不仅可以启动自己程序内的 Activity，还可以启动其他程序的 Activity，诸如打电话、发短信和调用网页等系统内置功能的调用都可以通过隐式 Intent 实现，常用的系统隐式 Intent 的动作如表 3.5 所示。

表 3.5　常用系统隐式 Intent 的动作

Intent 类的系统动作常量名	动作的常量值	功能描述
ACTION_MAIN	android.intent.action.MAIN	App 启动时的入口
ACTION_VIEW	android.intent.action.VIEW	显示数据
ACTION_EDIT	android.intent.action.EDIT	显示可编辑的数据

续表

Intent 类的系统动作常量名	动作的常量值	功能描述
ACTION_SEND	android.intent.action.SEND	分享内容
ACTION_CALL	android.intent.action.CALL	直接拨号
ACTION_DIAL	android.intent.action.DIAL	准备拨号
ACTION_SENDTO	android.intent.action.SENDTO	发送短信
ACTION_ANSWER	android.intent.action.ANSWER	接听电话
ACTION_SEARCH	android.intent.action.SEARCH	导航栏 SearchView 的搜索动作

下面通过一个案例讲解它们的应用。

3.3.3 隐式 Intent 案例

【案例 3-4】使用隐式 Intent 实现打电话、发短信和打开网页等系统功能，界面如图 3.19 所示。

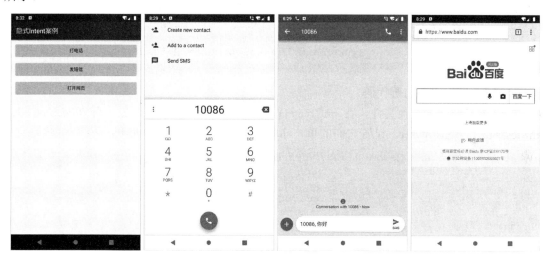

图 3.19　主界面、拨打电话、发送短信和打开网页的界面

步骤 1：创建项目

启动 Android Studio，创建名为 D0304_ImplicitIntent 的项目，将包名改为 com.example.implicit.intent，选择 Empty Activity，单击 Finish 按钮，等待系统构建完成。

步骤 2：完成 Activity 的界面布局

打开 activity_main.xml 布局文件，参照图 3.19 所示的主界面布局，增加三个按钮的布局，分别是打电话、发短信和打开网页，设置按钮的 id，具体代码如下：

```
1.  <?xml version="1.0" encoding="utf-8"?>
2.  <androidx.constraintlayout.widget.ConstraintLayout
3.      xmlns:android="http://schemas.android.com/apk/res/android"
4.      xmlns:app="http://schemas.android.com/apk/res-auto"
5.      xmlns:tools="http://schemas.android.com/tools"
6.      android:layout_width="match_parent"
7.      android:layout_height="match_parent"
8.      tools:context=".MainActivity">
9.      <Button
```

```xml
10.        android:id="@+id/btn_call"
11.        android:layout_width="match_parent"
12.        android:layout_height="wrap_content"
13.        android:layout_margin="10dp"
14.        android:text="打电话"
15.        app:layout_constraintLeft_toLeftOf="parent"
16.        app:layout_constraintRight_toRightOf="parent"
17.        app:layout_constraintTop_toTopOf="parent" />
18.    <Button
19.        android:id="@+id/btn_message"
20.        android:layout_width="match_parent"
21.        android:layout_height="wrap_content"
22.        android:layout_margin="10dp"
23.        android:text="发短信"
24.        app:layout_constraintLeft_toLeftOf="parent"
25.        app:layout_constraintRight_toRightOf="parent"
26.        app:layout_constraintTop_toBottomOf="@+id/btn_call" />
27.    <Button
28.        android:id="@+id/btn_browser"
29.        android:layout_width="match_parent"
30.        android:layout_height="wrap_content"
31.        android:layout_margin="10dp"
32.        android:text="打开网页"
33.        app:layout_constraintLeft_toLeftOf="parent"
34.        app:layout_constraintRight_toRightOf="parent"
35.        app:layout_constraintTop_toBottomOf="@+id/btn_message" />
36. </androidx.constraintlayout.widget.ConstraintLayout>
```

步骤3：编写事件监听代码

打开 MainActivity 类文件，在 onCreate()方法中初始化三个按钮对象，分别为 btnCall、btnMessage 和 btnBrowser，设置它们的单击事件的监听器。具体代码如下：

```java
1.  public class MainActivity extends AppCompatActivity implements View.OnClickListener {
2.      @Override
3.      protected void onCreate(Bundle savedInstanceState) {
4.          super.onCreate(savedInstanceState);
5.          setContentView(R.layout.activity_main);
6.          // 初始化控件
7.          Button btnCall = findViewById(R.id.btn_call);
8.          Button btnMessage = findViewById(R.id.btn_message);
9.          Button btnBrowser = findViewById(R.id.btn_browser);
10.         // 设置单击事件监听器
11.         btnCall.setOnClickListener(this);
12.         btnMessage.setOnClickListener(this);
13.         btnBrowser.setOnClickListener(this);
14.     }
15.     @Override
16.     public void onClick(View v) {
17.         // 实现按钮功能
18.     }
19. }
```

上述代码中，MainActivity 实现了 OnClickListener 接口，需要重写它的 onClick()方法，使得按钮在 MainActivity 对象上设置监听器。

步骤4：编写功能代码

在 MainActivity 类中定义 callPhone()方法实现打电话的功能，在此方法中创建 action 为 Intent.ACTION_DIAL 的隐式 Intent 对象。然后调用 Intent 的 setData()方法设置拨打电话的 data

数据，拨打给 10086 号码的 data 数据为"tel:10086"，setData()的参数是一个 Uri 对象，需要通过 Uri.parse()方法将此字符串转为 Uri 对象。

发短信功能由 sendMessage()方法实现，它的 action 为 Intent.ACTION_SENDTO，发送给 10086 号码的 data 数据为"smsto:10086"，并将它转为 Uri 对象；发送的信息使用 Intent 的 putExtra()方法发送额外数据。

打开网页功能的实现方法是 openBrowser()，action 为 Intent.ACTION_VIEW，data 数据就是网址，同样需要将字符串转为 Uri 对象。

具体代码如下：

```
@Override
public void onClick(View v) {
    switch (v.getId()) {
        case R.id.btn_call:
            callPhone();
            break;
        case R.id.btn_message:
            sendMessage();
            break;
        case R.id.btn_browser:
            openBrowser();
            break;
    }
}
// 打电话
private void callPhone() {
    Intent intent = new Intent(Intent.ACTION_DIAL);
    intent.setData(Uri.parse("tel:10086"));
    startActivity(intent);
}
// 发短信
private void sendMessage() {
    Intent intent = new Intent(Intent.ACTION_SENDTO);
    intent.setData(Uri.parse("smsto:10086"));
    intent.putExtra("sms_body","10086, 你好");
    startActivity(intent);
}
// 打开百度的主页
private void openBrowser() {
    Intent intent = new Intent(Intent.ACTION_VIEW);
    intent.setData(Uri.parse("https://www.baidu.com"));
    startActivity(intent);
}
```

步骤 5：运行项目

接下来运行项目，分别单击屏幕中的三个按钮，得到如图 3.19 所示的三个界面。

3.4 Activity 的数据传递

Android 应用程序一般由多个 Activity 组成，这些 Activity 之间不仅可以相互跳转，而且也可以传递数据，以及数据回传。本节将针对 Activity 之间的数据传递进行详细讲解。

3.4.1 Intent 数据传递

Android 提供 Intent 来实现 Activity 之间的跳转，Activity 之间的数据传递也是通过 Intent 实现的。Intent 提供多个 putExtra()重载方法传递基本类型、数组、字符串等类型的数据。putExtra()方法包含两个参数，第一个参数是字符串类型的键值，第二个参数是传递的数据，相当于 Intent 对象使用了键值对传递数据。接下来通过一个案例讲解 Intent 传递数据。

【案例 3-5】创建 ActivityDataDemo 项目，实现 Intent 传递简单类型、集合和对象等数据。

步骤 1：创建项目

启动 Android Studio，创建名为 D0305_ActivityData 的项目，选择 No Activity，将包名改为 com.example.activity.data，单击 Finish 按钮，等待系统构建完成。

步骤 2：创建 FirstActivity 和 SecondActivity

创建 FirstActivity，勾选 Generate a Layout File 和 Launcher Activity，单击 Finish 按钮完成。按照同样的方法创建 SecondActivity，只是不勾选 Launcher Activity。接下来编写界面布局，如图 3.20、图 3.21 所示。

图 3.20　FirstActivity 界面布局　　　　　图 3.21　SecondActivity 界面布局

步骤 3：编写事件监听代码

在 FirstActivity 类中初始化 Button 对象，通过 FirstActivity 类实现 OnClickListener 接口，设置 Button 在 FirstActivity 对象上单击事件监听器。具体代码如下：

```
1.  public class FirstActivity extends AppCompatActivity
2.          implements View.OnClickListener {
3.      @Override
4.      protected void onCreate(Bundle savedInstanceState) {
5.          super.onCreate(savedInstanceState);
6.          setContentView(R.layout.activity_first);
7.          // 初始化按钮对象
8.          Button btnData = findViewById(R.id.btn_data);
9.          Button btnList = findViewById(R.id.btn_list);
10.         Button btnObject = findViewById(R.id.btn_object);
11.         Button btnBundle = findViewById(R.id.btn_bundle);
12.         Button btnReturn = findViewById(R.id.btn_return);
13.         // 设置按钮的监听器
```

```
14.         btnData.setOnClickListener(this);
15.         btnList.setOnClickListener(this);
16.         btnObject.setOnClickListener(this);
17.         btnBundle.setOnClickListener(this);
18.         btnReturn.setOnClickListener(this);
19.     }
20.     @Override
21.     public void onClick(View v) {
22.         // 实现按钮功能
23.     }
24. }
```

步骤 4：Intent 传递数据

（1）Intent 直接传递简单类型数据

FirstActivity 类中的 btnData 按钮用于处理传递字符串数据，创建 Intent 对象，调用 Intent 的 putExtra()方法直接传递字符串数据，示例代码如下：

```
1. public void onClick(View v) {
2.     Intent intent = new Intent(FirstActivity.this, SecondActivity.class);
3.     switch (v.getId()) {
4.         case R.id.btn_data:
5.             intent.putExtra("data", "Activity传递字符串");
6.             startActivity(intent);
7.             break;
8.     }
9. }
```

SecondActivity 在接收数据时的代码如下：

```
1.  protected void onCreate(Bundle savedInstanceState) {
2.      super.onCreate(savedInstanceState);
3.      setContentView(R.layout.activity_second);
4.      // 初始化控件
5.      TextView tvData = findViewById(R.id.tv_data);
6.      // 接收 Intent 传递的字符串
7.      Intent intent = getIntent();
8.      String data = intent.getStringExtra("data");
9.      if (data != null) {
10.         tvData.setText("获取的字符串为：" + data);
11.     }
12. }
```

上述代码首先在 SecondActivity 中调用 getIntent()方法获取 Intent 对象，调用 Intent 对象的 getStringExtra()方法获取传递的字符串数据，这个方法需要传入传递数据时的键值。根据数据类型的不同，Intent 提供了多个 getXxxExtra()方法获取数据，其中的 Xxx 代表数据类型，如 getIntExtra()用于获取整型值、getStringArrayListExtra()用于获取字符串集合等，以此类推。

（2）Intent 传递 ArrayList 集合对象

在 FirstActivity 类的 onClick()方法中添加获取集合数据的按钮处理代码，调用 Intent 的 putIntegerArrayListExtra()方法传递 ArrayList 类型的整型数据 datas，示例代码如下：

```
1. case R.id.btn_list:
2.     ArrayList<Integer> datas = new ArrayList<>();
3.     datas.add(85);
4.     datas.add(90);
5.     datas.add(78);
6.     intent.putIntegerArrayListExtra("list", datas);
7.     startActivity(intent);
8.     break;
```

SecondActivity 调用 Intent 的 getIntegerArrayListExtra()方法获取 ArrayList 整型数据,示例代码如下:

```
1.  ArrayList<Integer> datas = intent.getIntegerArrayListExtra("list");
2.  if (datas != null) {
3.      tvData.setText("获取的列表数据为: " + datas.toString());
4.  }
```

（3）Intent 传递 Object 对象数据

可以直接使用 putExtra()方法传递 Object 对象,被传递的对象必须实现 Serialiable 接口或 Parcelable 接口,被传递对象 User 类的示例代码如下:

```
1.  public class User implements Serializable {
2.      private String name;
3.      private int age;
4.      // 有参构造方法
5.      public User(String name, int age) {
6.          this.name = name;
7.          this.age = age;
8.      }
9.      // 省略 getter/setter 方法
10. }
```

定义一个 User 对象,调用 Intent 的 putExtra()方法传递对象,示例代码如下:

```
1.  case R.id.btn_object:
2.      User user = new User("王晓", 20);
3.      intent.putExtra("object", user);
4.      startActivity(intent);
5.      break;
```

SecondActivity 在接收数据时,调用 Intent 的 getSerializableExtra()方法获取 Object 类型的对象,使用强制类型转换将 Object 转为 User 对象,示例代码如下:

```
1.  User user = (User) intent.getSerializableExtra("object");
2.  if (user != null) {
3.      tvData.setText("获取的对象数据为: " + user.toString());
4.  }
```

（4）Bundle 对象传递数据

根据 Android 的官方建议,Intent 传递的数据量大小应限制在 KB 级别,传递大型数据量可能会导致 TransactionTooLargeException 异常,传递数据量较大的信息应使用 Bundle 对象。Bundle 对象设置数据的方法为 putXxx(),参数也采用键值对的形式,其中 Xxx 代表数据类型,如 String 代表字符串、Int 代表整型等,调用 Intent 的 putExtras()方法设置 Bundle 类型数据,示例代码如下:

```
1.  case R.id.btn_bundle:
2.      Bundle bundle = new Bundle();
3.      bundle.putString("username", "王晓");
4.      bundle.putInt("age", 20);
5.      intent.putExtras(bundle);
6.      startActivity(intent);
7.      break;
```

SecondActivity 在接收数据时,调用 Intent 的 getExtras()方法获取 Bundle 对象,然后再调用 getString()、getInt()等方法获取传递的数据,参数就是传递的键值。示例代码如下:

```
1.  Bundle bundle = intent.getExtras();
2.  if(bundle != null) {
3.      String name = bundle.getString("username");
4.      int age = bundle.getInt("age");
5.      if(name != null) {
6.          tvData.setText("获取的bundle数据为: " + name + ", " + age);
7.      }
8.  }
```

步骤 4：运行项目

运行项目，单击按钮传递数据的结果如图 3.22 所示。

图 3.22　各种数据类型传值的运行结果

3.4.2　Activity 的数据回传

假设如下场景：一个应用程序的主界面是登录界面，首次运行必须先注册，用户希望注册成功后返回登录界面，同时将注册的用户名填充到登录界面的用户名输入框中。之前讲解的 startActivity()方法只能启动新 Activity，当返回原 Activity 时无法传回任何信息，为此，Activity 提供了 startActivityForResult()方法实现这个场景的需求。

startActivityForResult()方法有两个参数，第一个参数是 Intent，第二个参数是用于回调时判断数据来源的请求码，接下来完成【案例 3-5】中"传值并返回"按钮的功能。

步骤 1：编写 FirstActivity 事件处理代码

在 FirstActivity 类中编写 btn_return 按钮的处理代码，调用 startActivityForResult()方法传递字符串数据，设置请求码为整数值 1，也可以设置为任意整数值，示例代码如下：

```
1.  case R.id.btn_return:
2.      intent.putExtra("data", "传递字符串");
3.      startActivityForResult(intent, 1);
4.      break;
```

步骤 2：编写 SecondActivity 类的事件监听

SecondActivity 界面上的"返回"按钮完成关闭自己的同时，设置传递的数据信息，调用 setResult()方法返回 FirstActivity 界面，它的第一个参数是返回码，设置为 RESULT_OK 表示正常返回，第二个参数为包含回传信息的 Intent 对象。该按钮的事件监听及处理代码如下：

```
1.  Button btnBack = findViewById(R.id.btn_back);
2.  btnBack.setOnClickListener(new View.OnClickListener() {
3.      @Override
4.      public void onClick(View v) {
5.          Intent intent = new Intent();
6.          intent.putExtra("data", "返回FirstActivity");
7.          setResult(RESULT_OK, intent);
8.          finish();
9.      }
10. });
```

步骤 3：接收返回数据

FirstActivity 类需要重写 Activity 的 onActivityResult()方法以接收 SecondActivity 传回的数据，该方法有三个参数，第一个参数是发出的请求码，第二个参数是 SecondActivity 的返回码，第三个参数是 SecondActivity 返回的 Intent 对象，示例代码如下：

```
1.  @Override
2.  protected void onActivityResult(int requestCode, int resultCode,
                                    Intent data) {
3.      super.onActivityResult(requestCode, resultCode, data);
4.      TextView tvData = findViewById(R.id.tv_data);
5.      // 根据请求码 requestCode 进行判断
6.      if (requestCode == 1) {
7.          if (resultCode == RESULT_OK && data != null) {
8.              // 通过 Intent 对象获取返回的数据
9.              String returnData = data.getStringExtra("data");
10.             // 判断数据是否为 null
11.             if (returnData != null) {
12.                 tvData.setText("返回的数据为: " + returnData);
13.             }
14.         }
15.     }
16. }
```

注意：onActivityResult()方法中的 requestCode 必须与 startActivityForResult()方法中的 requestCode 的值相同，否则 onActivityResult()方法就不能获取 SecondActivity 传回的数据。

步骤 4：运行项目

接下来运行项目，传递数据并返回的运行结果如图 3.23 所示。

图 3.23　传值并返回的运行结果

至此，详细讲解了 Activity 之间的数据传递的方法，这种方法非常常用，如调用系统相机、通过相册获取照片、获取系统通信录等，读者应在充分理解的基础上根据场景进行灵活应用。

3.5 Activity 启动模式

针对不同的使用场景，Activity 有 4 种启动模式，分别是 standard、singleTop、singleTask 和 singleInstance，可以通过 AndroidManifest.xml 的<activity>标签的 Android:launchMode 进行设置。下面以案例 3-2 中的 3 个 Activity 为例，讲解这 4 种启动模式。

1. standard 模式

standard 模式是 Activity 启动的默认模式，这种模式下系统每次启动 Activity 时都会创建一个新的实例压入任务栈，并不关心这个 Activity 是否已存在。案例 3-2 中的 Activity1、Activity2 和 Activity3 的启动模式为 standard，它们的启动顺序为 Activity1->Activity2->Activity3，则它们的任务栈模型如图 3.24 所示。出栈时，位于栈顶的 Activity3 最先出栈。

2. singleTop 模式

如果 Activity1 已经位于栈顶，在 standard 模式下再次启动 Activity1 依旧会在栈顶创建一个新实例，不能直接复用，这种情况显然不太合理。singleTop 模式可以解决这个问题，它会判断新启动的 Activity 是否位于栈顶，如果位于栈顶则直接复用，否则创建新实例，任务栈模型如图 3.25 所示。

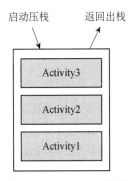

图 3.24　standard 模式

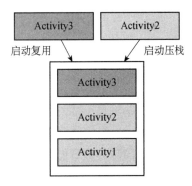

图 3.25　singleTop 模式

3. singleTask 模式

尽管 singleTop 模式可以有效解决栈顶重复创建 Activity 的问题，但如果 Activity 不位于栈顶，依旧会重复创建 Activity 的实例。第 3 种模式 singleTask 可以让 Activity 在整个应用程序的上下文中都只存在一个实例。当启动模式指定为 singleTask，则每次 Activity 启动时都会检查返回栈是否存在该 Activity 的实例，如果存在则直接复用，并且将该 Activity 上面的所有 Activity 都出栈，否则创建新实例，任务栈模型如图 3.26 所示。

4. singleInstance 模式

singleInstance 是 4 种模式中最特殊的，指定为 singleInstance 模式的 Activity 在首次启动

时会创建一个新的返回栈管理这个 Activity，如果该 Activity 的实例已经存在则直接复用，也就是说，在整个 Android 系统中只存在一个该 Activity 实例，可以看作是加强版的 singleTask 模式，任务栈模型如图 3.27 所示。

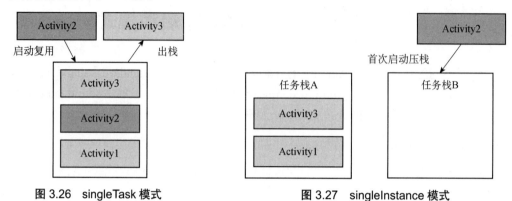

图 3.26 singleTask 模式　　　　　图 3.27 singleInstance 模式

综上所述，4 种启动模式之间的区别和适用场景归纳如表 3.6 所示。

表 3.6　4 种启动模式的区别和适用场景

启动模式	实例个数	每次生成新实例	存储的任务栈	适用场景
标准 standard	多个	是	与发送 Intent 的发送者在同一任务栈	普通界面
栈顶复用 singleTop	栈顶 1 个，不在栈顶多个	取决于是否在栈顶	与发送 Intent 的发送者在同一任务栈	通知等
栈内复用 singleTask	1 个	取决于是否在目标栈	取决于 taskAffinity	主页等
单例 singleInstance	1 个	不是	新建任务栈	闹钟等

3.6　Fragment

平板电脑与手机最大的区别在于屏幕的大小，平板电脑的屏幕大小一般在 7～10 英寸，有的甚至达到 12 英寸，这使得一些在手机上布局美观的界面在平板电脑上会存在比例失调等诸多问题，为了同时兼顾智能手机和平板电脑的开发，Android 3.0 引入了 Fragment 的概念，接下来将详细讲解 Fragment 的基本概念及应用。

3.6.1　Fragment 简介

Fragment 的中文含义是"片段"或"碎片"，这表明 Fragment 是嵌在 Activity 中使用的 UI 片段，一个 Activity 中可以组合多个 Fragment，多个 Activity 可重复使用某个 Fragment。Fragment 可以视为 Activity 的模块化组成部分，它具有自己的生命周期，能处理用户事件，并能在 Activity 运行时动态添加或移除，Android 和 Fragment 的关系如图 3.28 所示。

综上所述，Fragment 具备以下优势：
- 模块化 Modularity：无须将所有代码写在 Activity 中，而是分别写在各自的 Fragment 中。
- 重用性 Reuasbility：多个 Activity 可以重用一个 Fragment。
- 适配性 Adaptability：根据硬件的屏幕尺寸、屏幕方向，实现不同的布局。

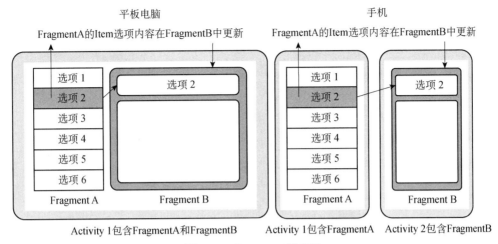

图 3.28 Fragment 的应用

从图 3.28 可以看出，平板电脑横屏时，FragmentA 和 FragmentB 同时嵌入到 Activity1 中；而在手机或平板电脑竖屏时，FragmentA 嵌入 Activity1，Fragment2 需要嵌入 Activity2，这样的布局既合理又美观。

3.6.2　使用 Fragment

创建 Fragment 的方法与 Activity 类似，自定义 Fragment 类继承自 androidx.fragment.app.Fragment，示例代码如下：

```
1.  public class TitleFragment extends Fragment {
2.      @Override
3.      public View onCreateView(LayoutInflater inflater, ViewGroup container,
4.                              Bundle savedInstanceState) {
5.          View view = inflater.inflate(R.layout.fragment_title,
                                         container, false);
6.          return view;
7.      }
8.  }
```

上述代码重写了 Fragment 的 onCreateView()方法，该方法的功能与 Activity 的 onCreate()方法类似，其中，调用 LayoutInflater 的 inflater()方法将布局文件 fragment_title.xml 动态加载到 Fragment 中。

Activity 有两种添加 Fragment 的方式，除了上述动态加载的方式，还可以通过在 Activity 的布局中添加<fragment>标签进行加载，下面使用一个案例演示这两种加载方式。

【案例 3-6】创建 Fragment 项目，演示 Fragment 的两种加载方式。两个 Fragment 界面如图 3.29 所示，运行结果如图 3.30 所示。

步骤 1：创建项目

启动 Android Studio，创建名为 D0306_Fragment 的项目，选择 Empty Activity 项，将包名改为 com.example.fragment，单击 Finish 按钮，等待项目构建完成。

步骤 2：创建 Fragment

右击 com.example.fragment，选择 New->Fragment->Fragment（Blank），打开创建 Fragment 对话框，填写 Fragment Name，分别创建名为 TitleFragment 和 ContentFragment 的 Fragment，如图 3.31 所示。

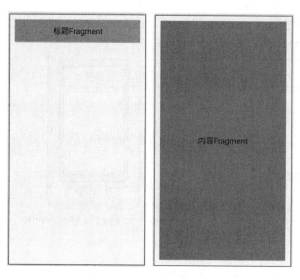

图 3.29　Fragment 布局　　　　　图 3.30　动态加载 Fragment 的运行结果

图 3.31　创建 Fragment 对话框

根据向导生成的 TitleFragment 类的代码比较复杂，代码如下所示：

```
1.  public class TitleFragment extends Fragment {
2.      private static final String ARG_PARAM1 = "param1";
3.      private static final String ARG_PARAM2 = "param2";
4.      private String mparam1;
5.      private String mparam2;
6.      public TitleFragment() {
7.          // Required empty public constructor
8.      }
9.      public static TitleFragment newInstance(String param1, String param2) {
10.         TitleFragment fragment = new TitleFragment();
11.         Bundle args = new Bundle();
12.         args.putString(ARG_PARAM1, param1);
13.         args.putString(ARG_PARAM2, param2);
14.         fragment.setArguments(args);
```

```
15.         return fragment;
16.     }
17.     @Override
18.     public void onCreate(Bundle savedInstanceState) {
19.         super.onCreate(savedInstanceState);
20.         if (getArguments() != null) {
21.             mParam1 = getArguments().getString(ARG_PARAM1);
22.             mParam2 = getArguments().getString(ARG_PARAM2);
23.         }
24.     }
25.     @Override
26.     public View onCreateView(LayoutInflater inflater, ViewGroup container,
27.                              Bundle savedInstanceState) {
28.         return inflater.inflate(R.layout.fragment_title, container, false);
29.     }
30. }
```

其中，

- 第2~3行定义了两个字符串常用，用于标识传递参数的键值。
- 第4~5行定义了两个参数mparam1、mparam2用于从Activity向Fragment传递数据。
- 第9~16行使用静态工厂方法newInstance()创建TitleFragment对象，使用Bundle对象设置两个参数，使用Fragment的setArguments()方法设置Bundle进行传递。
- 第17~24行重写了Fragment的onCreate()方法，使用Fragment的getArguments()方法接收传递的参数值。
- 第25~29行重写了Fragment的onCreateView()方法动态加载TitleFragment的布局。
- 如果无须传递参数给Fragment，可以删除与传递参数相关的代码。

步骤3：完成Fragment的界面布局

系统创建的Fragment默认使用FrameLayout帧布局，简单设置两个Fragment的内容和样式，如图3.29所示。此布局非常简单，使用TextView展示文本，此处不再展示XML布局代码。

步骤4：Activity添加Fragment

创建的Fragment必须添加到Activity中才能使用。Activity添加Fragment有静态和动态两种方式，下面一一讲解。

1. 静态加载

在Activity的布局文件中使用<fragment>标签静态添加Fragment，必须指定android:name属性，其属性值为Fragment的全限定名；tools:layout属性用于设置Fragment的布局，属性值为layout目录的fragment的名称；app:layout_constraintVertical_weight属性是ConstraintLayout布局设置纵向权重的属性。MainActivity加载TitleFragment和ContentFragment的示例代码如下：

```
1.  <?xml version="1.0" encoding="utf-8"?>
2.  <androidx.constraintlayout.widget.ConstraintLayout
3.      xmlns:android="http://schemas.android.com/apk/res/android"
4.      xmlns:app="http://schemas.android.com/apk/res-auto"
5.      xmlns:tools="http://schemas.android.com/tools"
6.      android:layout_width="match_parent"
7.      android:layout_height="match_parent"
8.      tools:context=".MainActivity">
9.      <fragment
10.         android:id="@+id/fragment_title"
11.         android:name="com.example.fragment.TitleFragment"
12.         android:layout_width="match_parent"
13.         android:layout_height="0dp"
14.         app:layout_constraintVertical_weight="1"
```

```
15.        app:layout_constraintLeft_toLeftOf="parent"
16.        app:layout_constraintRight_toRightOf="parent"
17.        app:layout_constraintTop_toTopOf="parent"
18.        app:layout_constraintBottom_toTopOf="@+id/fragment_content"
19.        tools:layout="@layout/fragment_title" />
20.    <fragment
21.        android:id="@+id/fragment_content"
22.        android:name="com.example.fragment.ContentFragment"
23.        android:layout_width="match_parent"
24.        android:layout_height="0dp"
25.        app:layout_constraintVertical_weight="7"
26.        app:layout_constraintLeft_toLeftOf="parent"
27.        app:layout_constraintRight_toRightOf="parent"
28.        app:layout_constraintTop_toBottomOf="@+id/fragment_title"
29.        app:layout_constraintBottom_toBottomOf="parent"
30.        tools:layout="@layout/fragment_content" />
31. </androidx.constraintlayout.widget.ConstraintLayout>
```

2. 动态加载

实际项目使用静态加载的场景很少，绝大多数是根据具体情况在程序运行时动态添加 Fragment 的，这样做的好处是可定制 Fragment。需要注意的是，若需要在运行时添加或替换 Fragment，则 Activity 布局必须有一个装载 Fragment 的布局容器。

首先，修改 activity_main.xml 的布局，使用两个 FrameLayout 布局替代原有的 <fragment> 标签，并删除 Fragment 相关的属性，具体代码如下：

```
1.  <?xml version="1.0" encoding="utf-8"?>
2.  <androidx.constraintlayout.widget.ConstraintLayout
3.      xmlns:android="http://schemas.android.com/apk/res/android"
4.      xmlns:app="http://schemas.android.com/apk/res-auto"
5.      xmlns:tools="http://schemas.android.com/tools"
6.      android:layout_width="match_parent"
7.      android:layout_height="match_parent"
8.      tools:context=".MainActivity">
9.      <FrameLayout
10.         android:id="@+id/fragment_title"
11.         android:layout_width="match_parent"
12.         android:layout_height="0dp"
13.         android:orientation="vertical"
14.         app:layout_constraintBottom_toTopOf="@+id/fragment_content"
15.         app:layout_constraintLeft_toLeftOf="parent"
16.         app:layout_constraintRight_toRightOf="parent"
17.         app:layout_constraintTop_toTopOf="parent"
18.         app:layout_constraintVertical_weight="1" />
19.     <FrameLayout
20.         android:id="@+id/fragment_content"
21.         android:layout_width="match_parent"
22.         android:layout_height="0dp"
23.         android:orientation="vertical"
24.         app:layout_constraintBottom_toBottomOf="parent"
25.         app:layout_constraintLeft_toLeftOf="parent"
26.         app:layout_constraintRight_toRightOf="parent"
27.         app:layout_constraintTop_toBottomOf="@+id/fragment_title"
28.         app:layout_constraintVertical_weight="5" />
29. </androidx.constraintlayout.widget.ConstraintLayout>
```

然后，在 MainActivity 类中使用 FragmentManager 对象管理 Fragment 的添加、移除、替换及执行其他事务，最终调用 commit() 方法提交。

Fragment 的添加、移除、替换这几种行为之间的区别介绍如下。

- add()方法：将 Fragment 添加到 Activity 容器中，可调用 hide()或 show()隐藏和显示。
- replace()方法：清除 Activity 容器中的 View 对象，并不是 Fragment 实例，不能使用 hide()或 show()隐藏和显示。
- remove()方法：移除 Fragment 实例，在使用之前使用 isAdded()方法进行判断。
- 当按下"返回"按钮进行回滚时，add()仅需要恢复原有的 Fragment，无须重新加载；而 replace()则需要重新创建原有的 Fragment。
- add()添加相同的 Fragment 会报 IllegalStateException 异常，而 replace()则不会。
- replace()用于不需要重新访问当前 Fragment 或者内存受限的情况。

Fragment 添加到 Activity 容器中的具体步骤如下：

（1）调用 getSupportFragmentManager()方法获取 FragmentManager 对象，它是 Fragment 的管理器。

（2）调用 FragmentManager 的 beginTransaction 开启 FragmentTransaction 事务。

（3）调用 FragmentTransaction 的 replace()或 add()向 Activity 的布局容器加载 Fragment，需要传入加载 Fragment 的容器的 id 和 Fragment 对象。

（4）调用 FragmentTransaction 的 addToBackStack()方法将 Fragment 添加到回退栈。

（5）调用 FragmentTransaction 的 commit()方法提交事务。

本案例将此过程封装成 replaceFragment()方法，此方法有两个参数，第一个参数是添加 Fragment 的 Activity 布局容器的 id，第二个参数是 Fragment 对象，代码如下：

```
1.   private void replaceFragment(int containerId, Fragment fragment) {
2.       // 获取 FragmentManager 对象
3.       FragmentManager manager = getSupportFragmentManager();
4.       // 开启事务
5.       FragmentTransaction transaction = manager.beginTransaction();
6.       // 向容器加载 fragment
7.       transaction.replace(containerId, fragment);
8.       // 将 fragment 加入回退栈，使得回滚时直接恢复，而无须重新创建
9.       transaction.addToBackStack(null);
10.      // 提交事务
11.      transaction.commit();
12.  }
```

在 MainActivity 类的 onCreate()方法中调用此方法加载 TitleFragment 和 ContentFragment，代码如下：

```
1.   protected void onCreate(Bundle savedInstanceState) {
2.       super.onCreate(savedInstanceState);
3.       setContentView(R.layout.activity_main);
4.
5.       replaceFragment(R.id.fragment_title, TitleFragment.newInstance());
6.       replaceFragment(R.id.fragment_content, ContentFragment.
                     newInstance("param1"));
7.   }
```

上述代码中的 ContentFragment 接收了 MainActivity 传递的参数，下一小节将详细讲解 Activity 和 Fragment 之间的数据传递。

3.6.3 Fragment 与 Activity 的交互

尽管 Fragment 是嵌在 Activity 中的，但是它们依旧各自存在不同的类中，实际开发中经

常会需要相互间进行通信。

1. 使用实例方法

FragmentManager 类提供了 findFragmentById()方法获取 Fragment 实例，该方法的参数就是 Activity 布局中的 Fragment 的 id，获取 Fragment 对象之后，就能调用 Fragment 中的 public 权限的方法了，代码如下：

```
1.  TitleFragment titleFragment = (TitleFragment) getSupportFragmentManager()
2.                  .findFragmentById(R.id.fragment_title);
```

Fragment 类提供了 getActivity()方法获取关联的 Activity 对象，如在 MainActivity 中加载 ContentFragment 对象后，在 ContentFragment 类中就可以用以下代码获取 MainActivity 对象，从而就能调用 MainActivity 类的方法了。另外，getActivity()方法也能获取 Fragment 的 Context 对象。

```
1.  MainActivity activity = (MainActivity) getActivity();
```

2. 使用 Bundle 传递数据

Activity 向 Fragment 传递数据，可以通过调用 setArguments()方法将 Bundle 对象传递给 Fragment，示例代码如下：

```
1.  // 创建 Fragment 对象
2.  public static ContentFragment newInstance(String param) {
3.      ContentFragment fragment = new ContentFragment();
4.      // 使用 Bundle 对象装载数据
5.      Bundle args = new Bundle();
6.      args.putString("param", param);
7.      // 传递 bundle 对象
8.      fragment.setArguments(args);
9.      return fragment;
10. }
```

Fragment 调用 getArguments()方法接收 Activity 传递的数据，示例代码如下：

```
1.  @Override
2.  public void onCreate(Bundle savedInstanceState) {
3.      super.onCreate(savedInstanceState);
4.      // 接收 Activity 传递的数据
5.      Bundle bundle = getArguments();
6.      if (bundle != null) {
7.          mParam = bundle.getString("param");
8.      }
9.  }
```

3. 使用 Listener 监听器

Fragment 通过事件监听与 Activity 进行通信。首先在 Fragment 类中定义事件监听的监听器接口，Activity 类实现该监听器。Fragment 在其 onAttach()生命周期方法中捕获接口实现，然后调用接口方法，触发监听方法与 Activity 通信。示例代码如下。

（1）在 ContentFragment 类中定义接口，并在 onAttach()方法中捕获：

```
1.  public class ContentFragment extends Fragment {
2.      // 定义接口对象
3.      private OnItemSelectedListener callback;
```

```
4.      // 捕获接口实现
5.      @Override
6.      public void onAttach(@NonNull Context context) {
7.          super.onAttach(context);
8.          try {
9.              // 获取接口对象
10.             this.callback = (OnItemSelectedListener) getActivity();
11.         } catch (ClassCastException e) {
12.             throw new ClassCastException(getActivity().toString()
13.                     + " must implement OnHeadlineSelectedListener");
14.         }
15.     }
16.     // 接口对象的set()方法
17.     public void setOnItemSelectedListener(OnItemSelectedListener callback) {
18.         this.callback = callback;
19.     }
20.     // 定义接口
21.     public interface OnItemSelectedListener {
22.         public void onContentSelected(int position);
23.     }
24. }
```

（2）父 MainActivity 类实现该接口，重写 onItemSelected()方法，添加 FragmentManager 的 OnAttach 监听器，重写 onAttachFragment()方法，设置 ContentFragment 的监听器，实现选择 item 的事件监听。示例代码如下：

```
1.  public class MainActivity extends AppCompatActivity
2.          implements ContentFragment.OnItemSelectedListener {
3.      private FragmentManager manager;
4.      @Override
5.      protected void onCreate(Bundle savedInstanceState) {
6.          ...
7.          // 设置 Fragment 的监听器
8.          manager.addFragmentOnAttachListener(new FragmentOnAttachListener() {
9.              @Override
10.             public void onAttachFragment(@NonNull FragmentManager
                                            fragmentManager,
11.                                         @NonNull Fragment fragment) {
12.                 if (fragment instanceof ContentFragment) {
13.                     ContentFragment contentFragment =
                                            (ContentFragment) fragment;
14.                     contentFragment.setOnItemSelectedListener
                                            (MainActivity.this);
15.                 }
16.             }
17.         });
18.     }
19.     // 实现接口方法
20.     @Override
21.     public void onContentSelected(int position) {
22.         Toast.makeText(this, String.format(Locale.CHINA,
                            "单击了第%d个", position),
23.                 Toast.LENGTH_SHORT).show();
24.     }
25.     ...
26. }
```

（3）ContentFragment 类触发事件，使用回调方法将事件传递给父 Activity。

```
1.  @Override
2.  public View onCreateView(LayoutInflater inflater, ViewGroup container,
3.                           Bundle savedInstanceState) {
4.      View view = inflater.inflate(R.layout.fragment_content,
                                     container, false);
5.      TextView tvContent = view.findViewById(R.id.tv_content);
6.      if(mParam != null) {
7.          tvContent.setText(mParam);
8.      }
9.      tvContent.setOnClickListener(new View.OnClickListener() {
10.         @Override
11.         public void onClick(View view) {
12.             // 调用回调方法
13.             callback.onContentSelected(1);
14.         }
15.     });
16.     return view;
17. }
```

3.6.4 Fragment 新特性

Fragment 现在以 Androidx Fragments 的形式成为 Jetpack 组件的一部分（注：Jetpack 将在第 9 章详细讲解），随着 Fragment 1.3.0-rc02 的发布，FragmentManager 内部进行了重构，它实现了 FragmentResultOwner 接口，这意味着 FragmentManager 可以充当 Fragment 之间数据传递的存储中介的角色，提供了 Fragment 之间传递数据的新方式，再也不用通过 Fragment 直接引用彼此实现数据传递，需要在 app/build.gradle 文件中添加 Fragment 1.3.0 的依赖实现此特性。

```
1.  dependencies {
2.      implementation "androidx.fragment:fragment:1.3.0-rc02"
3.  }
```

例如，将上例中的 ContentFragment 的数据传递给 TitleFragment，可以在 TitleFragment 调用 FragmentManager 的 setFragmentResultListener()中设置结果监听器，示例代码的第 8～15 行设置了监听器，其中的参数包括请求 key、Fragment 对象和 FragmentResultListener 监听器接口对象，并在 onFragmentResult()方法中接收数据，具体代码如下：

```
1.  @Override
2.  public View onCreateView(LayoutInflater inflater, ViewGroup container,
3.                           Bundle savedInstanceState) {
4.      View view = inflater.inflate(R.layout.fragment_title, container, false);
5.      final TextView tvTitle = view.findViewById(R.id.tv_title);
6.
7.      // 设置结果监听器，接收数据
8.      getParentFragmentManager().setFragmentResultListener("key", this,
9.              new FragmentResultListener() {
10.                 @Override
11.                 public void onFragmentResult(String key, @NonNull
                                                 Bundle bundle) {
12.                     String result = bundle.getString("title");
13.                     tvTitle.setText(result);
14.                 }
15.             });
16.     return view;
17. }
```

ContentFragment 使用相同的 key 调用 setFragmentResult()方法设置 Bundle 数据，如下第 16～18 行代码：

```
1.  @Override
2.  public View onCreateView(LayoutInflater inflater, ViewGroup container,
3.                           Bundle savedInstanceState) {
4.      View view = inflater.inflate(R.layout.fragment_content,
                                     container, false);
5.      TextView tvContent = view.findViewById(R.id.tv_content);
6.      if (mParam != null) {
7.          tvContent.setText(mParam);
8.      }
9.      tvContent.setOnClickListener(new View.OnClickListener() {
10.         @Override
11.         public void onClick(View view) {
12.             // 调用回调方法
13.             callback.onContentSelected(1);
14. 
15.             // 生成数据传递给 TitleFragment
16.             Bundle result = new Bundle();
17.             result.putString("title", "内容标题");
18.             getParentFragmentManager().setFragmentResult("key", result);
19.         }
20.     });
21.     return view;
22. }
```

综上所述，借助 FragmentManager 实现两个 Fragment 之间传递数据的流程如图 3.32 所示。

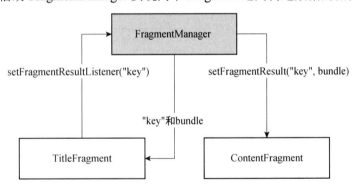

图 3.32　ContentFragment 传递数据给 TitleFragment

3.6.5　Fragment 的生命周期

通过 3.1.3 小节的学习，我们已经了解 Activity 的生命周期有运行、暂停、停止和销毁 4 种状态。Fragment 的生命周期不仅与 Activity 类似，而且由于 Fragment 必须关联 Activity 而存在，因此 Activity 的生命周期直接影响到 Fragment 的生命周期，如图 3.33 所示。

当 Activity 创建时，与之关联的 Fragment 也被创建，并处于启动状态，Activity 调用 Fragment 的 onAttach()方法使得 Fragment 附属于 Activity，然后 Activity 通过调用 onAttachFragment()方法接收到 Fragment 的引用，Fragment 调用 onCreate()方法进行创建，接着调用 onCreateView()方法构建界面视图，当 Activity 包含多个 Fragment 时，这个过程就会进行多遍，确保 Fragment 都被创建。

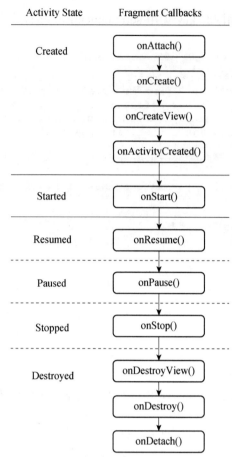

图 3.33 Activity 生命周期对 Fragment
生命周期的影响

当 Activity 处于运行状态时，可以动态添加、删除或替换 Fragment，当 Fragment 被添加时处于启动状态，被删除时则处于销毁状态。

当 Activity 暂停时，该 Activity 关联的所有 Fragment 都会被暂停，当 Activity 被销毁时，Activity 中所有的 Fragment 也会被销毁。

Fragment 有与 Activity 相同的生命周期方法，但比 Activity 多出几个方法，说明如下。

- onAttach()：Fragment 与 Activity 关联时调用。
- onCreateView()：Fragment 创建界面视图时调用。
- onActivityCreated()：Fragment 关联的 Activity 已经创建完成时调用。
- onDestroyView()：Fragment 的视图被销毁时调用。
- onDetach()：Fragment 与 Activity 解除关联时调用。

Fragment 生命周期与 Activity 非常相似，可以参考 Activity 生命周期案例自行创建项目体验 Fragment 生命周期的执行过程，此处不再赘述。

在理解 Fragment 生命周期的基础上，下面通过一组图（见图 3.34～图 3.37）分别展示 Fragment 的添加、替换方式及回退操作的生命周期的调用过程，帮助大家更深地理解 add()、replace()方式。

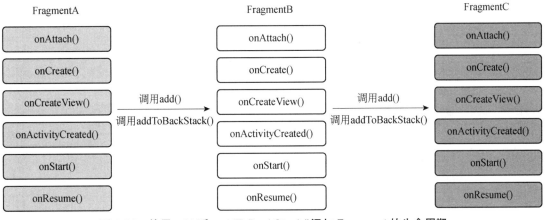

图 3.34 使用 add() 和 addToBackStack() 添加 Fragment 的生命周期

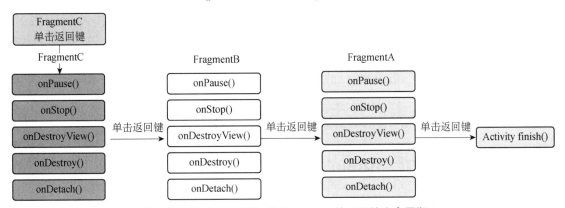

图 3.35 使用 add() 添加的 Fragment 的回退的生命周期

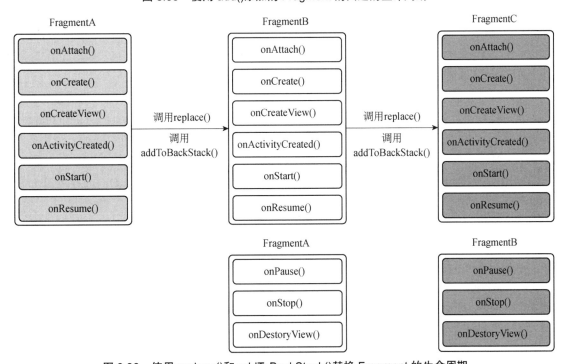

图 3.36 使用 replace() 和 addToBackStack() 替换 Fragment 的生命周期

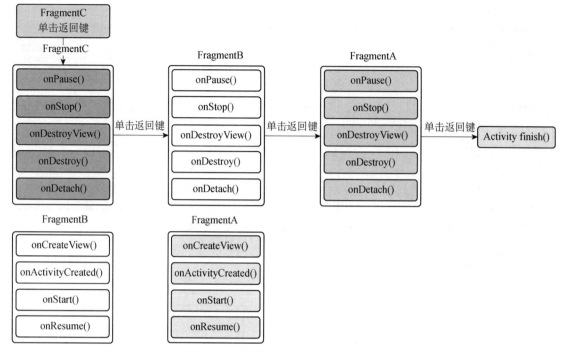

图 3.37 使用 replace()替换的 Fragment 的回退的生命周期

3.6.6 DialogFragment 对话框

应用程序与用户交互还可以使用弹出对话框完成，Android 的对话框实现有以下几种方式。
- Dialog：自定义对话框类继承 Dialog 类。
- AlertDialog：Android 原生提供的对话框。
- PopupWindow：弹出悬浮框实现对话框，它可以在指定位置显示。

以上三种方法都有一个共同的问题：没有与 Activity 的生命周期进行绑定，当屏幕切换时会消失，如果处理不当很可能引发异常。因此，Google 在 Android 3.0 中引入了 DialogFragment 来替代它们。DialogFragment 是一种特殊的 Fragment，用于在 Activity 界面中展示一个模态对话框，主要用于展示警告框、输入框等。DialogFragment 与 Fragment 有基本一致的生命周期，便于 Activity 更好地控制管理它，也解决了屏幕切换 Dialog 消失的问题。DialogFragment 还会随着屏幕切换自动调整对话框大小。

有两种创建 DialogFragment 的方法。
- 重写 onCreateDialog 方法：用于创建替代传统 Dialog 对话框的场景，布局简单且功能单一，利用 AlertDialog 或 Dialog 创建，不适用于多线程情况。
- 重写 onCreateView 方法：用于创建复杂内容弹窗或全屏展示效果的场景，布局和功能较复杂，使用自定义布局，适用于网络请求等异步操作。

下面通过一个案例讲解 DialogFragment 的使用。

【案例 3-7】使用 DialogFragment 实现登录、退出对话框，实现对话框的数据回传，主界面及对话框运行结果如图 3.38 所示。

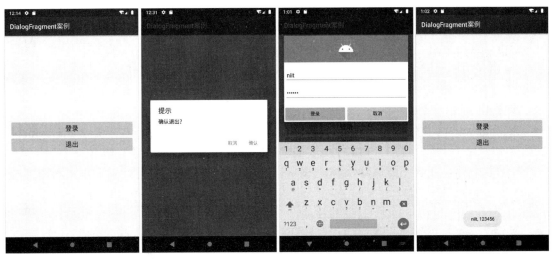

图 3.38 主界面、DialogFragment 的对话框、登录对话框和数据回传的效果图

步骤 1：创建项目

启动 Android Studio，创建名为 D0307_DialogFragment 的项目，选择 Empty Activity，将包名改为 com.example.dialogfragment，单击 Finish 按钮，等待项目构建完成。

步骤 2：完成 MainActivity 的布局和按钮的事件监听

按照图 3.38 所示的界面布局，在 activity_main.xml 中增加两个 Button，在 MainActivity 类中完成两个按钮的事件监听的代码。

步骤 3：创建自定义对话框类

在包 com.example.dialogfragment 下，创建 LogoutDialog 类继承 DialogFragment，重写 onCreateDialog()方法，在该方法中创建自定义 AlertDialog 对象，设置对话框的标题、消息内容、按钮等信息，示例代码如下：

```
1.   public class LogoutDialog extends DialogFragment {
2.       @NonNull
3.       @Override
4.       public Dialog onCreateDialog(@Nullable Bundle savedInstanceState) {
5.           return new AlertDialog.Builder(getActivity())
                                        // 使用 Builder 构造对话框
6.                   .setTitle("提示")    // 设置标题
7.                   .setMessage("确认退出？")  // 设置消息
8.                   .setPositiveButton("确认", new DialogInterface.
                                        OnClickListener() {
9.                       // 确认按钮的事件处理
10.                      @Override
11.                      public void onClick(DialogInterface
                                        dialogInterface, int i) {
12.                          dialogInterface.dismiss();
13.                          LogoutDialog.this.getActivity().finish();
14.                      }
15.                  })
16.                  .setNegativeButton("取消", new DialogInterface.
                                        OnClickListener() {
17.                      // 取消按钮的事件处理
18.                      @Override
19.                      public void onClick(DialogInterface
                                        dialogInterface, int i) {
```

```
20.                         dialogInterface.dismiss();
21.                     }
22.                 })
23.                 .create();    // 创建对话框
24.     }
25. }
```

步骤 4：显示对话框

在 MainActivity 类中初始化两个按钮控件，btnLogout 的 onClick 用于显示 LogoutFragment，参见第 10~16 行代码，第 13~14 行用于创建退出的对话框并显示，详情如下：

```
1. public class MainActivity extends AppCompatActivity {
2.     @Override
3.     protected void onCreate(Bundle savedInstanceState) {
4.         super.onCreate(savedInstanceState);
5.         setContentView(R.layout.activity_main);
6.         // 初始化控件对象
7.         Button btnLogout = findViewById(R.id.btn_logout);
8.         Button btnLogin = findViewById(R.id.btn_login);
9.         // 设置退出按钮的事件监听器，显示 LogoutDialog 对话框
10.        btnLogout.setOnClickListener(new View.OnClickListener() {
11.            @Override
12.            public void onClick(View v) {
13.                LogoutDialog logoutDialog = new LogoutDialog();
14.                logoutDialog.show(getSupportFragmentManager(), "退出");
15.            }
16.        });
17.        // 设置登录按钮的事件监听器，处理单击事件
18.        btnLogin.setOnClickListener(new View.OnClickListener() {
19.            @Override
20.            public void onClick(View v) {
21.
22.            }
23.        });
24.    }
25. }
```

步骤 5：运行项目

单击"运行"按钮，在主界面单击"退出"按钮，弹出图 3.38 的第 2 个界面所示的警告框，单击"确认"按钮可关闭主界面。

步骤 6：创建登录对话框的布局

在 res/layout 目录中创建如图 3.38 所示的布局文件 dialog_login.xml，在布局文件中用一个 ImageView 显示图标，用两个 EditText 分别表示登录用户名和密码，代码如下：

```
1. <LinearLayout xmlns:android="http://schemas.android.com/apk/res/android"
2.     android:orientation="vertical"
3.     android:layout_width="wrap_content"
4.     android:layout_height="wrap_content">
5.     <ImageView
6.         android:src="@mipmap/ic_launcher_round"
7.         android:layout_width="match_parent"
8.         android:layout_height="80dp"
9.         android:scaleType="center"
10.        android:background="#3DDC84"
11.        android:contentDescription="@string/app_name" />
12.    <EditText
13.        android:id="@+id/et_username"
14.        android:inputType="textEmailAddress"
```

```
15.         android:layout_width="match_parent"
16.         android:layout_height="wrap_content"
17.         android:layout_margin="8dp"
18.         android:hint="请输入用户名" />
19.     <EditText
20.         android:id="@+id/et_password"
21.         android:inputType="textPassword"
22.         android:layout_width="match_parent"
23.         android:layout_height="wrap_content"
24.         android:layout_margin="8dp"
25.         android:hint="请输入密码"/>
26. </LinearLayout>
```

步骤 7. 创建自定义对话框类

在 com.example.dialogfragment 包下创建 LoginDialog 类继承自 DialogFragment,重写 onCreateView()方法,此方法与 Fragment 的 onCreateView()的含义相同,都使用 LayoutInflater 对象加载布局,示例代码如下:

```
1. public class LoginDialog extends DialogFragment {
2.     @Override
3.     public View onCreateView(@NonNull LayoutInflater inflater,
4.                              @Nullable ViewGroup container,
5.                              @Nullable Bundle savedInstanceState) {
6.         View view = inflater.inflate(R.layout.dialog_login,
                                         container, false);
7.         return view;
8.     }
9. }
```

步骤 8:数据传回 Activity

当用户在对话框中输入用户名和密码后,如何将数据传回 Activity 呢?答案是也采用 Fragment 定义回调接口的方法,在对话框中定义 LoginInputListener 接口,通过该接口将事件传回给对话框的宿主 Activity。示例代码如下:

```
1.  public class LoginDialog extends DialogFragment {
2.      private LoginInputListener listener;
3.      @Override
4.      public View onCreateView(@NonNull LayoutInflater inflater,
5.                               @Nullable ViewGroup container,
6.                               @Nullable Bundle savedInstanceState) {
7.          View view = inflater.inflate(R.layout.dialog_login,
                                          container, false);
8.          return view;
9.      }
10.     @Override
11.     public void onAttach(@NonNull Context context) {
12.         super.onAttach(context);
13.         // 验证宿主 Activity 实现了回调接口
14.         try {
15.             // 实例化 LoginInputListener,将事件对象传给 Activity
16.             listener = (LoginInputListener) context;
17.         } catch (ClassCastException e) {
18.             // 如果 Activity 未实现接口,则抛出异常
19.             throw new ClassCastException(getActivity().toString()
20.                     + " 必须实现 LoginInputListener");
21.         }
22.     }
23.     // 登录信息的监听回调接口
```

```
24.    public interface LoginInputListener {
25.        public void onDialogPositiveClick(String username,
                                             String password);
26.        public void onDialogNegativeClick(DialogFragment dialog);
27.    }
28. }
```

上述代码的第 24~27 行定义了 LoginInputListener 接口，定义的两个方法用于将事件传回给宿主 Activity，其中，onDialogPositiveClick()对应对话框登录按钮事件，将用户名和密码传回 Activity；onDialogNegativeClick()对应取消按钮事件。第 14~25 行代码重写 onAttach()方法实例化 LoginInputListener 接口对象。

接下来，对话框的宿主 Activity 通过实现 LoginInputListener 接口接收对话框的事件，示例代码如下：

```
1.  public class MainActivity extends AppCompatActivity
2.          implements LoginDialog.LoginInputListener {
3.      @Override
4.      protected void onCreate(Bundle savedInstanceState) {
5.          super.onCreate(savedInstanceState);
6.          setContentView(R.layout.activity_main);
7.          // 初始化控件对象
8.          Button btnLogout = findViewById(R.id.btn_logout);
9.          Button btnLogin = findViewById(R.id.btn_login);
10.         // 设置退出按钮的事件监听器，显示 LogoutDialog 对话框
11.         btnLogout.setOnClickListener(new View.OnClickListener() {
12.             @Override
13.             public void onClick(View v) {
14.                 LogoutDialog logoutDialog = new LogoutDialog();
15.                 logoutDialog.show(getSupportFragmentManager(), "退出");
16.             }
17.         });
18.         // 设置登录按钮的事件监听器，显示 LoginDialog 对话框
19.         btnLogin.setOnClickListener(new View.OnClickListener() {
20.             @Override
21.             public void onClick(View v) {
22.                 LoginDialog loginDialog = new LoginDialog();
23.                 loginDialog.show(getSupportFragmentManager(), "登录");
24.             }
25.         });
26.     }
27.     @Override
28.     public void onDialogPositiveClick(String username, String password) {
29.     }
30.     @Override
31.     public void onDialogNegativeClick(DialogFragment dialog) {
32.     }
33. }
```

宿主 Activity 实现了 LoginInputListener 接口，对话框即可通过接口回调方法向 Activity 传递单击事件，示例代码如下：

```
1.  @Override
2.  public View onCreateView(@NonNull LayoutInflater inflater,
3.                  @Nullable ViewGroup container,
4.                  @Nullable Bundle savedInstanceState) {
5.      View view = inflater.inflate(R.layout.dialog_login, container, false);
6.      final EditText etUsername = view.findViewById(R.id.et_username);
7.      final EditText etPassword = view.findViewById(R.id.et_password);
```

```
8.      Button btnLogin = view.findViewById(R.id.btn_login);
9.      // 登录按钮的事件监听处理
10.     btnLogin.setOnClickListener(new View.OnClickListener() {
11.         @Override
12.         public void onClick(View v) {
13.             // 获取输入的用户名和密码
14.             String username = etUsername.getText().toString();
15.             String password = etPassword.getText().toString();
16.             // 触发 MainActivity 的确定按钮的事件
17.             listener.onDialogPositiveClick(username, password);
18.             // 关闭登录对话框
19.             LoginDialog.this.dismiss();
20.         }
21.     });
22.     Button btnCancel = view.findViewById(R.id.btn_cancel);
23.     // 取消按钮的事件监听处理
24.     btnCancel.setOnClickListener(new View.OnClickListener() {
25.         @Override
26.         public void onClick(View v) {
27.             // 触发 MainActivity 的取消按钮的事件
28.             listener.onDialogNegativeClick(LoginDialog.this);
29.             // 关闭登录对话框
30.             LoginDialog.this.dismiss();
31.         }
32.     });
33.     return view;
34. }
```

第 17、28 行分别将登录、取消事件传回宿主 Activity 对象，两个按钮的处理代码如下：

```
1.  // 登录事件处理
2.  @Override
3.  public void onDialogPositiveClick(String username, String password) {
4.      Toast.makeText(this, username + ", " + password, Toast.
                      LENGTH_SHORT).show();
5.  }
6.  // 取消事件处理
7.  @Override
8.  public void onDialogNegativeClick(DialogFragment dialog) {
9.      Toast.makeText(this, "取消", Toast.LENGTH_SHORT).show();
10. }
```

步骤 9：运行项目

完成以上编码后，运行项目，单击"登录"按钮进入登录对话框，输入用户名、密码，单击"登录"按钮关闭对话框，在主界面中显示用户名和密码，运行结果如图 3.38 所示。

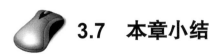

3.7 本章小结

本章讲解了 Activity 和 Fragment 两个非常重要的组件，首先介绍了 Activity 的基础知识、创建 Activity 的方法及 Activity 的生命周期；接着讲解了 Activity 之间的数据传递，以及 Activity 的启动模式；然后讲解了 Fragment 的基本概念，以及 Fragment 的应用和它的生命周期；最后讲解了 Google 推荐使用的对话框 DialogFragment 的基本使用方法。通过本章的学习，读者可以对 Android 的 Activity、Fragment 组件有了较为深入的了解。

习 题

一、选择题

1. 下列选项中,属于当前 Activity 被其他 Activity 覆盖时调用的方法的是（   ）。
   A. onCreate()   B. onResume()   C. onPause()   D. onDestroy()
2. 下列选项中,属于开启 Activity 方法的是（   ）。
   A. goToActivity()   B. goActivity()   C. startActivity()   D. 以上方法都对
3. 下列选项中,属于没有明确指定组件名的 Intent 类型的是（   ）。
   A. IntentFilter   B. 显式 Intent   C. 隐式 Intent   D. Intent
4. 下面关于 Activity 生命周期状态的描述中,正确的是（   ）。
   A. Activity 的运行状态很短暂
   B. Activity 在暂停状态时用户对它操作没有响应
   C. Activity 会停留在销毁的状态
   D. 处于暂停状态的 Activity 对用户来说是不可见的
5. 下列关于 Fragment 的描述中,正确的是（   ）。
   A. Fragment 不需要添加到 Activity 中也可以单独显示界面
   B. 只能在布局中添加 Fragment
   C. 只能在 Java 代码中添加 Fragment
   D. 可以通过 getFragmentManager()方法获取 FragmentManager 实例
6. 下列选项中,属于 Fragment 和 Activity 建立关联时调用的生命周期方法是（   ）。
   A. onActivityCreate()   B. onDetach()
   C. onActivityCreate()   D. onAttach()

二、简答题

1. 简述 Activity 生命周期的几种状态,它们各有什么特点。
2. 简述显式 Intent 和隐式 Intent 的区别。
3. 简述 Activity 的 4 种启动模式及其特点。
4. 简述 Fragment 生命周期的方法及调用时机。
5. 简述在 Activity 中动态加载 Fragment 的步骤。

三、编程题

1. 编写一个数据传递的程序,在第一个界面中输入两个数字,在第二个界面中会显示这两个数字的和。
2. 创建项目演示 Fragment 的生命周期方法的执行顺序。
3. 使用对话框的方式完成第 1 题。

第 4 章 Android 高级界面设计

Google 发布 Android 5.0 时提出了 Material Design 设计理念，发布了一整套设计指南。Material Design 设计理念让 Android 界面在体验上更加新鲜和简洁，同时能够更有效激发应用开发者的创作热情，使其带来更加卓越的应用界面。Google 同时推出了一系列实现 Material Design 效果的 Android 控件库——Android Design Support Library，如 FloatingActionButton、RecyclerView、CardView、CoordinatorLayout 等控件。本章将详细讲解这些高级界面布局、界面组件的应用场景和使用方法。

本章学习要点：

- 掌握 Android 高级界面布局的使用方法
- 掌握 Android 高级界面组件的应用场景和使用方法
- 理解 Material Design 设计理念
- 了解 Android 自定义 view 的使用方法

 ## 4.1 Material Design

Material Design 的中文名称为质感设计，是 Google 在 2014 年 I/O 大会推出的全新的设计语言，Material Design 不像 Android 以往采用的 Holo 风格那样深沉，增加或修改了阴影动画功能，使其更加跳动和富有活力。Google 认为，这种设计语言能为手机、平板电脑、台式机和其他智能设备提供更一致、更美观的外观体验。

Material Design 是针对可视化、交互性、动画特效及多屏幕适应的全面设计。使用 Material Design 要求 Android SDK 版本在 21 及以上，即 Android Lollipop/5.0 以上。Material Design 支持各种新的动画效果，颜色更加鲜艳，界面效果更加丰富。Google 希望 Android 开发者掌握这个新框架，从而让 Android 应用程序拥有统一的外观，就像苹果向开发者提出的设计原则一样。

在设计上，不能简单地将 Material Design 归纳为平面化设计，但也不能归纳为拟物化设计，毕竟它所使用的各种图案和形状并非是对现实实体的模拟，而是按照 Google 对数字世界的理解，以色彩、图案、形状进行视觉信息的划分。

Material Design 在设计上并没有完全抛弃 Google 过去在设计上取得的成果，比如它会继续使用阴影效果，就像在 Android 4.0 的下拉菜单底部的淡淡的阴影效果。

不过，比起过去的 Android Design 来说，Material Design 更有自己的目标，它不仅仅为了好看，还要让不同设备的屏幕表现出一致、美观的视觉体验及交互。为了统一设备应用程序的界面和交互，让用户得到连贯的体验，Material Design 不再让像素处于同一个平面，而是让它们按照规则处于空间当中，具备不同的维度。

随着 Android 5.0 的发布，Material Design 不仅成为 App 设计的标准，也广泛应用于网页设计中。它的设计理念让 Android 界面更加鲜活、简洁，也有效激发了 App 开发者的创作热情，带来更加卓越的应用界面。Google 为此推出了一系列实现 Material Design 理念的控件库 Android Design Support Library，包括：Floating Action Button、RecyclerView、CardView、Toolbar、Snackbar、TabLayout、CoordinatorLayout 等控件和布局，本章接下来的内容将结合实例，一一介绍它们的特点及使用方法。

Android Support Design 使用非常方便，只需在 app/build.gradle 文件中添加依赖包即可。

 ## 4.2 高级 UI 布局

4.2.1 ConstraintLayout

在实际开发过程中，经常会遇到一些复杂的 UI 布局，可能会出现布局嵌套过多的问题。嵌套越多，设备绘制视图所需的时间和计算功耗也就越多，页面响应越慢，用户体验越差。约束布局 ConstraintLayout 的推出就是为了解决布局嵌套过多带来的性能问题，用更灵活的方式定位和调整界面组件，相对 RelativeLayout 来说，其性能更好、布局也更灵活。ConstraintLayout 继承自 ViewGroup，可以在 Android SDK API 9 以上的 Android 系统中使用。

约束是 ConstraintLayout 的核心概念，也就是控件之间的位置关系，通过添加水平或垂直约束条件进行定位。每个约束条件都表示与其他控件、布局或隐形引导线之间连接或对齐方式，均定义了 View 在纵轴或者横轴上的位置；因此每个 View 在每个轴上都必须至少有一个约束条件，没有约束的 View 只会显示在屏幕的左上角。

ConstraintLayout 布局首先要明确约束控件、被约束控件分别是哪个 View，然后遵守以下规则创建约束条件：

- 每个 View 都必须至少有两个约束条件，即一个水平约束条件、一个垂直约束条件。
- 只能在共用同一平面的约束手柄与定位点之间创建约束条件。也就是说，View 的垂直平面（左侧和右侧）只能约束在另一个垂直平面上，而基准线则只能约束到其他基准线上。
- 每个约束句柄只能用于一个约束条件，但可以在同一定位点上创建多个约束条件。

从 Android Studio 2.2 起，应用程序默认的根布局都采用 ConstraintLayout，在最新的 Android 官方开发文档也推荐使用它，接下来详细讲解 ConstraintLayout 的使用方法。

【案例 4-1】ConstraintLayout 布局的使用。

步骤 1：创建项目

启动 Android Studio，创建名为 D0401_ConstraintLayout 的项目，选择 Empty Activity，将包名改为 com.example.layout，单击 Finish 按钮，等待项目构建完成。

步骤 2：添加依赖

在 app/build.gradle 文件中添加 ConstraintLayout 的依赖库：

```
1.  dependencies {
2.      implementation 'androidx.constraintlayout:constraintlayout:2.0.4'
3.  }
```

步骤 3：设置 ConstraintLayout 的定位
下面通过创建不同的布局文件讲解不同的定位方式。
（1）相对定位

在 res/layout 目录中创建 relative_position.xml，添加相对定位的两个按钮，设置相关的属性，如图 4.1 所示。相对定位是部件对于另一个位置的约束，如需将控件 B 放在控件 A 的右侧，设置控件 B 的 layout_constraintLeft_toRightOf 属性，代码如下：

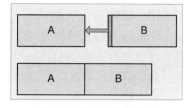

图 4.1　相对定位示意图

```
1.  <?xml version="1.0" encoding="utf-8"?>
2.  <androidx.constraintlayout.widget.ConstraintLayout
3.      xmlns:android="http://schemas.android.com/apk/res/android"
4.      xmlns:app="http://schemas.android.com/apk/res-auto"
5.      android:layout_width="match_parent"
6.      android:layout_height="match_parent">
7.      <Button
8.          android:id="@+id/btn_a"
9.          android:layout_width="wrap_content"
10.         android:layout_height="wrap_content"
11.         android:text="Button_A"
12.         app:layout_constraintLeft_toLeftOf="parent"
13.         app:layout_constraintTop_toTopOf="parent" />
14.     <Button
15.         android:id="@+id/btn_b"
16.         android:layout_width="wrap_content"
17.         android:layout_height="wrap_content"
18.         android:text="Button_B"
19.         app:layout_constraintLeft_toRightOf="@+id/btn_a"
20.         app:layout_constraintTop_toTopOf="@+id/btn_a" />
21. </androidx.constraintlayout.widget.ConstraintLayout>
```

从以上代码可以看出，通过设置 layout_constraintLeft_toRightOf 属性，可以使 Button_B 在 Button_A 的右侧。同理，可以根据开发需求，使用相对定位的其他属性，灵活实现组件的相对定位。图 4.2 描述了组件约束代表的含义，常用的属性如表 4.1 所示，使用时需要给它们赋一个其他组件的 id 值或者 parent。

表 4.1　ConstraintLayout 相对定位的属性

属性名称	功能描述
layout_constraintLeft_toLeftOf	控件的左边与另外一个控件的左边对齐
layout_constraintLeft_toRightOf	控件的左边与另外一个控件的右边对齐
layout_constraintRight_toLeftOf	控件的右边与另外一个控件的左边对齐
layout_constraintRight_toRightOf	控件的右边与另外一个控件的右边对齐
layout_constraintTop_toTopOf	控件的上边与另外一个控件的上边对齐
layout_constraintTop_toBottomOf	控件的上边与另外一个控件的底部对齐
layout_constraintBaseline_toBaselineOf	控件间的文本内容基准线对齐
layout_constraintStart_toEndOf	控件的起始边与另外一个控件的尾部对齐
layout_constraintStart_toStartOf	控件的起始边与另外一个控件的起始边对齐
layout_constraintEnd_toStartOf	控件的尾部与另外一个控件的起始边对齐
layout_constraintEnd_toEndOf	控件的尾部与另外一个控件的尾部对齐

（2）Margin 外边距

Margin 是指组件间的外边间距，如图 4.3 所示。设置 Margin 继续使用属性 layout_margin*，需要注意的是，要使 Margin 生效，必须具有对应方向的 layout_constraint*，否则 Margin 不生效。可以通过以下属性设置一个控件相对另一个控件的外边距，属性值必须大于或者等于 0。

- android:layout_marginStart。
- android:layout_marginEnd。
- android:layout_marginLeft。
- android:layout_marginTop。
- android:layout_marginRight。
- android:layout_marginBottom。

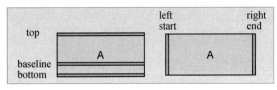

图 4.2　相对定位约束示意图

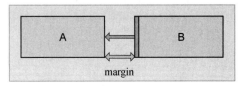

图 4.3　相对定位 Margin 示意图

（3）Gone widget

当一个相对的控件隐藏时，即控件可见性被设置为 gone 的时候，ConstraintLayout 也可以设置一个不同的边距，这些属性是 RelativeLayout 布局所没有的：

- layout_goneMarginStart。
- layout_goneMarginEnd。
- layout_goneMarginLeft。
- layout_goneMarginTop。
- layout_goneMarginRight。
- layout_goneMarginBottom。

（4）居中和偏移

在 RelativeLayout 布局中，把控件放在布局中间的方法是把 layout_centerInParent 设为 true，而在 ConstraintLayout 布局中的写法是：

```
1.  app:layout_constraintBottom_toBottomOf="parent"
2.  app:layout_constraintLeft_toLeftOf="parent"
3.  app:layout_constraintRight_toRightOf="parent"
4.  app:layout_constraintTop_toTopOf="parent"
```

此外，还可以设置 bias 属性，表示子控件相对父控件的位置偏移，如图 4.4 所示。

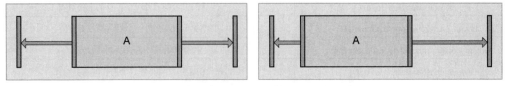

图 4.4　子控件相对父控件的位置偏移示意图

ConstraintLayout 提供的偏移属性：

- layout_constraintHorizontal_bias：水平偏移比例。
- layout_constraintVertical_bias：垂直偏移比例。

例如：需要实现水平偏移，给 Button_A 的 layout_constraintHorizontal_bias 赋值范围为 0～1，如果赋值为 0，则 Button_A 在布局的最左侧，如果赋值为 1，则 Button_A 在布局的最右侧，如果赋值为 0.5，则水平居中，如果赋值为 0.3，则向左侧偏移。同理，也可以设置垂直偏移。以水平向左偏移为例，界面布局为 bias_layout.xml，代码如下：

```
1.  <?xml version="1.0" encoding="utf-8"?>
2.  <androidx.constraintlayout.widget.ConstraintLayout
3.      xmlns:android="http://schemas.android.com/apk/res/android"
4.      xmlns:app="http://schemas.android.com/apk/res-auto"
5.      xmlns:tools="http://schemas.android.com/tools"
6.      android:layout_width="match_parent"
7.      android:layout_height="match_parent"
8.      tools:context=".MainActivity">
9.      <Button
10.         android:id="@+id/btn_a"
11.         android:layout_width="wrap_content"
12.         android:layout_height="wrap_content"
13.         android:text="Button_A"
14.         app:layout_constraintHorizontal_bias="0.3"
15.         app:layout_constraintLeft_toLeftOf="parent"
16.         app:layout_constraintRight_toRightOf="parent"
17.         app:layout_constraintTop_toTopOf="parent" />
18. </androidx.constraintlayout.widget.ConstraintLayout>
```

运行效果如图 4.5 所示。

（5）角度定位

角度定位是指可以用一个角度和一个距离来约束两个控件的中心，相关的属性有三个：

● layout_constraintCircle：引用圆心控件的 id。

● layout_constraintCircleRadius：到圆心控件中心的距离。

● layout_constraintCircleAngle：当前控件相对圆心控件的角度（以度为单位，从 0 到 360）。

创建 angle_positon.xml 布局文件，第 23～25 行代码实现 Button_B 在 Button_A 的顺时针 45 度位置，详细代码如下，界面效果如图 4.6 所示。

图 4.5　水平向左偏移效果图

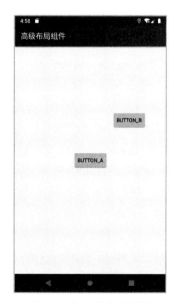
图 4.6　角度定位效果图

```xml
1.  <?xml version="1.0" encoding="utf-8"?>
2.  <androidx.constraintlayout.widget.ConstraintLayout
3.      xmlns:android="http://schemas.android.com/apk/res/android"
4.      xmlns:app="http://schemas.android.com/apk/res-auto"
5.      xmlns:tools="http://schemas.android.com/tools"
6.      android:layout_width="match_parent"
7.      android:layout_height="match_parent"
8.      tools:context=".MainActivity">
9.      <Button
10.         android:id="@+id/btn_a"
11.         android:layout_width="wrap_content"
12.         android:layout_height="wrap_content"
13.         android:text="Button_A"
14.         app:layout_constraintBottom_toBottomOf="parent"
15.         app:layout_constraintLeft_toLeftOf="parent"
16.         app:layout_constraintRight_toRightOf="parent"
17.         app:layout_constraintTop_toTopOf="parent" />
18.     <Button
19.         android:id="@+id/btn_b"
20.         android:layout_width="wrap_content"
21.         android:layout_height="wrap_content"
22.         android:text="Button_B"
23.         app:layout_constraintCircle="@+id/btn_a"
24.         app:layout_constraintCircleAngle="45"
25.         app:layout_constraintCircleRadius="150dp" />
26. </androidx.constraintlayout.widget.ConstraintLayout>
```

（6）尺寸约束

尺寸约束是指给控件设置尺寸大小，有以下 3 种方式：

● 使用指定的尺寸。

● 使用 wrap_content，让控件自己计算大小。当控件的高度或宽度为 wrap_content 时，可以使用下列属性来控制最大、最小的高度或宽度。

◇ android:minWidth：最小的宽度。

◇ android:minHeight：最小的高度。

◇ android:maxWidth：最大的宽度。

◇ android:maxHeight：最大的高度。

● 使用 0dp(match_constraint)。官方不推荐在 ConstraintLayout 布局中使用 match_parent，可设为 0dp，配合约束代替 match_parent。

（7）链式约束

两个控件之间的相互约束可以通过链式约束实现，链式约束中的第一个控件称为"链头"，针对链头设置 chainStyle 属性控制链式约束的样式，它有水平和垂直两种类型，分别为 layout_constraintHorizontal_chainStyle 和 layout_constraintVertical_chainStyle，有三个取值。

● spread：展开样式（首尾控件不固定，所有控件平均展开），也是默认值。

● spread_inside：内部展开样式（首尾控件固定，首尾中间的控件平均展开）。

● packed：打包样式（所有控件将连接在一起，可设置 bias 属性确定链式布局的位置）。

接下来创建 chain_positon.xml 布局文件，设置三个按钮之间的链式约束，界面效果如图 4.7 所示，三个界面的 layout_constraintHorizontal_chainStyle 依次为 spread、spread_inside 和 packed，示例代码如下。

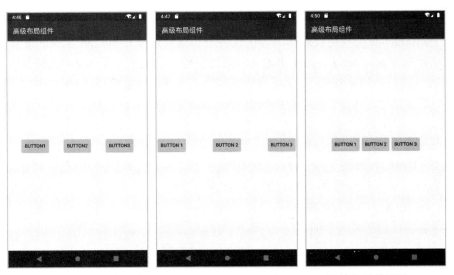

图 4.7　链式布局的 spread、spread_inside 和 packed 值的布局效果图

4.2.2　CoordinatorLayout

目前，折叠式布局在 App 中非常常见，它可以使界面效果更加丰富、灵动，极大地提高了用户体验。其中 CoordinatorLayout 和 AppBarLayout 的配合使用可以带来滑动联动的界面效果。

CoordinatorLayout 控件的作用就像它的命名，即协调者布局，官方文档称 CoordinatorLayout 是一个加强版 FrameLayout，其功能就是作为顶层的装饰布局及协调多个子 View 之间的交互，实现多种 Material Design 中提到的菜单侧滑、折叠、滑动删除 UI 元素等效果，这些效果主要包括：

- 让 FloatingActionButton 上下滑动，为 Snackbar 留出空间。
- 扩展或者缩小 Toolbar 或者头部，让主内容区域有更多的空间。
- 与 AppBarLayout 配合使用，控制某个 View 的扩展或收缩，以及显示的大小比例，包括视差效果动画、响应滚动事件等。

AppBarLayout 继承自 LinearLayout，是一个 Vertical 方向的 LinearLayout，其子 View 通过 setScrollFlags（int）或者在 XML 中的 app:layout_scrollFlags 属性设置它的滑动手势，它的取值包括：

- scroll：子 View 会跟随滚动事件一起滚动，相当于添加到 ScrollView 头部。
- exitUntilCollapsed：当 ScrollView 滑到顶部，就将子 View 折叠起来。
- enterAlways：只要屏幕下滑，View 就会立即拉下来。
- enterAlwaysCollapsed：当 ScrollView 滑到底，再将子 View 展开。
- snap：该属性使控件变得有弹性，如果控件下拉了 75%的高度，就会自动展开，如果只有 25%的区域显示，就会反弹回去关闭。

由于 AppBarLayout 设置了默认的 Behavior，因此它的根布局必须是 CoordinatorLayout，否则没有效果。而 AppBarLayout 下方的滑动控件，比如 RecyclerView、NestedScrollView 也必须在布局文件中设置 layout_behavior 属性实现与 AppBarLayout 绑定，如：app:layout_behavior= "@string/appbar_scrolling_view_behavior"，该属性值指向的是 com.google.android.material.appbar.

AppBarLayout$ScrollingViewBehavior，也就是 material 包下的 ScrollingViewBehavior 类的名称。ScrollingViewBehavior 类利用代理和组合模式将一些布局过程及 Nested Scrolling 过程展示出来，让开发者为 CoordinatorLayout 添加各种效果插件，实现多个子 View 之间的联动交互，因此，我们也可以扩展 Behavior 类实现自定义的交互效果。如果要在子 View 中使用 Behavior 进行控制，那么这个子 View 必须是 CoordinatorLayout 的直接子 View，否则就不能产生交互效果。

Coordinatorlayout 与 AppbarLayout 配合使用，可实现图片上下滚动折叠，通过设置 CollapsingToolbarLayout 的 app:layout_scrollFlags 属性为 scroll 和 exitUntilCollapsed 的组合在布局折叠后 Toolbar 保持不动的效果，如图 4.8 所示。在【案例 4-1】中创建布局文件 coordinator_layout.xml，实现代码如下：

```xml
1.  <?xml version="1.0" encoding="utf-8"?>
2.  <androidx.coordinatorlayout.widget.CoordinatorLayout
3.      xmlns:android="http://schemas.android.com/apk/res/android"
4.      xmlns:app="http://schemas.android.com/apk/res-auto"
5.      android:layout_width="match_parent"
6.      android:layout_height="match_parent">
7.      <com.google.android.material.appbar.AppBarLayout
8.          android:id="@+id/app_bar"
9.          android:layout_width="match_parent"
10.         android:layout_height="wrap_content">
11.         <com.google.android.material.appbar.CollapsingToolbarLayout
12.             android:id="@+id/collapsing_toolbar"
13.             android:layout_width="match_parent"
14.             android:layout_height="match_parent"
15.             android:fitsSystemWindows="true"
16.             app:layout_scrollFlags="scroll|exitUntilCollapsed">
17.             <ImageView
18.                 android:layout_width="match_parent"
19.                 android:layout_height="wrap_content"
20.                 android:fitsSystemWindows="true"
21.                 android:scaleType="centerInside"
22.                 android:src="@drawable/ic_android"
23.                 app:layout_collapseMode="parallax" />
24.             <androidx.appcompat.widget.Toolbar
25.                 android:id="@+id/toolbar"
26.                 android:layout_width="match_parent"
27.                 android:layout_height="?attr/actionBarSize"
28.                 app:layout_collapseMode="pin">
29.                 <TextView
30.                     android:layout_width="wrap_content"
31.                     android:layout_height="wrap_content"
32.                     android:text="折叠布局"
33.                     android:textColor="#FFFFFF"
34.                     android:textSize="24sp" />
35.             </androidx.appcompat.widget.Toolbar>
36.         </com.google.android.material.appbar.CollapsingToolbarLayout>
37.     </com.google.android.material.appbar.AppBarLayout>
38. </androidx.coordinatorlayout.widget.CoordinatorLayout>
```

上述第 11 行代码使用的 CollapsingToolbarLayout 的作用是提供一个可折叠的 Toolbar，它继承自 FrameLayout，用来实现子布局内不同元素响应滚动细节的布局，CollapsingToolBarLayout 必须是 AppBarLayout 的直接子 View 才有折叠效果。CollapsingToolBarLayout 的常用布局是包含一个 ImageView 和一个 Toolbar。

图 4.8 CoordinatorLayout 与 AppbarLayout 配合使用效果图

CollapsingToolbarLayout 常用属性说明如表 4.2 所示。

表 4.2 CollapsingToolbarLayout 常用属性说明

属性	含义	取值
app:title	设置标题内容	字符串
app:collapsedTitleGravity	设置收缩后 Title 的位置	left、right、top、bottom、center 等
app:expandedTitleGravity	设置扩展后 Title 的位置	取值同上
app:contentScrim	设置折叠时 Toolbar 的颜色	默认为 colorPrimary 的颜色
app:statusBarScrim	设置折叠时状态栏的颜色	默认为 colorPrimaryDark 的颜色
app:layout_scrollFlags	设置滑动组件与手势之间的关系	Scroll、exitUntilCollapsed、snap 等
app:layout_collapseParallaxMultiplier	设置视差滚动因子	值的范围：[0.0, 1.0]，值越大视差越大
app:layout_collapseMode	设置子 View 的折叠效果	parallax：视差模式，折叠时有视差折叠效果 pin：固定模式，折叠时 Toolbar 固定在顶端

使用 CollapsingToolbarLayout 实现折叠效果，需要注意的是：

● AppBarLayout 的高度必须固定。

● CollapsingToolbarLayout 的子视图设置 layout_collapseMode 属性。

● 关联悬浮视图设置 app:layout_anchor、app:layout_anchorGravity 属性。

通过以上详细的讲解，要达到 Material Design 风格的界面效果，需要多个布局的相互配合，用示意图（见图 4.9）说明它们之间的布局关系，可参考此示意图进行界面的嵌套设计。

4.2.3 TabLayout

在开发 Android 应用程序的过程中，经常需要用到切换界

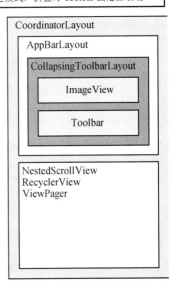

图 4.9 布局嵌套示意图

面的场景，通过一排 Tab 按钮切换不同的界面进行横向导航。Android 使用 TabLayout 布局实现 Tab 的切换，它是 Android Support 库的控件 android.support.design.widget.TabLayout，升级到 AndroidX 后，TabLayout 迁移到 material 包，在 com.google.android.material.tabs 包内。在开发过程中 TabLayout 一般结合 ViewPager 和 Fragment 来实现滑动的标签选择器。

TabLayout 的常用属性包括：
- app:tabGravity：Tab 对齐方式，取值为 fill 或者 center。
- app:tabIndicatorColor：Tab 指示符颜色，用来标识当前所选中的 Tab。
- app:tabMode：Tab 模式，取值为 fixed 或者 scrollable。fixed 表示固定个数，而 scrollable 表示 Tab 选项可以横向滚动。
- app:tabIndicatorFullWidth：选中指示器的宽度是否与文本相同，取值为 true 或 false。
- app:tabSelectedTextColor：选中的 Tab 选项上的文本颜色。
- app:tabTextColor：其他未被选中的 Tab 选项上的文本颜色。

应用 TabLayout 布局有以下两种方式：
① 通过 TabLayout 的 addTab()方法添加 Tab 实例或者直接使用 TabItem 控件。
② 使用 ViewPager 和 TabLayout 一站式管理 Tab，无须像第一种方式那样手动添加 Tab。

【案例 4-2】TabLayout 布局实现 Tab 切换，两种加载方式的界面效果如图 4.10、图 4.11 所示。

图 4.10 TabItem 布局的 Tab 界面

图 4.11 addTab()动态添加的 Tab 界面

步骤 1：创建项目

启动 Android Studio，创建名为 D0402_TabLayout 的项目，选择 Empty Activity，将包名改为 com.example.tablayout，单击 Finish 按钮，等待项目构建完成。

步骤 2：添加依赖库

在 app/build.gradle 文件中添加 TabLayout 的依赖库。

```
1.  dependencies {
2.      implementation 'com.google.android.material:material:1.2.1'
3.  }
```

步骤 3：设置布局

在 activity_main.xml 布局文件中添加 TabLayout 布局，分别用两种方式进行界面布局。

第一种方式实现图 4.10、图 4.11 所示的界面功能

(1) 在布局中添加 TabLayout,根据需要设置相应的属性。

```
1.  <com.google.android.material.tabs.TabLayout
2.      android:id="@+id/tab_layout"
3.      android:layout_width="match_parent"
4.      android:layout_height="wrap_content"
5.      app:layout_constraintLeft_toLeftOf="parent"
6.      app:layout_constraintRight_toRightOf="parent"
7.      app:layout_constraintTop_toTopOf="parent"
8.      app:tabGravity="fill"
9.      app:tabIndicatorColor="@color/colorAccent"
10.     app:tabIndicatorFullWidth="false"
11.     app:tabMaxWidth="0dp"
12.     app:tabSelectedTextColor="@color/colorPrimary"
13.     app:tabTextColor="@color/colorPrimary">
14. </com.google.android.material.tabs.TabLayout>
```

(2) 直接使用 TabItem 控件添加 TabItem,如下所示:

```
1.  <?xml version="1.0" encoding="utf-8"?>
2.  <androidx.constraintlayout.widget.ConstraintLayout
3.      xmlns:android="http://schemas.android.com/apk/res/android"
4.      xmlns:app="http://schemas.android.com/apk/res-auto"
5.      xmlns:tools="http://schemas.android.com/tools"
6.      android:layout_width="match_parent"
7.      android:layout_height="match_parent"
8.      tools:context=".MainActivity">
9.      <com.google.android.material.tabs.TabLayout
10.         android:id="@+id/tab_layout"
11.         android:layout_width="match_parent"
12.         android:layout_height="wrap_content"
13.         app:layout_constraintLeft_toLeftOf="parent"
14.         app:layout_constraintRight_toRightOf="parent"
15.         app:layout_constraintTop_toTopOf="parent"
16.         app:tabGravity="fill"
17.         app:tabIndicatorColor="@color/colorAccent"
18.         app:tabIndicatorFullWidth="false"
19.         app:tabMaxWidth="0dp"
20.         app:tabSelectedTextColor="@color/colorPrimary"
21.         app:tabTextColor="@color/colorPrimary">
22.         <com.google.android.material.tabs.TabItem
23.             android:layout_width="wrap_content"
24.             android:layout_height="wrap_content"
25.             android:id="@+id/tab_item1"
26.             android:text="热点"/>
27.         <com.google.android.material.tabs.TabItem
28.             android:layout_width="wrap_content"
29.             android:layout_height="wrap_content"
30.             android:id="@+id/tab_item2"
31.             android:text="新闻"/>
32.     </com.google.android.material.tabs.TabLayout>
33. </androidx.constraintlayout.widget.ConstraintLayout>
```

或者在 MainActivity 类的 onCreate()方法中通过 TabLayout 的 addTab()方法添加 Tab 实例:

```
1.  @Override
2.  protected void onCreate(Bundle savedInstanceState) {
3.      super.onCreate(savedInstanceState);
4.      setContentView(R.layout.activity_main);
5.      // 使用 addTab()方法添加 Tab 实例
```

```
6.     TabLayout tabLayout = findViewById(R.id.tab_layout);
7.     for (int i = 0; i < 4; i++) {
8.         tabLayout.addTab(tabLayout.newTab().setText("TAB_" + i));
9.     }
10. }
```

（3）为 TabLayout 添加选择 Tab 的 OnTabSelected 事件的监听器，当用户选中某个 Tab 后弹出 Toast 的消息提示，代码如下：

```
1.  tabLayout.addOnTabSelectedListener(new TabLayout.OnTabSelectedListener() {
2.      @Override
3.      public void onTabSelected(TabLayout.Tab tab) {
4.          Toast.makeText(MainActivity.this, tab.getText(),
                       Toast.LENGTH_LONG).show();
5.      }
6.      @Override
7.      public void onTabUnselected(TabLayout.Tab tab) {
8.
9.      }
10.     @Override
11.     public void onTabReselected(TabLayout.Tab tab) {
12.
13.     }
14. });
```

第二种方式实现如图 4.12 所示的界面功能

图 4.12 ViewPager 和 TabLayout 协同管理 Tab 的界面效果

（1）在左侧导航的包名上单击右键，选择 New->Activity->Empty Activity，使用向导创建名为 ViewPagerActivity 的 Activity，然后打开它的布局文件，在根布局下分别添加 TabLayout 布局和 ViewPager 控件，设置相关的属性，代码如下所示：

```
1.  <?xml version="1.0" encoding="utf-8"?>
2.  <androidx.constraintlayout.widget.ConstraintLayout
3.      xmlns:android="http://schemas.android.com/apk/res/android"
4.      xmlns:app="http://schemas.android.com/apk/res-auto"
5.      android:layout_width="match_parent"
6.      android:layout_height="match_parent">
7.      <com.google.android.material.tabs.TabLayout
8.          android:id="@+id/tab_layout"
9.          android:layout_width="match_parent"
```

```
10.        android:layout_height="wrap_content"
11.        app:layout_constraintLeft_toLeftOf="parent"
12.        app:layout_constraintRight_toRightOf="parent"
13.        app:layout_constraintTop_toTopOf="parent"
14.        app:tabGravity="fill"
15.        app:tabIndicatorColor="@color/colorAccent"
16.        app:tabIndicatorFullWidth="false"
17.        app:tabMaxWidth="0dp"
18.        app:tabSelectedTextColor="@color/colorPrimary"
19.        app:tabTextColor="@color/colorPrimary" />
20.    <androidx.viewpager.widget.ViewPager
21.        android:id="@+id/view_pager"
22.        android:layout_width="match_parent"
23.        android:layout_height="0dp" />
24.        app:layout_constraintBottom_toBottom="parent"
25.        app:layout_constraintTop_toBottomOf="@+id/tab_layout"
26. </androidx.constraintlayout.widget.ConstraintLayout>
```

（2）单击右键，然后在弹出的快捷菜单中选择 New->Fragment->Fragment（Blank），使用向导创建名为 TabFragment 的 Fragment，打开 fragment_tab.xml 文件，为 FrameLayout 和 TextView 添加 id 属性，第 13～15 行用于设置文本的字体、颜色和大小，代码如下：

```
1.  <?xml version="1.0" encoding="utf-8"?>
2.  <FrameLayout xmlns:android="http://schemas.android.com/apk/res/android"
3.      xmlns:tools="http://schemas.android.com/tools"
4.      android:id="@+id/frame_layout"
5.      android:layout_width="match_parent"
6.      android:layout_height="match_parent"
7.      tools:context=".TabFragment">
8.      <TextView
9.          android:id="@+id/tv_fragment"
10.         android:layout_width="match_parent"
11.         android:layout_height="match_parent"
12.         android:layout_margin="16dp"
13.         android:text="@string/hello_blank_fragment"
14.         android:textStyle="bold"
15.         android:textSize="24sp" />
16. </FrameLayout>
```

（3）修改 TabFragment 类，删除不必要的注释和代码，给 Fragment 传递一个参数，用于设置 TextView 的内容，在 onCreateView()中完成布局文件的注册、文本框添加文本及设置背景色，代码如下：

```
1.  public class TabFragment extends Fragment {
2.      private static final String ARG_PARAM = "param1";
3.      // 定义 TextView 显示的文本和界面的背景色 id 变量
4.      private String message;
5.      public TabFragment() {
6.          // Required empty public constructor
7.      }
8.      // 创建 TabFragment 对象，使用 setArguments()传递两个参数
9.      public static TabFragment newInstance(String param) {
10.         TabFragment fragment = new TabFragment();
11.         Bundle args = new Bundle();
12.         args.putString(ARG_PARAM, param);
13.         fragment.setArguments(args);
14.         return fragment;
15.     }
16.     @Override
17.     public void onCreate(Bundle savedInstanceState) {
```

```
18.         super.onCreate(savedInstanceState);
19.         // 获取传递的两个参数的值
20.         Bundle bundle = getArguments();
21.         if (bundle != null) {
22.             message = bundle.getString(ARG_PARAM);
23.         }
24.     }
25.     @Override
26.     public View onCreateView(LayoutInflater inflater, ViewGroup
                                    container,
27.                              Bundle savedInstanceState) {
28.         // 注册 Fragment 的布局文件
29.         final View view = inflater.inflate(R.layout.fragment_tab,
                                        container, false);
30.         // 初始化布局上的控件
31.         TextView tvMessage = view.findViewById(R.id.tv_fragment);
32.         // 设置文本和背景色
33.         if (!message.isEmpty()) {
34.            tvMessage.setText(message);
35.         }
36.         // 返回 view
37.         return view;
38.     }
39. }
```

（4）在 ViewPagerActivity 类中定义 initFragments()方法，初始化 3 个 Fragment，代码如下：

```
1.  // 初始化 fragments 结合
2.  private List<TabFragment> fragments;
3.  private void initFragments() {
4.      fragments = new ArrayList<>();
5.      fragments.add(TabFragment.newInstance("Fragment 热点"));
6.      fragments.add(TabFragment.newInstance("Fragment 新闻"));
7.      fragments.add(TabFragment.newInstance("Fragment 体育"));
8.  }
```

（5）在 ViewPagerActivity 类中创建用于 ViewPager 加载 3 个 Fragment 的 Adapter 的内部类，该 Adapter 继承自 FragmentStatePagerAdapter，重写 getItem()、getCount()和 getPageTitle() 三个方法，它们分别表示获取特定位置的 Fragment、Fragment 的格式及 ViewPager 的每个 Tab 标题，代码如下：

```
1.  private List<String> tabTitles = Arrays.asList("热点", "新闻", "体育");
2.  public class ViewPagerAdapter extends FragmentStatePagerAdapter {
3.      public ViewPagerAdapter(@NonNull FragmentManager fm, int behavior) {
4.          super(fm, behavior);
5.      }
6.      // 返回 position 位置的 Fragment
7.      @NonNull
8.      @Override
9.      public Fragment getItem(int position) {
10.         return fragments.get(position);
11.     }
12.     // 获取 tab 的个数
13.     @Override
14.     public int getCount() {
15.         return fragments.size();
16.     }
17.     // 获取每页的标题
18.     @Nullable
```

```
19.     @Override
20.     public CharSequence getPageTitle(int position) {
21.         return tabTitles.get(position);
22.     }
23. }
```

（6）在 ViewPagerActivity 类中定义 initView()方法用于初始化布局上的控件，代码如下：

```
1. // 初始化 ViewPager 等控件
2. private void initView() {
3.     TabLayout tabLayout = findViewById(R.id.tab_layout);
4.     ViewPager viewPager = findViewById(R.id.view_pager);
5.     ViewPagerAdapter adapter = new ViewPagerAdapter
                                    (getSupportFragmentManager());
6.     viewPager.setAdapter(adapter);
7.     tabLayout.setupWithViewPager(viewPager);
8. }
```

（7）在 ViewPagerActivity 类的 onCreate()方法中调用这两个方法，完成初始化工作，代码如下：

```
1. @Override
2. protected void onCreate(Bundle savedInstanceState) {
3.     super.onCreate(savedInstanceState);
4.     setContentView(R.layout.activity_viewpager);
5.     initFragments();
6.     initView();
7. }
```

（8）在 initView()方法中添加 ViewPager 的 PageChange 事件和 TabLayout 的 TabSelected 事件的监听器，代码如下：

```
1. // 设置监听器
2. viewPager.addOnPageChangeListener(
3.         new TabLayout.TabLayoutOnPageChangeListener(tabLayout));
4. tabLayout.addOnTabSelectedListener(new TabLayout.OnTabSelectedListener() {
5.     @Override
6.     public void onTabSelected(TabLayout.Tab tab) {
7.         viewPager.setCurrentItem(tab.getPosition());
8.     }
9.     @Override
10.    public void onTabUnselected(TabLayout.Tab tab) {
11.    }
12.    @Override
13.    public void onTabReselected(TabLayout.Tab tab) {
14.    }
15. });
```

（9）修改 AndroidManifest.xml 配置清单文件，使得 ViewPagerActivity 为启动界面，然后启动模拟器运行，效果如图 4.12 所示。

```
1. <activity android:name=".ViewPagerActivity" >
2.     <intent-filter>
3.         <action android:name="android.intent.action.MAIN" />
4.         <category android:name="android.intent.category.LAUNCHER" />
5.     </intent-filter>
6. </activity>
```

4.2.4 DrawerLayout

手机上的侧滑菜单很常见，如单击 QQ 左上角的头像就会打开一个侧滑的菜单，Google 官方提供的 DrawerLayout 布局可以很方便地实现侧滑菜单的效果，也称为"抽屉式导航栏"。抽屉式导航栏是显示应用主导航菜单的界面面板，DrawerLayout 结合 NavigationView 共同实现这个侧滑导航。当用户触摸应用栏中的抽屉式导航栏图标或用户从屏幕的左边缘滑动手指时显示出来。下面通过一个案例讲解 DrawerLayout 的使用。

【案例 4-3】DrawerLayout 结合 NavigationView 实现左侧导航菜单，以及菜单项的事件处理。界面效果如图 4.13～图 4.15 所示。

图 4.13 启动界面

图 4.14 弹出的菜单界面

图 4.15 单击菜单项后的界面

步骤 1：创建项目

启动 Android Studio，创建名为 D0403_DrawerLayout 的项目，将包名改为 com.example.layout，选择 Empty Activity，单击 Finish 按钮，等待项目构建完成。

步骤 2：添加依赖库

在 app/build.gradle 文件中添加 DrawerLayout 的依赖库：

```
1.  dependencies {
2.      implementation 'com.google.android.material:material:1.2.1'
3.  }
```

步骤 3：添加样式

values 目录下的 style.xml 文件用于定义界面的主题和样式，在此文件中添加工具条标题和图标的颜色，代码如下：

```
1.  <!--修改图标颜色-->
2.  <style name="AppTheme.ToolBar" parent="AppTheme">
3.      <item name="colorControlNormal">@android:color/white</item>
4.  </style>
```

步骤 4：设计界面布局

首先，打开 activity_main.xml 布局文件，修改根布局为 CoordinatorLayout，再添加工具栏

Toolbar 布局，设置 Toolbar 的 id、背景色、主题、导航图标及标题的文本和颜色，具体代码如下：

```
1.  <com.google.android.material.appbar.AppBarLayout
2.      android:id="@+id/appbar_layout"
3.      android:layout_width="match_parent"
4.      android:layout_height="wrap_content">
5.      <androidx.appcompat.widget.Toolbar
6.          android:id="@+id/toolbar"
7.          android:layout_width="match_parent"
8.          android:layout_height="?attr/actionBarSize"
9.          android:background="?attr/colorPrimary"
10.         android:theme="@style/AppTheme.ToolBar"
11.         app:navigationIcon="@drawable/ic_back"
12.         app:title="@string/app_name"
13.         app:titleTextColor="@android:color/white" />
14. </com.google.android.material.appbar.AppBarLayout>
```

然后，添加 DrawerLayout 布局，该布局一般包含两个子布局：内容布局和侧滑菜单布局，侧滑菜单使用 com.google.android.material.navigation 包的 NavigationView 控件实现，NavigationView 是 Material 库中的导航组件，它的 headerLayout 属性用于设置侧滑菜单布局的头部，menu 属性用于设置侧滑菜单布局头部下的菜单项，layout_gravity 用于设置滑出方向，start 为从左侧滑出，end 为从右侧滑出，具体代码如下：

```
1.  <androidx.drawerlayout.widget.DrawerLayout
2.      android:id="@+id/drawer_layout"
3.      android:layout_width="match_parent"
4.      android:layout_height="match_parent"
5.      android:layout_marginTop="?attr/actionBarSize">
6.      <!-- 内容区 -->
7.      <LinearLayout
8.          android:layout_width="match_parent"
9.          android:layout_height="wrap_content"
10.         android:gravity="center"
11.         android:orientation="vertical">
12.         <TextView
13.             android:id="@+id/tv_content"
14.             android:layout_width="wrap_content"
15.             android:layout_height="wrap_content"
16.             android:padding="20dp"
17.             android:text="内容区"
18.             android:textSize="20sp" />
19.     </LinearLayout>
20.     <!-- 左侧菜单 -->
21.     <com.google.android.material.navigation.NavigationView
22.         android:id="@+id/nav_view"
23.         android:layout_width="280dp"
24.         android:layout_height="match_parent"
25.         android:layout_gravity="start"
26.         android:fitsSystemWindows="true"
27.         app:headerLayout="@layout/drawer_header"
28.         app:menu="@menu/nav_menu" />
29. </androidx.drawerlayout.widget.DrawerLayout>
```

接着，在 res/layout 目录上右击，选择 New->Layout Resource File，创建抽屉菜单头部的布局文件 drawer_header.xml，此布局代码比较简单，直接查看代码，不再赘述。

```
1.  <?xml version="1.0" encoding="utf-8"?>
2.  <LinearLayout xmlns:android="http://schemas.android.com/apk/res/android"
3.      xmlns:app="http://schemas.android.com/apk/res-auto"
4.      android:layout_width="match_parent"
5.      android:layout_height="180dp"
6.      android:background="?attr/colorPrimary"
7.      android:gravity="bottom"
8.      android:orientation="vertical"
9.      android:padding="16dp"
10.     android:theme="@style/ThemeOverlay.AppCompat.Dark">
11.     <ImageView
12.         android:id="@+id/iv_header_image"
13.         android:layout_width="wrap_content"
14.         android:layout_height="wrap_content"
15.         app:srcCompat="@mipmap/ic_launcher_round" />
16.     <TextView
17.         android:id="@+id/tv_user"
18.         android:layout_width="wrap_content"
19.         android:layout_height="wrap_content"
20.         android:paddingTop="8dp"
21.         android:text="Android Studio"
22.         android:textAppearance="@style/TextAppearance.AppCompat.Body1" />
23.     <TextView
24.         android:id="@+id/tv_mail"
25.         android:layout_width="wrap_content"
26.         android:layout_height="wrap_content"
27.         android:text="android.studio@android.com"
28.         android:textSize="14sp" />
29. </LinearLayout>
```

步骤 5：设计菜单

在 res 目录上右击，选择 New->Android Resource File，资源类型选择 Menu，创建菜单文件 nav_menu.xml，使用 tool:showIn 属性指定包含此文件的布局，便于查看它嵌入到父级布局中的显示，系统在编译时会将其移除。<group>标签的 checkableBehavior 属性用于指定菜单项的选择行为，取值有 none（不可选）、single（单选）和 all（多选）。菜单项使用 id、icon 和 title 等属性设置它的 id、图标和标题，详细代码如下：

```
1.  <?xml version="1.0" encoding="utf-8"?>
2.  <menu xmlns:android="http://schemas.android.com/apk/res/android"
3.      xmlns:tools="http://schemas.android.com/tools"
4.      tools:showIn="nav_view">
5.      <group android:checkableBehavior="single">
6.          <item
7.              android:id="@+id/nav_collect"
8.              android:icon="@drawable/ic_menu_camera"
9.              android:title="收藏"/>
10.         <item
11.             android:id="@+id/nav_manage"
12.             android:icon="@drawable/ic_menu_tools"
13.             android:title="工具"/>
14.         <item
15.             android:id="@+id/nav_about"
16.             android:icon="@drawable/ic_menu_about"
17.             android:title="关于"/>
18.     </group>
19.     <item android:title="Communicate">
20.         <menu>
21.             <item
22.                 android:id="@+id/nav_share"
```

```
23.                android:icon="@drawable/ic_menu_share"
24.                android:title="分享"/>
25.            <item
26.                android:id="@+id/nav_send"
27.                android:icon="@drawable/ic_menu_logout"
28.                android:title="退出"/>
29.        </menu>
30.    </item>
31. </menu>
```

步骤6：编写代码

首先打开 MainActivity 类文件完成工具栏 Toolbar、抽屉布局 DrawerLayout、菜单导航 NavigationView 和内容区 TextView 等控件的初始化，代码如下：

```
1.  private TextView tvContent;
2.  @Override
3.  protected void onCreate(Bundle savedInstanceState) {
4.      super.onCreate(savedInstanceState);
5.      setContentView(R.layout.activity_main);
6.
7.      // 初始化控件对象
8.      Toolbar toolbar = findViewById(R.id.toolbar);
9.      DrawerLayout mainLayout = findViewById(R.id.drawer_layout);
10.     NavigationView navMenu = findViewById(R.id.nav_view);
11.     tvContent = findViewById(R.id.tv_content);
12. }
```

然后，设置标题文本，使用 ActionBarDrawerToggle 类建立 Toolbar 与 DrawerLayout 的关联，实现 Toolbar 的左上角按钮在抽屉打开和关闭时的图标切换及动画效果，ActionBarDrawerToggle 类实现了 DrawerLayout.DrawerListener 接口，它实现了带有动画效果改变的 android.R.id.home 图标、DrawerLayout 的打开和隐藏及监听打开和隐藏等功能，因此可以将它作为 DrawerLayout 的事件监听器的参数，监听抽屉侧滑打开和隐藏事件，实现联动。

```
1.  @Override
2.  protected void onCreate(Bundle savedInstanceState) {
3.      ...
4.      // Toolbar 设置标题
5.      toolbar.setTitle("DrawerLayout");
6.
7.      // 设置Toolbar 与 DrawerLayer 关联
8.      ActionBarDrawerToggle toggle = new ActionBarDrawerToggle
                                            (this, mainLayout,
9.              toolbar, R.string.open, R.string.close);
10.     //初始化状态
11.     toggle.syncState();
12.     mainLayout.addDrawerListener(toggle);
13. }
```

最后，设置菜单项的单击事件 NavigationItemSelectedListener 的监听器，菜单项的事件处理通过重写 onNavigationItemSelectedListener()方法实现。

```
1.  protected void onCreate(Bundle savedInstanceState) {
2.      ...
3.      // 设置抽屉菜单项的监听
4.      navMenu.setNavigationItemSelectedListener(
5.              new NavigationView.OnNavigationItemSelectedListener() {
6.          @Override
7.          public boolean onNavigationItemSelected(@NonNull MenuItem item) {
```

```
8.              tvContent.setText("单击了 " + item.getTitle() + " 菜单项");
9.              return false;
10.         }
11.     });
12. }
```

步骤 7：运行项目

在 AndroidManifest.xml 配置清单文件中，取消系统设置的默认工具栏，设置 MainActivity 的 theme 属性为 Theme.MaterialComponents.DayNight.NoActionBar，然后运行，通过侧滑、单击工具栏的图标打开抽屉菜单，单击菜单项检查是否在内容区显示设置的文本内容，得到如图 4.13 所示的界面。

多种 Material Design 风格的布局能极大地提升界面的美观和用户体验，读者应在实际开发中多加练习。接下来将详细讲解 Material Design 控件的使用。

4.3　高级 UI 组件

4.3.1　RecyclerView

Google 自 Android 5.0 推出了 RecyclerView 控件，它是列表控件 ListView 的高级版本，也是手机上最常用的布局控件之一，用于以列表形式展现数据，它能创建带有 Material Design 风格的复杂列表，因此，本书也就不再讲解 ListView 控件。

RecyclerView 控件和 ListView 有很多相似的地方，但是 RecyclerView 比 ListView 更加灵活，它将布局、绘制、数据绑定等都拆分成不同的类进行管理，高度解耦。RecyclerView 列表中的视图还提供了一种插拔式的体验，通过扩展 RecyclerView.ViewHolder 定义列表项 item 实例，负责显示一个带有视图的项，扩展性非常强。

RecyclerView 针对列表项 item 的增删提供了默认的动画效果，但开发者也可根据实际开发需要，通过 RecyclerView.ItemAnimator 类进行扩展，调用 RecyclerView.setItemAnimator() 方法实现自定义动画。

使用 RecyclerView 控件涉及的相关类的描述如下。

● RecyclerView.Adapter：数据适配器，继承自 RecyclerView.Adapetr 类，用于绑定数据项和列表项的 item 布局，为每个 item 布局绑定数据。

● RecyclerView.LayoutManager：布局管理器，设置每个列表项 item 在 RecyclerView 中的布局，控制它们的显示或隐藏。当列表项 item 重用或者回收时，LayoutManger 都会向 Adapter 请求新的数据替换原有数据，当需要显示大型数据集时，这种回收重用机制能有效提高性能，避免创建大量的列表项对象或者频繁调用 findViewById()方法。RecyclerView 提供了以下三种内置的 LayoutManager，也可以继承 RecyclerView.LayoutManager 实现自定义布局。

◇ LinearLayoutManager：线性布局，横向或者纵向排列列表项。

◇ GridLayoutManager：表格布局，以表格形式展示列表项。

◇ StaggeredGridLayoutManager：流式布局，类似瀑布流效果。

● RecyclerView.ViewHolder：承载 Item 视图的子布局。

● RecyclerView.ItemDecoration：给每一项 Item 视图添加子 View，常用来画分割线。

● RecyclerView.ItemAnimator：负责处理数据添加或者数据删除时的动画效果。

使用 RecyclerView 控件的通用编码过程如下：

```
1.  RecyclerView recyclerView = findViewById(R.id.recyclerView);
2.  LinearLayoutManager layoutManager = new LinearLayoutManager(this);
3.  // 设置布局管理器
4.  recyclerView.setLayoutManager(layoutManager);
5.  // 设置为垂直布局，这也是默认的
6.  layoutManager.setOrientation(OrientationHelper.VERTICAL);
7.  // 设置 Adapter
8.  recyclerView.setAdapter(recycleAdapter);
9.   // 设置分隔线
10. recyclerView.addItemDecoration(new DividerGridItemDecoration(this));
11. // 设置增加或删除条目的动画
12. recyclerView.setItemAnimator(new DefaultItemAnimator());
```

在开发过程中，尤其需要注意 Adapter 类的编码，RecyclerView 的 Adapter 与 ListView 的 Adapter 有一定的区别，RecyclerView.Adapter 需要实现以下三个方法。

（1）onCreateViewHolder()：此方法通过 inflate()方法将列表项 item 布局编译为 View 对象，返回以这个对象为参数的 ViewHolder 对象。

（2）onBindViewHolder()：此方法主要将数据渲染到列表项的 ViewHolder 的 View 控件中。

（3）getItemCount()：此方法类似于 ListView 的 BaseAdapter 适配器的 getCount()方法，即数据的总条目。

了解了 RecyclerView 的基本概念之后，下面通过一个案例练习 RecyclerView 的使用。

【案例 4-4】使用 RecyclerView 显示水果列表，界面如图 4.16 所示。

图 4.16　RecyclerView 的界面布局及单击响应结果

步骤 1：创建项目

启动 Android Studio，创建名为 D0404_RecyclerView 的项目，将包名改为 com.example. recyclerview，选择 Empty Activity，单击 Finish 按钮，等待项目构建完成。

步骤 2：添加依赖库

在 app/build.gradle 文件中添加 RecyclerView 的依赖库。

```
1. dependencies {
2.     implementation 'com.google.android.material:material:1.2.1'
3. }
```

步骤 3：编写布局代码

在编写布局文件之前，修改收集的水果图片使其大小一致，然后将其复制到 res/drawable 目录下。

首先，打开 activity_main.xml 布局文件，添加 RecyclerView 布局，设置 id 属性，代码如下：

```xml
<?xml version="1.0" encoding="utf-8"?>
<androidx.constraintlayout.widget.ConstraintLayout
    xmlns:android="http://schemas.android.com/apk/res/android"
    xmlns:app="http://schemas.android.com/apk/res-auto"
    xmlns:tools="http://schemas.android.com/tools"
    android:layout_width="match_parent"
    android:layout_height="match_parent"
    tools:context=".MainActivity">
    <androidx.recyclerview.widget.RecyclerView
        android:id="@+id/rv_fruit"
        android:layout_width="match_parent"
        android:layout_height="match_parent"
        app:layout_constraintBottom_toBottomOf="parent"
        app:layout_constraintLeft_toLeftOf="parent"
        app:layout_constraintRight_toRightOf="parent"
        app:layout_constraintTop_toTopOf="parent" />
</androidx.constraintlayout.widget.ConstraintLayout>
```

然后，在 res/layout 目录上右击，选择 New->Layout Resource File，创建列表项的布局文件 item_fruit.xml，设置 ImageView 和 TextView 两个控件分别展示水果的图片和文本。使用 RecyclerView 时，item 的根布局的 layout_height 属性需设置为 wrap_content 或固定高度，如果设置为 match_parent 会造成一个 item 布局占据整个手机屏幕的现象。设置 android:gravity 属性为 center 使两个控件垂直居中，代码如下：

```xml
<?xml version="1.0" encoding="utf-8"?>
<androidx.constraintlayout.widget.ConstraintLayout
    xmlns:android="http://schemas.android.com/apk/res/android"
    xmlns:app="http://schemas.android.com/apk/res-auto"
    android:layout_width="match_parent"
    android:layout_height="80dp">
    <ImageView
        android:id="@+id/iv_fruit"
        android:layout_width="64dp"
        android:layout_height="64dp"
        android:layout_margin="8dp"
        android:gravity="center_vertical"
        android:scaleType="fitXY"
        app:layout_constraintEnd_toStartOf="@+id/tv_name"
        app:layout_constraintHorizontal_bias="0.35"
        app:layout_constraintStart_toStartOf="parent"
        app:layout_constraintTop_toTopOf="parent"
        app:layout_constraintVertical_chainStyle="packed" />
    <TextView
        android:id="@+id/tv_name"
        android:layout_width="0dp"
        android:layout_height="wrap_content"
        android:layout_marginStart="8dp"
        android:gravity="center_vertical"
        android:textSize="16sp"
        app:layout_constraintBottom_toBottomOf="parent"
        app:layout_constraintEnd_toEndOf="parent"
        app:layout_constraintHorizontal_bias="0.65"
        app:layout_constraintStart_toEndOf="@+id/iv_fruit"
```

```
30.            app:layout_constraintTop_toTopOf="parent" />
31. </androidx.constraintlayout.widget.ConstraintLayout>
```

步骤 4：编写功能代码

首先，在 java 包上右击，选择 New->Java Class，创建 Fruit 实体类，从列表项展示的内容可以得出，该类需要两个成员，即水果名称和图片 id，类的代码如下：

```
1.  public class Fruit {
2.      private String name;
3.      private int imageId;
4.      // 构造方法
5.      public Fruit(String name, int imageId) {
6.          this.name = name;
7.          this.imageId = imageId;
8.      }
9.      // 省略 getter/setter 方法
10. }
```

然后，打开 MainActivity 类文件，新建 initData()方法随机构造 50 个十种水果的随机集合；新建 initView()方法完成初始化 RecyclerView 对象，设置它的布局管理器、分割线和动画等功能，本案例使用 LinearLayoutManager 布局线性展示水果列表，最后在 onCreate()方法中调用这两个方法，完成初始化工作，代码如下：

```
1.  private void initView() {
2.      // 初始化对象
3.      RecyclerView rvFruit = findViewById(R.id.rv_fruit);
4.      // 设置布局管理器
5.      rvFruit.setLayoutManager(new LinearLayoutManager(this));
6.      // 设置分割线
7.      rvFruit.addItemDecoration(new DividerItemDecoration(this,
8.              DividerItemDecoration.VERTICAL));
9.      // 设置动画
10.     rvFruit.setItemAnimator(new DefaultItemAnimator());
11.     // 设置适配器
12.     final FruitAdapter adapter = new FruitAdapter(fruits);
13.     rvFruit.setAdapter(adapter);
14. }
15. private List<Fruit> fruits = new ArrayList<Fruit> ();
16. private List<Fruit> fruitList = new ArrayList<Fruit> () {
17.     {
18.         add(new Fruit("苹果", R.drawable.apple));
19.         add(new Fruit("香蕉", R.drawable.banana));
20.         add(new Fruit("橙子", R.drawable.orange));
21.         add(new Fruit("樱桃", R.drawable.cherry));
22.         add(new Fruit("芒果", R.drawable.mango));
23.         add(new Fruit("梨子", R.drawable.pear));
24.         add(new Fruit("草莓", R.drawable.strawberry));
25.         add(new Fruit("菠萝", R.drawable.pineapple));
26.         add(new Fruit("西瓜", R.drawable.watermelon));
27.         add(new Fruit("葡萄", R.drawable.grape));
28.     }
29. };
30. // 初始化数据
31. private void initData() {
32.     fruits.clear();
33.     final Random random = new Random();
34.     for (int i = 0; i < 50; i++) {
```

```
35.            fruits.add(fruitList.get(random.nextInt(fruitList.size())));
36.        }
37. }
```

步骤5：创建列表适配器类

在java包上右击，选择New->Class Name，创建继承自RecyclerView.Adapter的FruitAdapter类，重写三个方法，并创建内部类ViewHolder完成item_fruit.xml布局中控件的初始化，代码如下：

```
1.  public class FruitAdapter extends RecyclerView.Adapter
        <FruitAdapter.ViewHolder> {
2.      private final List<Fruit> fruits;
3.      // 构造方法，接收传递的数据集合
4.      public FruitAdapter(List<Fruit> fruits) {
5.          this.fruits = fruits;
6.      }
7.      @NonNull
8.      @Override
9.      public ViewHolder onCreateViewHolder(@NonNull ViewGroup parent,
                                            int viewType) {
10.         // 加载列表项布局
11.         View view = LayoutInflater.from(parent.getContext())
12.                 .inflate(R.layout.item_fruit, parent, false);
13.         return new ViewHolder(view);
14.     }
15.     @Override
16.     public void onBindViewHolder(@NonNull ViewHolder holder,
                                    int position) {
17.         // 给列表项控件赋值
18.         Fruit fruit = fruits.get(position);
19.         holder.tvName.setText(fruit.getName());
20.         holder.ivFruit.setImageResource(fruit.getImageId());
21.     }
22.     @Override
23.     public int getItemCount() {
24.         return fruits.size();
25.     }
26.     public static class ViewHolder extends RecyclerView.ViewHolder {
27.         TextView tvName;
28.         ImageView ivFruit;
29.         public ViewHolder(@NonNull View itemView) {
30.             super(itemView);
31.             tvName = itemView.findViewById(R.id.tv_name);
32.             ivFruit = itemView.findViewById(R.id.iv_fruit);
33.         }
34.     }
35. }
```

步骤6：编写列表项单击事件的处理代码

RecyclerView没有提供类似setOnItemClickListener()的事件监听器，设置单击事件有两种处理方法：一种是直接在onBindViewHolder()方法中对item绑定单击事件，但这种方式增加了程序的耦合，所以不推荐；第二种是通过在Adapter类中定义监听的回调接口实现，接下来仔细讲解编码过程。

（1）创建接口OnItemClickListener，定义onItemClick()方法，代码如下：

```
1.  public interface OnItemClickListener {
2.      void onItemClick(View view, int position);
3.  }
```

（2）在 Adapter 类中定义 listener 成员及 set 方法。

```
1.  public class FruitAdapter extends RecyclerView.Adapter<FruitAdapter.
        ViewHolder> {
2.      ...
3.      private OnItemClickListener listener;
4.      public void setItemClickListener(OnItemClickListener listener) {
5.          this.listener = listener;
6.      }
7.  }
```

（3）在 onBindViewHolder 中设置 ViewHolder 的 view 的单击事件监听器，调用实现类对象的 onItemClick()方法。注意：必须在整个 view 上设置，仅在某个控件上设置不会响应单击事件。

```
1.  public void onBindViewHolder(@NonNull ViewHolder holder, int position) {
2.      ...
3.      holder.itemView.setOnClickListener(new View.OnClickListener() {
4.          @Override
5.          public void onClick(View v) {
6.              if (listener != null) {
7.                  listener.onItemClick(v, position);
8.              }
9.          }
10.     });
11. }
```

（4）MainActivity 类定义的 Adapter 对象调用 setOnItemClickListener 方法传入 OnItemClickListener 的实现类对象。

```
1.  private void initView() {
2.      ...
3.      // 设置 item 单击事件监听
4.      adapter.setItemClickListener(new FruitAdapter.OnItemClickListener() {
5.          @Override
6.          public void onItemClick(View view, int position) {
7.              final Fruit fruit = fruits.get(position);
8.              Toast.makeText(MainActivity.this,
9.                      String.format(Locale.CHINA, "单击了第%d 行的%s",
10.                             Position + 1, fruit.getName()),
11.                     Toast.LENGTH_SHORT).show();
12.         }
13.     });
14. }
```

至此，就能使用 RecyclerView 展示列表数据，修改布局管理器就能以不同的方式展示数据，非常方便。RecyclerView 的 Adapter 的设计将布局、数据解耦合，使得逻辑更为清晰，编码也更简单。RecyclerView 的内部实现也使得列表数据的加载得到了优化，读者应在实际项目开发中多练习。

4.3.2 CardView

CardView 是 Google 官方发布的 Material Design 风格的卡片布局控件，以往需要自定义 Shape 实现卡片的圆角和阴影效果，现在这些效果只需设置 CardView 属性即可实现，开发者可以很方便地将布局做成卡片，自带有圆角、阴影等效果，让卡片更具有物理世界的真实体验。

CardView 继承自 FrameLayout 类，可以理解为带圆角阴影和水波纹效果的 FrameLayout 布局，常用属性如表 4.3 所示。

表 4.3 CardView 常用属性的含义描述

属性	属性含义
app:cardBackgroundColor	设置背景颜色
app:cardCornerRadius	设置圆角半径
app:cardElevation	设置 z 轴的阴影深度
app:cardMaxElevation	设置 z 轴的最大高度值
app:contentPadding	设置内容与边距的间隔
app:cardUseCompatPadding	设置 Android 5.0 及以上版本是否添加 padding，默认值为 false Android 5.0 以下版本默认添加 padding
app:cardPreventConrerOverlap	是否给 content 添加 padding 来阻止与圆角重叠，默认值为 true

图 4.17 CardView 展示列表界面

下面通过案例讲解 CardView 的应用。

【案例 4-4（续）】使用 CardView 展示水果列表，如图 4.17 所示。

步骤 1：添加依赖包

在 app/build.gradle 文件中添加 CardView 的依赖包。

```
1.  dependencies {
2.      implementation 'com.google.android.material:
                       material:1.2.1'
3.  }
```

步骤 2：添加列表项的卡片布局

打开 D0404_RecyclerView 项目，在 res/layout 目录上右击，选择 New->Layout Resouce File，创建列表项布局文件 item_fruit_card.xml，设置它的根布局为 CardView，设置卡片的圆角半径 cardCornerRadius、阴影 cardElevation 等属性值。在它的内部添加 ImageView 控件显示水果图片，TextView 显示水果名字，代码如下：

```
1.  <androidx.cardview.widget.CardView
2.      xmlns:android="http://schemas.android.com/apk/res/android"
3.      xmlns:app="http://schemas.android.com/apk/res-auto"
4.      android:layout_width="match_parent"
5.      android:layout_height="wrap_content"
6.      android:layout_margin="6dp"
7.      app:cardCornerRadius="8dp"
8.      app:cardElevation="8dp">
9.      <LinearLayout
10.         android:layout_width="match_parent"
11.         android:layout_height="wrap_content"
12.         android:orientation="vertical">
13.         <ImageView
14.             android:id="@+id/iv_fruit"
15.             android:layout_width="match_parent"
16.             android:layout_height="128dp"
17.             android:scaleType="centerCrop" />
18.         <TextView
19.             android:id="@+id/tv_name"
20.             android:layout_width="wrap_content"
21.             android:layout_height="wrap_content"
```

```
22.                android:layout_gravity="center_horizontal"
23.                android:layout_margin="8dp"
24.                android:textSize="20sp" />
25.        </LinearLayout>
26. </androidx.cardview.widget.CardView>
```

步骤 3 修改适配器类 FruitAdapter

打开 FruitAdapter 类文件，修改 onCreateViewHolder()方法，将布局文件修改为 item_fruit_card，代码如下：

```
1. public ViewHolder onCreateViewHolder(@NonNull ViewGroup parent,
                                        int viewType) {
2.     // 加载列表项布局
3.     View view = LayoutInflater.from(parent.getContext())
4.             .inflate(R.layout.item_fruit_card, parent, false);
5.     return new ViewHolder(view);
6. }
```

步骤 4：修改 RecyclerView 的布局管理器

打开 MainActivity 类文件，修改 initView()方法中设置布局管理器的代码，代码如下：

```
1. private void initView() {
2.     ...
3.     // 设置布局管理器(2 列的网格布局器)
4.     rvFruit.setLayoutManager(new GridLayoutManager(this, 2));
5. }
```

4.3.3 FloatingActionButton

FloatingActionButton 是一种浮动的圆形按钮，具有一些独特的动态效果，比如变形、弹出、位移等，代表着在当前页面上用户的特定操作，简称为 FAB。与普通按钮相比，FloatingActionButton 可以为 Android 应用程序带来更加绚丽的界面效果及用户体验。

FAB 继承自 ImageView，拥有 ImageView 的所有属性，常用属性如表 4.4 所列。

表 4.4 FAB 的常用属性描述

属性	含义	取值
app:backgroundTint	边框背景颜色	颜色值
app:tint	图标颜色，无法修改图片颜色	颜色值
app:rippleColor	单击时的涟漪颜色	颜色值
app:borderWidth	FAB 的边框大小	通常设为 0dp
app:elevation	默认状态的 Z 轴的阴影大小	默认值：6dp
app:pressedTranslationZ	单击状态的 Z 轴的偏移量	默认值：12dp
app:fabSize	FAB 的大小	auto、normal（56dp）、mini（40dp）
app:src	FAB 的图标	符合 Design 设计的图标大小为 24dp
app:layout_anchor	设置锚点	
app:layout_anchorGravity	设置相对锚点的位置	bottom、center、right、left、top 等

FAB 的常用方法包括：

- void show()：显示按钮。
- void hide()：隐藏按钮。

- boolean isShown()：返回显示状态。
- void setOnClickListener（OnClickListener）：设置单击监听器。

FAB 的默认填充色使用 styles.xml 样式的 theme 的 colorAccent，自定义它的颜色值即可改变 FAB 按钮颜色，如：设置为#FF4081 的 FAB 按钮颜色是玫红色。colorAccent 还会改变 EditText、RadioButton 及 CheckBox 等控件选中时的颜色。FAB 的默认背景色是 theme 的 colorControlHighlight，自定义它的颜色值即可改变 FAB 按钮单击后的背景色。

设置 theme 的 colorAccent 和 colorControlHighlight 会影响到整个 App 的颜色，通过设置 backgroundTint 和 rippleColor 属性值可以只改变 FAB 的填充色与背景色，示例代码如下：

```
1.  <android.support.design.widget.FloatingActionButton
2.      ....
3.      app:backgroundTint="@color/myColorAccent"
4.      app:rippleColor="@color/myColorHighlight"
5.  />
```

值得注意的是，在 Android 5.x 及以上版本的设备中显示的 FAB，如果没有出现阴影，需要设置 app:borderWidth="0dp"；如果按上述设置后，阴影出现了，但是会有矩形的边界，此时需要设置一个合适的 margin 的值。

官方推荐的最佳实践包括以下要素：
- 按照标准设定尺寸，不要轻易修改它的样式。
- 单击和按压的时候，加深焦点的颜色表示这是一个单击。
- 不要过度使用 FAB，应用于当前页面最主要的操作，每个页面最好只有一个 FAB。
- 推荐只使用一个 FAB，如果需要多个操作，可以单击后将它展开显示更多操作按钮。
- FAB 的颜色比较抢眼，最好对其设定一些积极的操作，如：创建、分享等，避免轻微和破坏性的操作，如删除等。

在【案例4-4】的主界面中添加 FAB 按钮，修改 activity_main.xml 文件，添加 FAB 的布局，设置背景色、图标、大小及锚点等属性，代码如下：

```
1.  <com.google.android.material.floatingactionbutton.FloatingActionButton
2.      android:id="@+id/fab_add"
3.      android:layout_width="wrap_content"
4.      android:layout_height="wrap_content"
5.      android:layout_margin="16dp"
6.      android:backgroundTint="#FF4081"
7.      android:elevation="5dp"
8.      android:src="@drawable/ic_add"
9.      app:backgroundTint="@android:color/white"
10.     app:borderWidth="0dp"
11.     app:fabSize="normal"
12.     app:tint="@android:color/white"
13.     app:layout_anchor="@+id/rv_fruit"
14.     app:layout_anchorGravity="bottom|end" />
```

为 FAB 设置监听事件，当用户单击该按钮时弹出 Snackbar 提示，代码如下：

```
1.  private void initView() {
2.      ...
3.      // 设置 FAB 的单击事件监听
4.      FloatingActionButton fab = findViewById(R.id.fab_add);
5.      fab.setOnClickListener(new View.OnClickListener() {
6.          @Override
7.          public void onClick(View view) {
8.              Snackbar.make(view, "添加一个水果信息", Snackbar.LENGTH_LONG)
```

```
9.                        .setAction("Action", null).show();
10.         }
11.     });
12. }
```

FAB 及单击后的界面效果如图 4.18 所示。

从图 4.18 可以看到，位于底部的 FAB 按钮会被弹出的 Snackbar 遮盖，解决方案是将 activity_main 的根布局改为 CoordinatorLayout，因为 CoordinatorLaout 能监听它的子 View 的事件以实现避免遮挡，虽然 Snackbar 不是 CoordinatorLayout 的子 View，但由于其 View 参数是 CoordinatorLayout 的子元素，也能实现此效果，单击后的界面如图 4.19 所示。

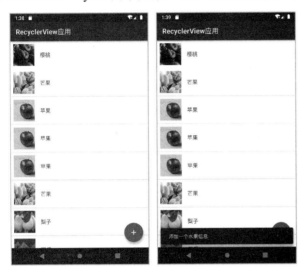

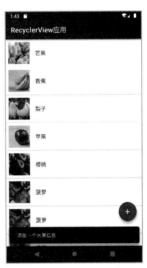

图 4.18　FAB 及单击的界面效果　　　　　　图 4.19　FAB 不被遮盖的界面效果

4.3.4　NavigationView

Google 在 Android 5.0 中推出的 NavigationView 是遵循 Material Design 设计风格的菜单导航类，用来规范侧滑菜单的基本样式，它包括头部布局 headerLayout 和内容菜单 menu 两部分，配合 DrawerLayout、Toolbar 实现侧滑抽屉式菜单功能，界面效果参见 4.2.4 小节的图 4.14，在 App 开发中经常用到，Google 官方也推荐使用。

NavigationView 的使用和普通的侧滑菜单基本相同，只需要在 DrawerLayout 中添加即可，它的布局设计及菜单项选择监听的编码实现在 4.2.4 小节的 DrawerLayout 案例中已经详细讲解，现归纳一下它的常用属性和方法，如表 4.5 所示。

表 4.5　NavigationView 的常用属性和方法描述

属性名称	含义描述
app:insetForeground="@android:color/transparent"	沉浸式展示
app:headerLayout="@layout/nav_header "	添加 Header 布局
app:menu="@menu/nav_menu"	添加标签 Item 的菜单
android:layout_gravity="start"	左侧拉出 NavigationView
方法名称	含义描述
addHeaderView(View view)	将视图添加为导航菜单的标题

方法名称	含义描述
removeHeaderView(View view)	删除已添加的标题视图
setItemBackgroundResource(int resId)	设置菜单项的背景
inflateMenu(int resId)	在此导航视图中添加菜单资源
setItemTextColor(ColorStateList textColor)	设置菜单项使用的文本颜色
setItemIconTintList(ColorStateList tint)	设置菜单项使用 Icon 的颜色
setNavigationItemSelectedListener(NavigationView listener)	设置菜单项选中的监听器

4.3.5 ViewPager

ViewPager 是 Android 的一个自带动画效果的视图滑动切换组件，可以通过手势滑动完成 View 的切换，主要用于 App 的引导页或图片轮播等。ViewPager 控件有如下特性：
- ViewPager 类继承自 ViewGroup 类，也就是一个控件容器，可以在其中添加其他控件。
- ViewPager 类需要一个 Adapter 适配器绑定视图，主要使用 PagerAdapter 适配器类。
- ViewPager 经常与 Fragment 共同使用。

ViewPager 类的常用方法描述如表 4.6 所示。

表 4.6 ViewPager 类的常用方法描述

方法名称	方法含义
setAdapter(PagerAdapter)	将 ViewPager 与适配器进行绑定
addOnPageChangeListener(ViewPager.OnPageChangeListener)	添加 Pager 滑动的监听器
removeOnPageChangeListener(ViewPager.OnPageChangeListener)	移除监听器
clearOnPageChangeListener()	清除监听器
setCurrentItem(int, boolean)	控制页面直接跳转到指定位置的界面
getCurrentItem()	获取当前页面的索引
setPageTransformer(boolean, ViewPager.PageTransformer)	设置滑动动画，如：翻转、渐进渐出
setOffscreenPageLimit(int)	设置当前页面左右两侧缓存的页面数量

创建继承自 PagerAdapter 的适配器类必须覆盖以下方法。
- getCount()：获取 ViewPager 一共有多少页面。
- isViewFromObject(View, Object)：确定页面 View 是否与 instantiateItem()方法返回的 key 对象相关联。
- instantiateItem(ViewGroup, int)：创建给定位置的页面，适配器将创建的 View 添加到给定的容器 container 中。
- destroyItem(ViewGroup,int, Object)：移除给定位置的页面。

PagerAdpater 类的通用实现代码如下所示，代码中使用的 views 是展示页面的结合：

```
1.    class MyPagerAdapter extends PagerAdapter {
2.        @Override
3.        public int getCount() {
4.            return views.size();
5.        }
6.        @Override
```

```
7.      public boolean isViewFromObject(@NonNull View view,
                                        @NonNull Object object) {
8.          return view == object;
9.      }
10.     @NonNull
11.     @Override
12.     public Object instantiateItem(@NonNull ViewGroup container,
                                      int position) {
13.         View view = views.get(position);
14.         TextView tvTitle = view.findViewById(R.id.tv_fragment);
15.         tvTitle.setText(tabTitles.get(position));
16.         container.addView(view);
17.         return view;
18.     }
19.     @Override
20.     public void destroyItem(@NonNull ViewGroup container,
21.                             int position, @NonNull Object object) {
22.         container.removeView(views.get(position));
23.     }
24. }
```

实际开发中一般不直接使用 PagerAdapter，而是使用它的两个子类 FragmentPagerAdapter 和 FragmentStatePagerAdapter。FragmentPagerAdapter 适配器会将所有 Fragment 都缓存在 FragmentManager 中，界面切换时 Fragment 不会被销毁，因此会占用大量内存，一般用于少数静态 Fragment 的场景。FragmentStatePagerAdapter 适配器用于界面较多的场景，因为它只缓存 Fragment 的状态，界面切换会销毁 Fragment，以节约内存。

在 Android 实际开发中，TabLayout + ViewPager + Fragment 是一个常见组合，能很好地实现分页滑动展示，4.2.3 节的 TabLayout 案例详细讲解了它们的组合应用，可以参考其中的 ViewPager 的代码实现。

4.3.6 Toolbar

Toolbar 是在 Android 5.0 中推出的 Material Design 风格的导航控件，它比 ActionBar 更具设计弹性，官方推荐使用 Toolbar 取代 ActionBar，作为 Android 客户端的导航栏。

Toolbar 与 ActionBar 不同之处在于，ActionBar 独立于布局之外，会固定在界面顶部，但 Toolbar 直接在布局文件中定义，可以在界面布局结构中依照需求任意配置，可以跟着 ScrollView 滚动，也可以与布局中的其他 View 交互、对滑动事件的响应等，调用 setSupportActionBar()或 setActionBar()方法使用 Toolbar 来取代 ActionBar。

除此之外，设计 Toolbar 时可以进行各种自定义，如设置导航栏图标、设置 App 的 logo、设置标题和子标题、添加一个或多个的自定义控件、添加 Action Menu。

【案例 4-5】实现如图 4.20 所示的 Toolbar 功能。

步骤 1：创建项目

启动 Android Studio，创建名为 D0405_Toolbar 的项目，选择 Empty Activity，将包名改为 com.example.toolbar，单击 Finish 按钮，等待项目构建完成。

图 4.20　实现以上功能的 Toolbar 效果图

步骤 2：添加工具栏

首先，打开 res/values/styles.xml，自定义样式继承自 appcompat 库的 NoActionBar 的主题背景，设置相关的颜色属性，如颜色、文本字体等。

```xml
1. <style name="NoActionBarTheme" parent="Theme.AppCompat.Light.NoActionBar">
2.     <item name="colorPrimary">@color/colorPrimary</item>
3.     <item name="colorPrimaryDark">@color/colorPrimaryDark</item>
4.     <item name="colorAccent">@color/colorAccent</item>
5.     <item name="android:textColorPrimary">@color/textColorPrimary</item>
6. </style>
```

然后，打开 AndroidManifest.xml 文件，修改<application>元素的样式主题为 NoActionBarTheme，替代原生 ActionBar 类提供的工具栏，代码如下：

```xml
1. <application
2.     android:theme="@style/NoActionBarTheme">
3. </application>
```

接着，在 Activity 的布局文件中添加 Toolbar 控件，它的高度设置建议参照 Material Design 规范，android:navigationIcon 用于设置左上角的导航图标，app:popupTheme 用于设置弹出菜单的样式，android:theme 用于设置工具栏的主题样式，代码如下：

```xml
1. <androidx.appcompat.widget.Toolbar
2.     android:id="@+id/toolbar"
3.     android:layout_width="match_parent"
4.     android:layout_height="?attr/actionBarSize"
5.     android:background="@color/colorPrimary"
6.     android:elevation="4dp"
7.     app:navigationIcon="@drawable/ic_menu"
8.     android:theme="@style/ThemeOverlay.AppCompat.ActionBar"
9.     app:popupTheme="@style/ThemeOverlay.AppCompat.Light" />
```

最后，在 Activity 类的 onCreate()方法中调用 setSupportActionBar()方法，将自定义 Toolbar 设为 Activity 的工具栏，就可以访问 ActionBar 类提供的各种方法，如：隐藏和显示工具栏，先调用 getSupportActionBar()方法获取 ActionBar 对象的引用，然后调用它的 hide()方法即可隐藏工具栏。

```java
1. @Override
2. protected void onCreate(Bundle savedInstanceState) {
3.     super.onCreate(savedInstanceState);
4.     setContentView(R.layout.activity_main);
5.     Toolbar toolbar = findViewById(R.id.toolbar);
6.     setSupportActionBar(toolbar);
7. }
```

步骤 3：添加和处理菜单操作

Toolbar 可以添加当前上下文的选项菜单，当菜单项无法全部显示时，就会显示到"溢出"菜单中，也可以通过设置 showAsAction 属性让菜单项始终显示到"溢出"菜单中。选项菜单的创建和菜单项事件响应操作参见 2.4 节的相关内容。app:actionProviderClass 用于设置菜单项的视图提供器。

```xml
1. <?xml version="1.0" encoding="utf-8"?>
2. <menu xmlns:android="http://schemas.android.com/apk/res/android"
3.     xmlns:app="http://schemas.android.com/apk/res-auto">
4.     <item android:id="@+id/action_share"
5.         android:title="分享"
6.         app:showAsAction="ifRoom"
7.         app:actionProviderClass="androidx.appcompat.widget.
                                    ShareActionProvider"/>
8.     <item
```

```
9.          android:id="@+id/action_settings"
10.         android:icon="@drawable/ic_settings"
11.         android:title="设置"
12.         app:showAsAction="never" />
13. </menu>
```

然后,Activity 类重写 onCreateOptionsMenu()加载菜单资源,重写 onOptionsItemSelected()响应菜单项的单击事件,代码如下:

```
1.  @Override
2.  public boolean onCreateOptionsMenu(Menu menu) {
3.      getMenuInflater().inflate(R.menu.main_menu, menu);
4.      return super.onCreateOptionsMenu(menu);
5.  }
6.  @Override
7.  public boolean onOptionsItemSelected(@NonNull MenuItem item) {
8.      Toast.makeText(this, "单击了" + item.getTitle(), Toast.LENGTH_
                      SHORT).show();
9.      return super.onOptionsItemSelected(item);
10. }
```

步骤 4:使用操作视图

操作视图是一种特殊的操作,能够在工具栏中提供丰富功能,如搜索功能,可以直接在工具栏中输入搜索文字,无须更改 Activity 或 Fragment。在菜单资源中创建相关的<item>元素,使用 actionViewClass 添加操作的控件类,使用 actionLayout 属性设置控件的布局资源,并将 showAsAction 属性设为"ifRoom|collapseActionView"或"never|collapseActionView",使得操作视图在工具栏中显示为图标,当用户单击时,则会展开填满工具栏,如图 4.21 所示,代码如下:

```
1.  <item
2.      android:id="@+id/action_search"
3.      android:icon="@drawable/ic_search"
4.      android:title="搜索"
5.      app:actionViewClass="android.widget.SearchView"
6.      app:showAsAction="ifRoom|collapseActionView" />
```

当单击图标时,操作视图便会展开执行搜索功能,通过在 Activity 类中调用 MenuItem 的 setOnActionExpandListener()设置监听器,其成员是实现了 MenuItem.OnActionExpandListener 监听器接口的类,代码如下:

```
1.  @Override
2.  public boolean onCreateOptionsMenu(Menu menu) {
3.      // 加载选项菜单
4.      getMenuInflater().inflate(R.menu.main_menu, menu);
5.      // 获取 Search 菜单项
6.      MenuItem searchItem = menu.findItem(R.id.action_search);
7.      // 设置展开和折叠的监听器
8.      searchItem.setOnActionExpandListener(new MenuItem.
                                      OnActionExpandListener() {
9.          @Override
10.         public boolean onMenuItemActionExpand(MenuItem item) {
11.             return true;
12.         }
13.         @Override
14.         public boolean onMenuItemActionCollapse(MenuItem item) {
15.             return true;
```

```
16.        }
17.    });
18.    return super.onCreateOptionsMenu(menu);
19. }
```

获取填写的搜索字符串，需要调用菜单项的 getActionView()获取 SearchView 操作视图对象，然后设置它的 onQueryTextListener 监听器，重写 onQueryTextSubmit()和 onQueryTextChange()两个方法，代码如下：

```
1.  @Override
2.  public boolean onCreateOptionsMenu(Menu menu) {
3.      ...
4.      // 获取 Search 菜单项的操作视图对象
5.      SearchView searchView = (SearchView) searchItem.getActionView();
6.      // 设置输入搜索文本的监听器
7.      searchView.setOnQueryTextListener(new SearchView.
                                          OnQueryTextListener() {
8.          @Override
9.          public boolean onQueryTextSubmit(String query) {
10.             Toast.makeText(MainActivity.this,
11.                 "搜索字符串为: " + query, Toast.LENGTH_SHORT).show();
12.             return false;
13.         }
14.         @Override
15.         public boolean onQueryTextChange(String newText) {
16.             return false;
17.         }
18.     });
19.     return super.onCreateOptionsMenu(menu);
20. }
```

步骤 5：运行项目

运行项目，单击"溢出"菜单标志，可以弹出菜单，单击"搜索"按钮，展开搜索框，填写搜索文本，使用 Toast 显示搜索文本的信息，运行效果如图 4.21 所示。

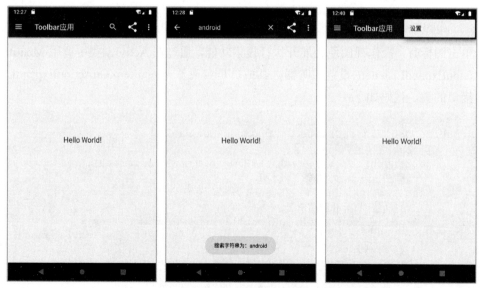

图 4.21 Toolbar 的界面、单击搜索按钮及溢出菜单的界面

4.4 自定义 View

尽管系统提供了丰富的控件，但在实际开发中，依旧会遇到系统提供的控件无法满足产品设计需求的情况，此时可以创建自定义 View 子类，控制 View 的外观和功能，从而满足开发需求。

创建自定义 View 控件之前，需要先了解 View 的绘制过程，整个绘制过程分为以下三个步骤。

1. Measure

在 Measure 操作中，测量操作的分发由 ViewGroup 完成，ViewGroup 通过对子节点进行遍历并分发测量操作，它的测量过程根据父节点的 MeasureSpec 及子节点的 LayoutParams 等信息计算子节点的宽高，最终形成父容器的宽高，具体的测量方式也可以通过重写 onMeasure() 方法设定。onMeasure() 方法的定义如下：

```
1.  protected void onMeasure(int widthMeasureSpec, int heightMeasureSpec){
2.
3.  }
```

2. Layout

Layout 过程用于确定每个 View 在父容器中的位置。Layout 过程是一个自上而下的过程，先设置父视图位置，再设置子视图，父视图位置一定程度上决定了子视图位置。也可以通过 onLayout() 方法来自定义具体的布局流程。Layout 过程是根据测量出的结果及对应的参数来确定每个控件应该显示的位置的。onLayout() 方法的定义如下：

```
1.  protected void onLayout(boolean changed, int left, int top,
                            int right,int bottom){
2.
3.  }
```

3. Draw

根据前两步所获得的具体布局参数，调用 onDraw() 方法对各个控件进行绘制，绘制的顺序为背景->控件内容->子控件绘制->绘制边缘及滚动条等。onDraw() 方法的定义如下：

```
1.  protected void onDraw(Canvas canvas){
2.
3.  }
```

以上的绘制流程说明控件的绘制由 ViewGroup 分发给子 View，各自完成自身的测量与布局操作，然后由根节点进行绘制，最终形成完整的界面，绘制过程的方法调用流程如图 4.22 所示。

按类型划分，自定义 View 的实现方式有以下 3 种。
- 组合控件：将一组 View 控件组合为一个新的控件，方便多处复用。
- 继承控件：继承系统原生 View 控件或 ViewGroup 控件，在原有的基础上进行扩展。
- 自绘控件：继承 View 或 ViewGroup，设计一个全新的控件或布局。

自绘控件的开发有一定的难点，此处不再讲解，接下来通过一个简单继承控件的案例进一步理解自定义 View 的创建过程。

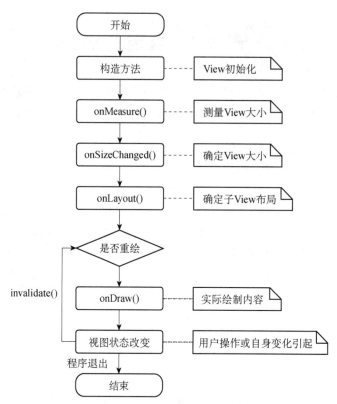

图 4.22　View 绘制的方法调用流程

【案例 4-6】创建验证码输入框的自定义 View，如图 4.23 所示。

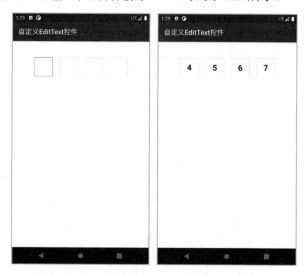

图 4.23　自定义验证码的 View 输入前后的界面图

步骤 1：创建项目

启动 Android Studio，创建名为 D0406_CustomView 的项目，将包名改为 com.example.customview，选择 Empty Activity，单击 Finish 按钮，等待项目构建完成。

步骤 2：创建自定义 View

在 com.example.customview 包上右击，选择 New->Java Class，创建 CodeEditText 类继承

自 androidx 包的 AppCompatEditText 类。首先初始化验证码格式、输入框大小、边框属性等基本属性，代码如下所示：

```java
1.  public class CodeEditText extends AppCompatEditText {
2.      private int textColor;                    // 验证码文本颜色
3.      private int maxLength = 4;                // 输入的最大长度
4.      private int strokeWidth;                  // 边框宽度
5.      private int strokeHeight;                 // 边框高度
6.      private int strokePadding = 20;           // 边框之间的距离
7.      private final Rect rect = new Rect();    // 用矩形保存方框的位置、大小信息
8.      private Drawable strokeDrawable;          // 方框背景
9.      private OnTextFinishListener finishListener;  // 输入结束监听器
10.     // 构造方法，初始化输入框大小、边框大小、间距及验证码个数等基本参数
11.     public CodeEditText(Context context, AttributeSet attrs) {
12.         super(context, attrs);
13.         TypedArray typedArray = context.obtainStyledAttributes(attrs,
14.             R.styleable.CodeEditText);
15.         int indexCount = typedArray.getIndexCount();
16.         for (int i = 0; i < indexCount; i++) {
17.             int index = typedArray.getIndex(i);
18.             if (index == R.styleable.CodeEditText_strokeHeight) {
19.                 this.strokeHeight = (int) typedArray.
                                    getDimension(index, 60);
20.             } else if (index == R.styleable.CodeEditText_strokeWidth) {
21.                 this.strokeWidth = (int) typedArray.getDimension
                                    (index, 60);
22.             } else if (index == R.styleable.CodeEditText_strokePadding) {
23.                 this.strokePadding = (int) typedArray.getDimension
                                    (index, 20);
24.             } else if (index == R.styleable.CodeEditText_
                                    strokeBackground) {
25.                 this.strokeDrawable = typedArray.getDrawable(index);
26.             } else if (index == R.styleable.CodeEditText_strokeLength) {
27.                 this.maxLength = typedArray.getInteger(index, 4);
28.             }
29.         }
30.         typedArray.recycle();
31.         if (strokeDrawable == null) {
32.             throw new NullPointerException("strokeDrawable 对象不能为空!");
33.         }
34.         setMaxLength(maxLength);              // 设置最大长度
35.         setLongClickable(false);              // 禁止长按操作
36.         setBackgroundColor(Color.TRANSPARENT); // 去掉背景颜色
37.         setCursorVisible(false);              // 不显示光标
38.     }
39.     // 设置最大长度数
40.     private void setMaxLength(int maxLength) {
41.         if (maxLength >= 0) {
42.             setFilters(new InputFilter[]{new InputFilter.LengthFilter
                                    (maxLength)});
43.         } else {
44.             setFilters(new InputFilter[0]);
45.         }
46.     }
47. }
```

上述代码第 11～29 行获取了布局中设置的属性集合，然后遍历设置输入框的大小、边距、背景及验证码个数等基本属性，其中使用的 styleable 资源定义在 values 目录的 attrs.xml 属性

资源文件中，代码如下：

```xml
1.  <?xml version="1.0" encoding="utf-8"?>
2.  <resources>
3.      <declare-styleable name="CodeEditText">
4.          <attr name="strokeLength" format="integer" />
5.          <attr name="strokeWidth" format="dimension" />
6.          <attr name="strokeHeight" format="dimension" />
7.          <attr name="strokePadding" format="dimension" />
8.          <attr name="strokeBackground" format="reference" />
9.      </declare-styleable>
10. </resources>
```

初始化属性参数之后，计算输入框的宽度，它的计算公式为：边框宽度*验证码数量+边框间距 *(验证码数量−1)，根据计算结果重写 onMeasure()方法进行测量布局。

```java
1.  @Override
2.  protected void onMeasure(int widthMeasureSpec, int heightMeasureSpec) {
3.      super.onMeasure(widthMeasureSpec, heightMeasureSpec);
4.      // 当前输入框的宽高信息
5.      int width = getMeasuredWidth();
6.      int height = getMeasuredHeight();
7.      int widthMode = MeasureSpec.getMode(widthMeasureSpec);
8.      int heightMode = MeasureSpec.getMode(heightMeasureSpec);
9.      // 判断高度是否小于推荐高度
10.     if (height < strokeHeight) {
11.         height = strokeHeight;
12.     }
13.     // 输入框宽度 = 边框宽度 * 数量 + 边框间距 *(数量-1)
14.     int recommendWidth = strokeWidth * maxLength + strokePadding *
                            (maxLength - 1);
15.     // 判断宽度是否小于推荐宽度
16.     if (width < recommendWidth) {
17.         width = recommendWidth;
18.     }
19.     widthMeasureSpec = MeasureSpec.makeMeasureSpec(width, widthMode);
20.     heightMeasureSpec = MeasureSpec.makeMeasureSpec(height, heightMode);
21.     setMeasuredDimension(widthMeasureSpec, heightMeasureSpec);
        // 设置测量布局
22. }
```

然后重写 onDraw()方法重绘输入框的背景和文本，代码如下：

```java
1.  @Override
2.  protected void onDraw(Canvas canvas) {
3.      super.onDraw(canvas);
4.      textColor = getCurrentTextColor();
5.      setTextColor(Color.TRANSPARENT);  // 将系统的文本颜色设为透明
6.      setTextColor(textColor);           // 重置文本颜色
7.      drawStrokeBackground(canvas);      // 重绘输入框
8.      drawText(canvas);   // 重绘文本
9.  }
```

drawStrokeBackground()方法首先保存画布上已有的元素，然后根据输入框的属性重绘，最后绘制拥有输入焦点的输入框的边框，代码如下：

```java
1.  // 绘制方框
2.  private void drawStrokeBackground(Canvas canvas) {
3.      // 下面绘制方框背景颜色，确定反馈位置
4.      rect.left = 0;
```

```
5.      rect.top = 0;
6.      rect.right = strokeWidth;
7.      rect.bottom = strokeHeight;
8.      int count = canvas.getSaveCount();   // 当前画布保存的状态
9.      canvas.save();   // 保存画布
10.     // 绘制每个输入框
11.     for (int i = 0; i < maxLength; i++) {
12.         strokeDrawable.setBounds(rect);   // 设置位置
13.         strokeDrawable.setState(new int[]{android.R.attr.state_
                         enabled});   //设置状态
14.         strokeDrawable.draw(canvas);    // 画到画布上
15.         // 确定下一个方框的位置
16.         float dx = rect.right + strokePadding;   // X坐标位置
17.         canvas.save();    // 保存画布
18.         canvas.translate(dx, 0);   // 移动画布到下一个位置
19.     }
20.     // 把画布还原到画反馈之前的状态,这样就还原到最初位置了
21.     canvas.restoreToCount(count);
22.     canvas.translate(0, 0);    // 画布归位
23.     // 绘制高亮状态的边框
24.     int activatedIndex = Math.max(0, getEditableText().length());
25.     rect.left = strokeWidth * activatedIndex + strokePadding *
                    activatedIndex;
26.     rect.right = rect.left + strokeWidth;
27.     strokeDrawable.setState(new int[]{android.R.attr.state_focused});
28.     strokeDrawable.setBounds(rect);
29.     strokeDrawable.draw(canvas);
30. }
```

drawText()方法的第 6～15 行代码的含义是根据验证码的长度,使用画笔对象 textPaint 设置颜色,获取文本框的大小,然后计算绘制文本的坐标,调用 canvas 对象的 drawText()方法绘制文本。代码如下:

```
1.  // 重绘文本
2.  private void drawText(Canvas canvas) {
3.      int count = canvas.getSaveCount();
4.      canvas.translate(0, 0);
5.      int length = getEditableText().length();
6.      for (int i = 0; i < length; i++) {
7.          String text = String.valueOf(getEditableText().charAt(i));
8.          TextPaint textPaint = getPaint();
9.          textPaint.setColor(textColor);
10.         textPaint.getTextBounds(text, 0, 1, rect);    // 获取文本大小
11.         // 计算(x, y) 坐标
12.         int x = strokeWidth/2+(strokeWidth + strokePadding) *
                    i - (rect.centerX());
13.         int y = canvas.getHeight() / 2 + rect.height() / 2;
14.         canvas.drawText(text, x, y, textPaint);
15.     }
16.     canvas.restoreToCount(count);
17. }
```

最后设置文本输入发生变化的监听器,监听当文本输入结束后关闭软键盘,调用文本输入结束的监听器 TextFinishListener 的 onTextFinish()方法,使得 CodeEditText()变为只读,代码如下:

```
1.  @Override
2.  protected void onTextChanged(CharSequence text, int start,
3.                               int lengthBefore, int lengthAfter) {
```

```
4.      super.onTextChanged(text, start, lengthBefore, lengthAfter);
5.      // 当前文本长度
6.      int textLength = getEditableText().length();
7.      if (textLength == maxLength) {
8.          hideSoftInput();
9.          if (finishListener != null) {
10.             finishListener.onTextFinish(getEditableText().
                                    toString(), maxLength);
11.         }
12.     }
13. }
14. // 关闭软键盘
15. public void hideSoftInput() {
16.     InputMethodManager imm = (InputMethodManager) getContext()
17.             .getSystemService(Context.INPUT_METHOD_SERVICE);
18.     if (imm != null)
19.         imm.hideSoftInputFromWindow(getWindowToken(),
20.             InputMethodManager.HIDE_NOT_ALWAYS);
21. }
22. // 设置输入完成监听
23. public void setOnTextFinishListener(OnTextFinishListener
                                    finishListener) {
24.     this.finishListener = finishListener;
25. }
```

步骤3：使用自定义View

完成验证码输入框自定义View之后，打开activity_main.xml，添加此控件，代码如下：

```
1.  <com.example.customview.CodeEditText
2.      android:id="@+id/et_code"
3.      android:layout_width="wrap_content"
4.      android:layout_height="wrap_content"
5.      android:layout_marginTop="48dp"
6.      android:inputType="number"
7.      android:singleLine="true"
8.      android:textSize="24sp"
9.      android:textStyle="bold"
10.     app:strokeBackground="@drawable/bg_code_edit"
11.     app:strokeHeight="56dp"
12.     app:strokeLength="4"
13.     app:strokePadding="20dp"
14.     app:strokeWidth="56dp" />
```

app:strokeBackground 用于设置输入框是否拥有焦点的选择器，当输入框拥有焦点时设置边框的颜色为橙色，否则为灰色，代码如下：

```
1.  <?xml version="1.0" encoding="utf-8"?>
2.  <selector xmlns:android="http://schemas.android.com/apk/res/android">
3.      <item android:state_focused="true">
4.          <shape>
5.              <stroke android:width="1dp" android:color="#FFFB9C00" />
6.              <corners android:radius="4dp" />
7.          </shape>
8.      </item>
9.      <item>
10.         <shape>
11.             <stroke android:width="1dp" android:color="#EEEEEE" />
12.             <corners android:radius="4dp" />
13.         </shape>
14.     </item>
15. </selector>
```

从上述代码可以看出，可以像使用普通控件一样使用自定义 View，设置各种属性，如设置控件宽和高的属性 layout_width 和 layout_heigh，设置每个输入框的宽高和边距 app:strokeHeight、app:strokeWidth、app:strokePadding，以及验证码的长度 app:strokeLength。但需要注意的是，自定义 View 在使用时务必使用全限定名，否则系统将无法找到这个 View。

在布局文件中加入自定义 View 的布局之后，就可以在 Activity 类中使用这个控件，并监听验证码输入完成后，将控件设置为不可编辑，代码如下：

```
1.  @Override
2.  protected void onCreate(Bundle savedInstanceState) {
3.      super.onCreate(savedInstanceState);
4.      setContentView(R.layout.activity_main);
5.
6.      CodeEditText codeEditText = findViewById(R.id.et_code);
7.      codeEditText.setOnTextFinishListener(new CodeEditText.
                                        OnTextFinishListener() {
8.          @Override
9.          public void onTextFinish(CharSequence text, int length) {
10.             codeEditText.setEnabled(false);
11.         }
12.     });
13. }
```

本节重点讲解了 Android 自定义控件的应用场景及使用方法，并结合自定义 View 的 Demo，使读者进一步了解如何使用自定义控件来满足实际开发需求。

4.5 本章小结

本章主要讲解了 Android 应用程序的高级界面设计，首先介绍了 Material Design 的基本概念，然后分别讲解了高级 UI 布局和高级 UI 组件的应用场景及使用方法，接着讲解了自定义 View 的使用方法，最后介绍了常用资源与样式的使用。希望通过本章的学习，读者对 Android 应用程序的界面设计与开发有更深的认识和了解，进一步强化 Android 界面开发的技能。

习　题

一、选择题

1. Material Design，中文名称为质感设计，是 Google 在 2014 年 I/O 大会上推出的全新的设计语言，使用 Material Design 需要 API（　　）以上。

　　A. 18　　　　　　　　B. 28　　　　　　　　C. 21　　　　　　　　D. 19

2. 在 ConstraintLayout 中，可以设置 bias 属性实现（　　）。

　　A. 子控件的透明度　　　　　　　　B. 子控件相对父控件的位置偏移

　　C. 子控件的角度定位　　　　　　　D. 子控件在 Z 轴方向的位置

3. 以下关于 RecyclerView 主要类的描述中错误的是（　　）。

　　A. RecyclerView.Adapter：可以托管数据集合，为每一项 Item 绑定数据

　　B. RecyclerView.LayoutManager：负责 Item 视图的布局显示管理

　　C. RecyclerView.ItemAnimator：负责处理数据添加或者数据删除时的动画效果

D. RecyclerView.ItemDecoration：负责承载 Item 视图的布局

4. 通常 View 的绘制可以分为 3 个过程，以下哪个不属于 View 的绘制过程（　　）。

A. Init　　　　　　B. Layout　　　　　　C. Draw　　　　　　D. Measure

5. 以下关于 DrawerLayout 的说法中错误的是（　　）。

A. DrawerLayout 中第一个子 View 为主要内容，即抽屉没有打开时显示的布局

B. DrawerLayout 中第二个子 View 是抽屉 view，即抽屉布局

C. 抽屉 view 的常用属性 layout_gravity="left"，表示抽屉始终展示在页面的左侧

D. 使用 DrawerLayout 时一般都需要有两个子 View

二、简答题

1. 简述 RecyclerView 提供的 3 种内置的 LayoutManager。

2. 简述 ConstraintLayout 的主要特性。

3. 简述 CardView 的常用属性。

三、编程题

1. 请使用 CoordinatorLayout 控件，协调 FloatingActionButton 和 Snackbar 的显示，当用户单击页面底部的 FloatingActionButton 时，弹出 Snackbar，提示"是否新增一条短信？"，同时 FloatingActionButton 上移，使 Snackbar 不会覆盖 FloatingActionButton。

2. 请综合使用本章所学 DrawerLayout、NavigationView 等高级控件开发新闻类 App 的主页面。

3. 在界面中添加一个 CardView，并在其中添加一张图片，页面中添加 3 个进度条，当用户拖动进度条时，分别改变卡片的圆角、阴影、图片间距，具体效果如图 4.24 所示。

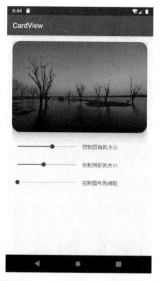

图 4.24　CardView 示例效果图

第 5 章 数据存储

任何应用程序都需要存储或读取数据，Android 应用程序也不例外，如是否记住用户登录信息、数据保存、图片存储等。Android 提供了应用程序私有、共享、偏好设置和数据库等不同应用场景的多种数据存储技术，这些存储技术可以将数据保存到不同的存储介质上，分别有 SharedPreferences、文件存储、SQLite 数据库、ContentProvider 和网络存储等 5 种方式，网络存储将在后续章节中讲解，本章重点讲解 Android 的前 4 种存储方式。

本章学习目标：

- 了解 5 种存储方式的含义及特点
- 掌握 SharedPrederences 的使用
- 掌握使用文件存储的方法
- 掌握 SQLite 数据库的应用，实现数据的增删改查功能
- 掌握 ContentProvider 在应用程序之间的数据访问方法
- 了解 Android 的第三方库的使用方法

 ## 5.1 SharedPreferences 存储

SharedPreferences 是 Android 提供的一种简单的数据存储类，类似 Web 开发中的 Cookie，它使用键值对的方式保存应用程序的一些简单配置信息，如用户登录的信息、播放音乐退出时的状态、设置选项等。SharedPreferences 支持多种不同数据类型的存储，也是 Android 中最简单的存储技术。本节详细讲解 SharedPreferences 的使用方法。

5.1.1 存储数据

SharedPreferences 是存储 key-value 键值对数据的 XML 文件，数据存储在手机内存的私有目录/data/data/<包名>/shared_prefs 中。存储数据首先需要获取 SharedPreferences 对象，Android 提供两种方法得到 SharedPreferences 对象。

1. Context 类的 getSharedPreferences()方法

此方法接收两个参数，第一个参数用于指定 SharedPreferences 文件的名称，如果指定的文件不存在则创建文件；第二个参数用于指定操作模式，目前只有 MODE_PRIVATE 模式可

选，表示只有当前的应用可以对这个文件进行读写。Android 4.2 版本之后，MODE_WORLD_WRITEABLE 和 MODE_WORLD_READABLE 模式都已被弃用。

2. Activity 类的 getPreferences()方法

此方法只有一个操作模式的参数，它自动将当前活动的类名当作 SharedPreferences 的文件名。获取 SharedPreferences 对象后就可以将数据存储到 SharedPreferences 文件中，具体步骤如下：

（1）调用 SharedPreferences 的 edit()方法获取 Editor 对象。SharedPreferences 对象本身只能获取数据，数据的存储和修改需要通过 SharedPreferences 的内部接口 Editor 实现。

（2）通过 Editor 对象存储 key-value 键值对数据。Editor 对象的 putXxx()方法用于保存键值对，其中 Xxx 表示 value 的不同数据类型，如：String 类型用 putString()方法，整型数据则使用 putInt()方法，以此类推。PutXxx()方法接收两个参数，第一个参数是 String 类型的 key 值，第二个参数是 value 值，value 的类型与 Xxx 指定的类型匹配。

（3）调用 apply()或 commit()方法提交数据，数据被存储在指定的 XML 文件中。

由于 getPreferences()方法不能自定义文件名，所以推荐使用 getSharedPreferences()方法，示例代码如下：

```
1.  SharedPreferences sp = getSharedPreferences("user.xml", MODE_PRIVATE);
2.  SharedPreferences.Editor edit = sp.edit();
3.  edit.putString("username", "安卓小绿人");
4.  edit.putInt("age", 20);
5.  edit.apply();
```

5.1.2 读取数据

SharedPreferences 的存储非常简单，读取也非常简单，只需使用 SharedPreferences 对象的 getXxx()方法即可获取，其中 Xxx 的含义与 putXxx()方法相同。示例代码如下：

```
1.  SharedPreferences sp = getSharedPreferences("user", MODE_PRIVATE);
2.  String username = sp.getString("username", "不存在");
3.  int age = sp.getInt("age", 12);
```

图 5.1 MainActivity 的界面布局

getXxx()方法需要两个参数，第一个参数是获取数据的 key 值，第二个参数是默认值，如果 key 值不存在，则返回默认值。例如，sp.getString("username", "不存在")，若 username 不存在则返回"不存在"。除了 getXxx()方法，SharedPreferences 还提供了 remove(String key)方法用于删除键为 key 值的 value 值，clear()方法用于清空文件中的所有数据。

下面通过一个具体案例介绍 SharedPreferences 的数据存储和读取功能。

【案例 5-1】使用 SharedPreferences 对象存储"记住我"的功能，界面如图 5.1 所示。

步骤 1：创建项目

启动 Android Studio，创建名为 D0501_SharedPreferences 的项目，选择 Empty Activity 项，将包名改为 com.example.

sharedpreferences，单击 Finish 按钮，等待项目构建完成。

步骤 2：完成 MainActivity 的界面布局

打开 activity_main.xml 布局文件，完成如图 5.1 所示的登录界面布局，具体代码如下：

```xml
1.  <?xml version="1.0" encoding="utf-8"?>
2.  <androidx.constraintlayout.widget.ConstraintLayout
3.      xmlns:android="http://schemas.android.com/apk/res/android"
4.      xmlns:app="http://schemas.android.com/apk/res-auto"
5.      xmlns:tools="http://schemas.android.com/tools"
6.      android:layout_width="match_parent"
7.      android:layout_height="match_parent"
8.      tools:context=".MainActivity">
9.      <ImageView
10.         android:id="@+id/iv_header"
11.         android:layout_width="wrap_content"
12.         android:layout_height="wrap_content"
13.         android:layout_margin="16dp"
14.         android:src="@mipmap/ic_launcher_round"
15.         app:layout_constraintLeft_toLeftOf="parent"
16.         app:layout_constraintRight_toRightOf="parent"
17.         app:layout_constraintTop_toTopOf="parent" />
18.     <EditText
19.         android:id="@+id/et_username"
20.         android:layout_width="match_parent"
21.         android:layout_height="wrap_content"
22.         android:layout_marginLeft="8dp"
23.         android:layout_marginTop="16dp"
24.         android:layout_marginRight="8dp"
25.         android:hint="请输入用户名"
26.         android:inputType="text"
27.         app:layout_constraintLeft_toLeftOf="parent"
28.         app:layout_constraintRight_toRightOf="parent"
29.         app:layout_constraintTop_toBottomOf="@+id/iv_header" />
30.     <EditText
31.         android:id="@+id/et_password"
32.         android:layout_width="match_parent"
33.         android:layout_height="wrap_content"
34.         android:layout_marginLeft="8dp"
35.         android:layout_marginTop="16dp"
36.         android:layout_marginRight="8dp"
37.         android:hint="请输入密码"
38.         android:inputType="textPassword"
39.         app:layout_constraintLeft_toLeftOf="parent"
40.         app:layout_constraintRight_toRightOf="parent"
41.         app:layout_constraintTop_toBottomOf="@+id/et_username" />
42.     <CheckBox
43.         android:id="@+id/cb_remember"
44.         android:layout_width="match_parent"
45.         android:layout_height="wrap_content"
46.         android:layout_marginLeft="8dp"
47.         android:layout_marginTop="16dp"
48.         android:layout_marginRight="8dp"
49.         android:text="记住我"
50.         android:textSize="20sp"
51.         app:layout_constraintLeft_toLeftOf="parent"
52.         app:layout_constraintRight_toRightOf="parent"
53.         app:layout_constraintTop_toBottomOf="@+id/et_password" />
54.     <Button
55.         android:id="@+id/btn_login"
56.         android:layout_width="match_parent"
```

```
57.        android:layout_height="wrap_content"
58.        android:layout_marginLeft="8dp"
59.        android:layout_marginTop="16dp"
60.        android:layout_marginRight="8dp"
61.        android:gravity="center_horizontal"
62.        android:orientation="horizontal"
63.        android:text="登    录"
64.        android:textSize="24sp"
65.        app:layout_constraintLeft_toLeftOf="parent"
66.        app:layout_constraintRight_toRightOf="parent"
67.        app:layout_constraintTop_toBottomOf="@+id/cb_remember" />
68. </androidx.constraintlayout.widget.ConstraintLayout>
```

步骤 3：编写控件初始化代码

打开 MainActivity 类文件，在 onCreate()方法中初始化"登录"按钮对象为 btnLogin、用户名和密码的 EditText 对象 etUsername 和 etPassword、CheckBox 对象 cbRemember 设置按钮的监听器及事件处理方法，具体代码如下：

```
1.  private void initView() {
2.      etUsername = findViewById(R.id.et_username);
3.      etPassword = findViewById(R.id.et_password);
4.      cbRemember = findViewById(R.id.cb_remember);
5.
6.      Button btnLogin = findViewById(R.id.btn_login);
7.      btnLogin.setOnClickListener(new View.OnClickListener() {
8.          @Override
9.          public void onClick(View v) {
10.
11.         }
12.     });
13. }
```

步骤 4：编写"记住我"功能的实现代码

打开 MainActivity 类文件，编写存储用户名和密码的 remember()方法、检查"记住我"按钮是否选中 checkRemember()方法及清空数据的方法 clear()，具体代码如下：

```
1.  private void remember(String username, String password) {
2.      SharedPreferences sp = getSharedPreferences("user", MODE_PRIVATE);
3.      SharedPreferences.Editor edit = sp.edit();
4.      edit.putString("username", username);
5.      edit.putString("password", password);
6.      edit.apply();
7.  }
8.  private void checkRemember() {
9.      SharedPreferences sp = getSharedPreferences("user", MODE_PRIVATE);
10.     etUsername.setText(sp.getString("username", ""));
11.     etPassword.setText(sp.getString("password", ""));
12. }
13. private void clear() {
14.     SharedPreferences sp = getSharedPreferences("user", MODE_PRIVATE);
15.     SharedPreferences.Editor edit = sp.edit();
16.     edit.remove("username");
17.     edit.clear();
18.     edit.apply();
19. }
```

在读取的过程中，需要注意 getString()方法的第一个参数的值务必与存入时的 key 值一致，否则就不能获取到值。调用 clear()方法清空键值对，务必调用 apply()方法确保提交。clear()

方法只清空 XML 文件中的键值对内容，并不会删除文件本身。

步骤 5：实现登录功能

```
1.  public void onClick(View v) {
2.      String username = etUsername.getText().toString();
3.      String password = etPassword.getText().toString();
4.
5.      if (!TextUtils.isEmpty(username) && !TextUtils.isEmpty(password)) {
6.          if(cbRemember.isChecked()) {
7.              remember(username, password);
8.          } else {
9.              clear();
10.         }
11.         Toast.makeText(MainActivity.this, "登录成功", Toast.LENGTH_
                        SHORT).show();
12.     } else {
13.         Toast.makeText(MainActivity.this, "用户名或密码不能为空",
14.             Toast.LENGTH_SHORT).show();
15.     }
16. }
```

第 6 行代码检查"记住我"是否选中，如果选中，则调用 remember()方法将数据存储到 user.xml 中，此文件存放在/data/data/com.example.sharedpreferences/shared_prefs 目录下，打开 Device File Explorer 找到此文件。如果没选中，则调用 clear()方法清空数据。生成的 user.xml 文件位置如图 5.2 所示。

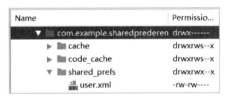

图 5.2　生成的 user.xml 文件位置

双击打开 user.xml 文件，存储的数据信息如下所示，输入的用户名和密码的数据已经存储在 user.xml 中。需要注意的是，密码类的敏感信息是不能明码存储在文件中的，这里只是案例，实际项目一般不保存密码信息。

```
1.  <?xml version='1.0' encoding='utf-8' standalone='yes'?>
2.  <map>
3.      <string name="password">123456</string>
4.      <string name="username">jack</string>
5.  </map>
```

步骤 6：重启项目，检查登录信息

关闭应用程序重新启动，在 onCreate()方法中获取 user.xml 中的登录信息，代码如下：

```
1.  protected void onCreate(Bundle savedInstanceState) {
2.      super.onCreate(savedInstanceState);
3.      setContentView(R.layout.activity_main);
4.
5.      initView();
6.      checkRemember();
7.  }
```

以上是 SharedPreferences 的存取数据的方法，它一般用于存储少量信息，大量数据需要使用文件存储，接下来进行具体讲解。

5.2 文件存储

文件存储是 Android 最常用的存储技术之一，用于存储一些简单的文本文件或二进制文件，它与 Java 中的文件存储类似，都是通过输入/输出（I/O）流把数据存储到文件中的。

5.2.1 文件存储简介

Android 的文件存储方式分为内部存储和外部存储，此分类是根据应用程序的安装目录进行划分的，不要与内置存储卡和外置存储卡混淆。Android 对开发者屏蔽了内置存储卡和外置 SD 卡的物理硬件。Android 的存储分类及存储位置如图 5.3 所示，内部存储和外部存储的实现方法不同，下面分别进行讲解。

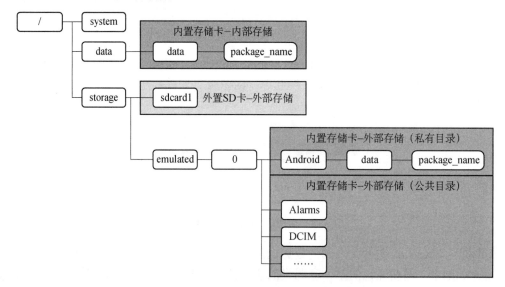

图 5.3　Android 的存储分类及存储位置

5.2.2 内部存储

Android 内部存储用于存储应用程序的私有文件，其他应用不能访问这些文件。系统为每个应用程序在/data/data/<包名>/xxx 下自动创建与之对应的目录，xxx 目录代表不同的存储类型，如：上一节讲解的 SharedPreferences 也属于内部存储，它使用 shared_pref 目录，文件存储在 files 目录下。当应用程序卸载后，内部存储的文件及目录都会被删除。另外，Android 4.4 以上版本的私有目录访问无须申请读写权限。

【案例 5-2】使用内部存储保存图 5.1 所示的用户登录信息。

步骤 1：创建项目

启动 Android Studio，创建名为 D0502_FileInternalStorage 的项目，将包名改为 com.example.internalstorage，按照默认步骤，等待构建完成。

步骤2：编写"记住我"功能，将数据写入文件

将数据存入内部存储，可以按以下步骤进行。

① 调用 Context 提供的 openFileOutput()方法获取 FileOutputStream 对象，此方法有两个参数，第一个参数是文件名称，第二个参数是文件的操作模式。应用程序运行后，系统会自动在应用程序目录下创建一个名为 files 的目录，用于保存私有文件。

② 调用 FileOutputStream 对象的 write()方法，将数据转为字节数组写入文件中。

③ 调用该对象的 close()方法，关闭输出流。

拷贝案例 5-1 的 activity_main.xml 的界面布局，打开 MainActivity 类文件，创建 remember()方法，将登录用户名和密码以字节形式写入 uesr.txt 文件，示例代码如下：

```
1.  private void remember(String username, String password) {
2.      try {
3.          FileOutputStream fos = openFileOutput("user.txt", Context.
                                                  MODE_PRIVATE);
4.          fos.write((username + ", " + password).getBytes());
5.          fos.close();
6.      }catch (IOException e) {
7.          e.printStackTrace();
8.      }
9.  }
```

openFileOutput()方法的第一个参数是文件名，第二个参数是操作模式，共有 4 种模式：MODE_PRIVATE 是创建内容模式，它的含义是创建文件并把它设为私有；MODE_APPEND 是添加内容模式，也就是说如果文件存在则追加内容，不存在则创建；MODE_WORLD_WRITEABLE 和 MODE_WORLD_READABLE 在 Android 4.2 版本之后被舍弃了。

user.txt 文件被创建在 files 目录下，如图 5.4 所示，文件内容为"jack, 123456"。

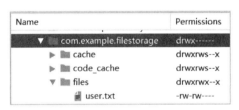

图 5.4 user.txt 文件位置

步骤3：读取文件中的数据

按照以下步骤从内部存储中读取 user.txt 文件中的内容。

① 调用 Context 提供的 openFileInput()方法并传入文件名为参数，它会返回一个 FileInputStream 对象。

② 调用 FileInputStream 对象的 read()方法读取字节数组。

③ 调用该对象的 close()方法，关闭输入流。

打开 MainActivity 文件，创建 checkRemember()方法读取文件数据，示例代码如下：

```
1.  private void checkRemember() {
2.      try {
3.          FileInputStream fis = openFileInput("user.txt");
4.          if(fis.available() == 0) {
5.              Toast.makeText(this, "文件内容为空", Toast.LENGTH_SHORT).show();
6.              return;
7.          }
8.          BufferedReader reader = new BufferedReader
```

```
                    (new InputStreamReader(fis));
9.         String[] datas = reader.readLine().split(",");
10.        if(datas.length > 1) {
11.            etUsername.setText(datas[0]);
12.            etPassword.setText(datas[1]);
13.        }
14.        fis.close();
15.    }catch (IOException e) {
16.        e.printStackTrace();
17.    }
18. }
```

第4~7行代码使用 FileInputStream 对象的 available()方法判断文件内容是否为空，如果为空则说明文件没内容，直接返回。第8~13行代码通过 BufferedReader 对象读取一行字符串，调用字符串的 split()方法将其转为字符串数组，然后将其填充到用户名和密码的文本框中。需注意的是，读取文件内容的方式必须与写入的方式一致，否则就无法获取内容。

如果未选中"记住我"，则在 user.txt 文件中填写空字符串即可，代码自行添加。

3. 存储缓存数据

如果只想暂时保留某些数据，则可以使用缓存进行保存。缓存数据的存储可以调用 getCacheDir()返回一个 File 对象，该对象指向应用程序私有的缓存目录，读写方式无须改变。当设备的内部存储空间不足时，Android 可能会删除这些缓存文件以回收空间。

5.2.3 外部存储

图 5.5 读取图片的界面图

Android 外部存储一般存储容量较大的文件，外部存储又分为私有存储和公有存储，私有存储用于存储应用程序中的容量较大的文件，公有存储则存储与其他应用共享的文件，如音频、视频、文档等。外部存储可以全局访问，但需要申请读写权限。当应用程序卸载后，私有存储的文件会被删除，而公有存储的文件则会被保留。接下来仍然通过一个案例来讲解外部存储的使用方法。

【案例 5-3】使用外部存储读写图片文件，界面如图 5.5 所示。

步骤 1：创建项目

启动 Android Studio，创建名为 D0503_FileExternalStorage 的项目，选择 Empty Activity，将包名改为 com.example.externalstorage，按照默认步骤，等待构建完成。

步骤 2：申请权限

由于外部存储的数据缺乏安全性，Android 系统规定：在访问外部存储时需要申请权限，否则程序在运行时会因为没有访问权限而崩溃。权限有两种申请方式：静态权限申请和动态权限申请。

（1）静态权限申请

静态权限申请适合 Android 6.0 以下的版本，该方式是在配置清单文件 AndroidManifest.xml 的<manifest>节点中添加<uses-permission>标签，申请外部存储读写权限的代码如下：

```
1. <?xml version="1.0" encoding="utf-8"?>
2. <manifest xmlns:android="http://schemas.android.com/apk/res/android"
3.     package="com.example.filestorage">
```

```
4.     <uses-permission android:name="android.permission.READ_
                                      EXTERNAL_STORAGE" />
5.     <uses-permission android:name="android.permission.WRITE_
                                      EXTERNAL_STORAGE" />
6.     <application...>
7. </manifest>
```

(2)动态权限申请

Android 从 6.0 版本开始改变了权限的管理模式,提供了 Permission 机制用来对应用程序执行的某些具体操作进行权限细分和访问控制,目的是保护 Android 用户的隐私。Permission 机制的权限按保护级别可分为正常权限、签名权限、危险权限和特殊权限。按照请求的时机可分为安装时请求和运行时请求。

Android 将权限分为几个保护级别,最重要的保护级别为正常权限和危险权限。正常权限对用户隐私或其他应用程序的操作几乎没有风险,应用程序只需要在配置清单文件中声明,系统在安装时自动授予该权限,如网络、蓝牙、Wi-Fi 等。危险权限指应用程序涉及的用户隐私的数据或资源的风险较大,危险权限一共有 9 组,分别为:日历、照相机、联系人、位置、麦克风、电话、传感器、短信和存储卡,如表 5.1 所示。

表 5.1 危险权限与权限组

权限组	权限
CALENDAR	READ_CALENDAR、WRITE_CALENDAR
CAMERA	CAMERA
CONTACTS	READ_CONTACTS、WRITE_CONTACTS GET_ACCOUNTS
LOCATION	ACCESS_FINE_LOCATION、ACCESS_COARSE_LOCATION
MICROPHONE	RECORD_AUDIO
PHONE	READ_PHONE_STATE、CALL_PHONE、READ_CALL_LOG、WRITE_CALL_LOG、ADD_VOICEMAIL、USE_SIP、PROCESS_OUTGOING_CALLS
SENSORS	BODY_SENSORS
SMS	SEND_SMS、RECEIVE_SMS、READ_SMS、RECEIVE_WAP_PUSH、RECEIVE_MMS
STORAGE	READ_EXTERNAL_STORAGE、WRITE_EXTERNAL_STORAGE

- 检查并请求权限

要使用危险权限,应用程序不仅需要在配置清单文件中声明,还必须在代码中动态申请,提示用户授予该权限。需要注意的是,不要在用户打开应用程序时检查或请求权限,而要等到用户选择或打开需要特定权限的功能时再检查或请求权限。例如,动态申请 SD 卡的写权限的示例代码如下:

```
1.  // 申请写 SD 的权限,要求 Android 的版本大于 6.0(Build.VERSION_CODES.M)
2.  if (Build.VERSION.SDK_INT >= Build.VERSION_CODES.M) {
3.      if (ContextCompat.checkSelfPermission(this,
4.              Manifest.permission.WRITE_EXTERNAL_STORAGE) !=
5.              PackageManager.PERMISSION_GRANTED) {
6.          ActivityCompat.requestPermissions(this,
7.                  new String[]{Manifest.permission.WRITE_EXTERNAL_
                                  STORAGE}, 1);
8.          return;
9.      }
10. }
```

图 5.6　申请权限对话框

首先判断 Android 的版本是否大于 6.0，然后调用 checkSelfPermission()方法判断申请的写权限是否已经授予，如果没有则调用 requestPermissions()方法申请权限。checkSelfPermission()方法有两个参数，第一个参数为 Context 上下文，第二个参数是申请的权限。requestPermissions()则需要三个参数，第一个参数也是 Context 上下文，第二个参数是申请的权限数组，第三个参数则是请求码。在添加权限检查的代码后运行程序，会显示一个对话框供用户选择是否授权，如图 5.6 所示。

● 处理权限请求响应

单击 ALLOW 按钮后，应用程序会调用申请权限的回调方法 onRequestPermissionsResult()处理权限请求响应，该方法有三个参数，第一个参数是请求权限时的请求码，第二个参数是权限的字符串数组，第三个参数是允许权限码的整型数组，如果被授予则它的第一个值就是常量 PackageManager.PERMISSION_GRANTED。示例代码如下：

```
1.  @Override
2.  public void onRequestPermissionsResult(int requestCode,
3.                                         @NonNull String[] permissions,
4.                                         @NonNull int[] grantResults) {
5.      super.onRequestPermissionsResult(requestCode, permissions,
                                          grantResults);
6.      if (grantResults.length == 0 ||
7.              grantResults[0] != PackageManager.PERMISSION_GRANTED) {
8.          return;
9.      }
10.     switch (requestCode) {
11.         case 1:  // 写SD卡
12.             saveToSD();
13.             break;
14.         case 2:  // 读SD卡
15.             readFromSD();
16.             break;
17.     }
18. }
```

● 写数据

外部存储的公共存储根目录为"/storage/emulated/0/"，调用 Environment 类的静态方法 getExternalStoragePublicDirectory()即可获取，该方法的参数就是不同类型的目录的字符串，不同类型文件的目录如图 5.7 所示。例如，使用 Environment.DIRECTORY_PICTURES 值为"Pictures"，该目录用于存储图片文件，便于 Android 的媒体扫描器能正确识别这些文件。

将图片存入 Pictures 目录的示例代码如下：

```
1.  private void saveToSD() {
2.      // 申请写SD的权限，要求Android的版本大于6.0
            (Build.VERSION_CODES.M)
3.      ...
4.      // 获取外部存储的Pictures目录，创建存储文件
```

图 5.7　外部存储的目录类型

```
5.          File path = Environment.getExternalStoragePublicDirectory(
6.                  Environment.DIRECTORY_PICTURES);
7.          File file = new File(path, MainActivity.FILE_NAME);
8.          try {
9.              file.createNewFile();
10.             // 获取 ImageView 的 Bitmap 图片对象
11.             BitmapDrawable drawable = (BitmapDrawable) ivPic.getDrawable();
12.             Bitmap bitmap = drawable.getBitmap();
13.             // 将 Bitmap 对象写入 SD 卡
14.             FileOutputStream fos = new FileOutputStream(file);
15.             bitmap.compress(Bitmap.CompressFormat.JPEG, 100, fos);
16.             // 关闭输出流
17.             fos.flush();
18.             fos.close();
19.         } catch (IOException e) {
20.             e.printStackTrace();
21.             Toast.makeText(MainActivity.this, "文件保存失败", Toast.LENGTH_
                        SHORT).show();
22.         }
23. }
```

运行程序，单击"存入 SD 卡"按钮，即可将 ImageView 显示的图片保存到 Pictures 目录，文件名为 demo.jpg，存储位置如图 5.7 所示。

● 读数据

从外部存储读取文件也需要动态申请权限，流程与写 SD 的权限申请类似，读取 SD 卡的权限的字符串是 Manifest.permission.READ_EXTERNAL_STORAGE。从外部存储目录读取图片文件的示例代码如下：

```
1.  private void readFromSD() {
2.          // 申请读 SD 的权限，要求 android 的版本大于 6.0(Build.VERSION_CODES.M)
3.          ...
4.          // 读取 SD 卡上的文件
5.          File path = Environment.getExternalStoragePublicDirectory(
6.                  Environment.DIRECTORY_PICTURES);
7.          File file = new File(path, MainActivity.FILE_NAME);
8.          try {
9.              // 创建 file 的文件输入流
10.             FileInputStream fis = new FileInputStream(file);
11.             // 将文件流写入 imageview
12.             ivPic.setImageBitmap(BitmapFactory.decodeStream(fis));
13.             // 关闭输入流
14.             fis.close();
15.         } catch (IOException e) {
16.             e.printStackTrace();
17.             Toast.makeText(MainActivity.this, "文件不存在或读取失败",
18.                     Toast.LENGTH_SHORT).show();
19.         }
20. }
```

如果应用程序已经被授予过读权限，则第 5～17 行代码是获取 Pictures 目录读取 demo.jpg 文件的代码段，使用 BitmapFactory 的 decodeStream()方法将输入字节流转为 Bitmap 对象，由 ImageView 控件显示。

Android 的文件存储底层依旧采用 Java 的 I/O 技术进行输入/输出流的读写，下面对内部存储、外部存储的目录结构、权限要求做一个总结，详细描述如表 5.2 所示。

表 5.2　Android 的存储特性总结

存储位置		路径	Android 版本	是否需要权限	访问方法	卸载后是否保留文件
内部存储		data/data/包名	所有	不需要	Context.getFilesDir() Context.getCacheDir()	不保留
外部存储	私有目录	Android/data/包名	API≥19	不需要	Context.getExternalFilesDir() Context.getExternalCacheDir()	不保留
			API＜19	需要		
	公有目录	DCIM Pictures Music Podcasts Alarms Download ……	API＜29	需要	Environment.getExternalStorageDirectory() Environment.getExternalStoragePublicDirectory()	保留
				不需要	Storage Access Framework	
			API≥29	需要	MediaStore API 读写其他应用的 Media 类型文件	
				不需要	MediaStore API 读写其他应用的 Media 类型文件 Storage Access Framework (访问其他应用非 Media 类型文件)	

5.3　SQLite 数据库存储

前两节讲解的数据存储技术只适用于数据量少的情况，对于结构化数据，Android 系统提供了内置的数据库系统 SQLite，它不仅支持标准的 SQL 语句，遵循数据库的 ACID 事务原则，而且运行速度快、占用资源少，特别适合在移动设备上使用。本节将详细讲解 SQLite 数据库系统的使用。

5.3.1　SQLite 数据库简介

SQLite 是 2000 年由 D.Richard Hipp 发布的开源的嵌入式数据库引擎，实现了自包容、无服务器、零配置及事务性。它是一个零配置的轻量级数据库，包括表在内的所有数据都存放在单个文件中，除 Android 外，许多开源项目也使用 SQLite。

SQLite 基本符合 SQL-92 标准，可以在所有主流的操作系统中运行，很容易地执行数据的增删改查操作。但由于移动设备的内存有限，SQLite 不能执行复杂的查询，不支持外键和左右连接，也不支持 Alter Table 的部分功能，尽管如此，它仍有很多突出的优点：轻量级；无须安装和管理配置；存储在单一文件中的数据库；支持数据库大小可达 2TB；没有额外依赖，独立性强；源代码完全开源；支持多种开发语言。

SQLite 采用动态数据类型，可以根据存入的值自动判断，它 5 种数据类型，详情如表 5.3 所示。虽然它支持的类型只有 5 种，但实际上它也接收 varchar、char、decimal 等类型，SQLite 在运算或保存时会将它们转换为对应的 5 种数据类型。

表 5.3　SQLite 支持的数据类型

数据类型	描述
NULL	空值
INTEGER	带符号的整数，根据值的大小存储在 1、2、3、4、6 或 8 字节中

数据类型	描 述
REAL	浮点值，存储为 8 字节的 IEEE 浮点数字
TEXT	文本字符串，使用数据库编码（UTF-8、UTF-16BE 或 UTF-16LE）存储
BLOB	二进制数据，根据它的输入进行存储

5.3.2 创建数据库

由于 SQLite 数据库并不需要像其他数据库那样进行身份验证，使得获得 SQLite Database 对象就像获取文件一样简单。Android 并不自动提供数据库，为了方便管理数据库，Android 提供了 SQLiteOpenHelper 类，借助该类使得创建和升级数据库变得非常简单。

SQLiteOpenHelper 是一个抽象类，它有两个抽象方法 onCreate()和 onUpgrade()，分别实现数据库的创建和升级。使用 SQLiteOpenHelper 类需要创建一个自定义类继承它，并重写这两个方法。示例代码如下所示：

```java
1.  public class DBHelper extends SQLiteOpenHelper {
2.      // 创建表的sql字符串
3.      private final static String CREATE_TABLE_STUDENT =
            "create table t_student(" +
4.              "_id integer primary key autoincrement, " +
5.              "name varchar(20), classmate varchar(30), age integer)";
6.      // 构造方法
7.      public DBHelper(@Nullable Context context) {
8.          super(context, "student.db", null, 1);
9.      }
10.     // 数据库第一次被创建时自动调用
11.     @Override
12.     public void onCreate(SQLiteDatabase db) {
13.         db.execSQL(DBHelper.CREATE_TABLE_STUDENT);
14.     }
15.     // 当数据库版本号增加时被自动调用
16.     @Override
17.     public void onUpgrade(SQLiteDatabase db, int oldVersion,
                              int newVersion) {
18.     }
19. }
```

上述代码中，第 3~7 行创建了两个 SQL 语句字符串，分别是创建表和删除表，第 9~11 行自定义类 DBHelper 的构造方法，通过 super()方法调用父类的构造方法，传入 4 个参数，分别为上下文对象、数据库名称、游标工厂（通常是 null）和数据库版本号。

第 13~16 行重写 onCreate()方法，用于初始化数据库表，当数据库第一次创建时调用，创建的数据库文件存储在"/data/data/包名/databases"目录中。第 18~22 行 onUpgrade()方法在 newVersion 的版本号增加时调用，否则不调用。需要注意的是，数据库的创建和更新都是在应用程序运行时自动调用的，无须由代码调用这两个方法。

onCreate()方法的参数 SQLiteDatabase 是一个数据库对象，即对应一个数据库文件。SQLiteOpenHelper 提供了两个实例方法获取 SQLiteDatabase 对象，getReadableDatabase()方法返回以只读方式打开数据库的对象，一般在表查询的时候使用；getWritableDatabase()方法返回以写入方式打开数据库的对象，一般用于表的增、删、改操作。

SQLiteDatabase 类提供了表的增删改查、执行 SQL 语句等常用方法，如表 5.4 所示。

表 5.4 SQLiteDatabase 常用方法

方法描述	方法定义
增加记录	insert(String table, String nullColumn, ContentValues values)
删除记录	delete(String table, String whereClause, String[] whereArgs)
修改记录	update(String table, ContentValues values, String whereClause, String[] whereArgs)
查询记录	query(String table, String[] columns, String selection, String[] selectionArgs, String groupBy, String having, String orderBy)
执行 SQL 语句	execSQL(String sql)
关闭数据库	close()

5.3.3 SQLite 数据库操作

完成创建数据库的自定义类之后，接下来详细讲解数据库表的增、删、改、查操作。以 onCreate()方法创建的 t_student 表为例，讲解如何进行表的基本操作。首先创建 StudentDao 类，创建 DBHelper 对象，通过构造方法引入 Context 上下文对象，示例代码如下：

```
1.  public class StudentDao {
2.      private DBHelper dbHelper;
3.      public StudentDao(Context context) {
4.          dbHelper = new DBHelper(context);
5.      }
6.  }
```

1. 添加数据

使用 SQLiteDatabase 对象的 insert()方法向 t_student 表中增加一条数据的示例代码如下：

```
1.  public void insert(String name, String classmate, int age) {
2.      // 打开数据库
3.      SQLiteDatabase db = dbHelper.getWritableDatabase();
4.      // 封装数据
5.      ContentValues values = new ContentValues();
6.      values.put("name", name);
7.      values.put("classmate", classmate);
8.      values.put("age", age);
9.      // 执行语句
10.     db.insert("t_student", null, values);
11.     // 关闭数据库
12.     db.close();
13. }
```

上述代码通过 getWritableDatabase()方法获得 SQLiteDatabase 对象，将 t_student 表需要的数据添加到 ContentValues 对象中，ContentValues 类与 Map 结构类似，以键值对的方式存储数据，它的 put()方法用于添加数据，其第一个参数是表的字段名，第二个参数是数据；最后调用 insert()方法将组装的数据添加到 t_student 表中。

insert()方法有 3 个参数，第一个参数是表名，第二个参数用于在未指定数据的情况下为空的列自动赋值 NULL 值，一般传入 null 即可，第三个参数是 ContentValues 对象。

如果你熟悉 insert 的 SQL 语句，也可以使用以下代码添加数据，与 Java 的 JDBC 方法非常类似。

```
1.  public void insert(String name, String classmate, int age) {
2.      SQLiteDatabase db = dbHelper.getWritableDatabase();
3.      String sql = "insert into t_student(name, classmate, age)
                     values(?, ?, ?)";
4.      db.execSQL(sql, new String[]{name, classmate, String.valueOf(age)});
5.      db.close();
6.  }
```

需要注意的是，SQLiteDatabase 对象使用结束后，务必调用 close()方法关闭数据库连接，否则会不断消耗内存，最终报数据库未关闭异常。

2. 删除数据

SQLiteDatabase 类的 delete()方法用于删除表中的数据，此方法接收 3 个参数，第一个参数与 insert()方法一样，也是表名，用于指定删除数据的表；第二个参数对应 SQL 语句的 where 部分，表示删除所有_id 为?的行，?是一个占位符，它的值由第三个参数提供的字符串数组按问号的顺序指定，不指定默认删除所有行。示例代码如下：

```
1.  public void delete(int _id) {
2.      SQLiteDatabase db = dbHelper.getWritableDatabase();
3.      db.delete("t_student", "_id=?", new String[]{String.valueOf(_id)});
4.      db.close();
5.  }
```

使用 SQL 语句删除数据的代码如下：

```
1.  String sql = "delete from student where _id=?";
2.  db.execSQL(sql, new String[]{String.valueOf(_id)});
```

3. 更新数据

SQLiteDatabase 类提供 update()方法用于更新数据，这个方法接收 4 个参数，第一个参数也是表名，第二个参数是 ContentValues 对象，第三、第四个参数是用于更新的条件和数据，不指定则默认更新所有行。示例代码如下：

```
1.  public void update(String name, String classmate, int age) {
2.      // 打开数据库
3.      SQLiteDatabase db = dbHelper.getWritableDatabase();
4.      // 封装数据
5.      ContentValues values = new ContentValues();
6.      values.put("name", name);
7.      values.put("classmate", classmate);
8.      values.put("age", age);
9.      // 执行语句
10.     db.update("t_student", values, "_id=?", new String[]{"1"});
11.     db.close();
12. }
```

上述代码的含义与添加、删除数据类似，就不再详细说明。使用 SQL 语句更新数据的代码如下：

```
1.  String sql = "update student set name=?, classmate=?, age=? where _id=?";
2.  db.execSQL(sql, new String[]{student.getName(), student.getClassmate(),
3.              String.valueOf(student.getAge()), String.valueOf
                        (student.get_id())});
```

4. 查询数据

查询数据是数据库增删改查操作中最复杂的操作，SQLiteDatabase 类提供 query()方法实现数据查询功能，这个方法返回描述数据集合的游标对象 Cursor，查询到的数据都从这个对象获取。这个方法的参数也比较多，最简单的也有 7 个参数，具体描述如表 5.5 所示。

表 5.5 query()的常用参数的含义

参数	含义	对应的 SQL 语句
table	查询的数据表	from table_name
columns	需要查询的字段，也就是列名	select column1, column2
selection	查询的子条件，相当于 select 的 where 部分，可使用占位符	where columm = value
selectionArges	对应 selection 的占位符的值	—
groupBy	指定需要 group by 的列	groub by column
having	指定需要 having 的列	having column
orderBy	指定需要 order by 的列	order by column

下面通过代码体验 query()方法的用法，示例代码如下：

```
1.   public void select(int _id) {
2.       // 打开数据库
3.       SQLiteDatabase db = dbHelper.getReadableDatabase();
4.       // 数据查询
5.       Cursor cursor = db.query("t_student", null, "_id=?",
6.               new String[]{String.valueOf(_id)}, null, null, null);
7.       // 获取查询结果
8.       if(cursor.moveToNext()) {
9.           String name = cursor.getString(cursor.getColumnIndex("name"));
10.          String classmate = cursor.getString(cursor.getColumnIndex
                                      ("classmate"));
11.          int age = cursor.getInt(cursor.getColumnIndex("age"));
12.      }
13.      // 关闭数据库
14.      cursor.close();
15.      db.close();
16.  }
```

上述代码中的第 8～12 行通过 Cursor 对象将数据库的查询结果转为 Java 变量的值。Cursor 对象的 moveToNext()方法实现移动游标指向下一条记录，moveToFirst()方法实现将游标指向第一条记录，getColumnIndex()方法实现获取某一列在表中对应的位置索引，然后调用 getXxx()方法获取该列的数据，其中 Xxx 代表数据类型，如 getString()获取字符串、getInt()获取整型值，以此类推。查询也可以直接使用 SQL 语句完成，代码如下：

```
1.   Cursor cursor = db.rawQuery("select * from t_student where _id=?",
2.           new String[]{String.valueOf(_id)});
```

需要注意的是，execSQL()方法只能执行 insert、delete 和 update 之类的 SQL 语句，select 语句需要用 rawQuery()方法执行。

下面通过一个完整的实例讲解 SQLite 数据库的增删改查操作。

【案例 5-4】使用数据库实现学生信息表的增删改查，界面布局如图 5.8 所示。插入和更新数据界面如图 5.9 所示。

图 5.8　主界面布局

图 5.9　插入和更新数据界面

步骤 1：创建项目

启动 Android Studio，创建名为 D0504_SQLite 的项目，选择 Empty Activity，将包名改为 com.example.sqlite，按照默认步骤，等待构建完成。

步骤 2：添加 RecyclerView 的依赖包

在 app/build.gradle 文件中添加 RecyclerView 的依赖。

```
1.  dependencies {
2.      implementation ' com.google.android.material:material:1.2.1'
3.  }
```

步骤 3：创建不同的功能包

创建如图 5.10 所示包名的 package，其中，activity 包存放 Activity 类，adapter 包存放列表的适配器，dao 包存放数据库表操作的封装，entity 包存放数据库表的实体类，utils 包存放 SQLite 数据库的工具类等。

图 5.10　项目的包结构

步骤 4：实现 MainActivity 的界面布局

打开 activity_main.xml 文件，创建如图 5.8 所示的界面布局，底部有 3 个按钮完成添加、修改和删除功能，按钮上面是学生的信息列表，使用 RecyclerView 实现。RecyclerView 列表中每个 item 的布局文件是 item_student.xml，使用线性布局，横向显示姓名、班级和年龄三个文本框。

步骤 5：实现 SQLite 数据库的工具类的编码

在 utils 包中创建 DBHelper 类继承 SQLiteOpenHelper 类，重写 onCreate()和 onUpgrade()方法，分别创建和升级数据库表 t_student。

```
1.  public class DBHelper extends SQLiteOpenHelper {
2.      // 创建表的sql字符串
3.      private final static String CREATE_TABLE_STUDENT =
            "create table t_student(" +
4.              "_id integer primary key autoincrement, " +
5.              "name varchar(20), classmate varchar(30), age integer)";
```

```
6.      // 删除表的sql字符串
7.      private final static String DROP_TABLE_STUDENT ="drop table if
                                                   exists t_student";
8.      // 构造方法
9.      public DBHelper(@Nullable Context context) {
10.         super(context, "student.db", null, 1);
11.     }
12.     // 数据库第一次被创建时自动调用
13.     @Override
14.     public void onCreate(SQLiteDatabase db) {
15.         db.execSQL(DBHelper.CREATE_TABLE_STUDENT);
16.     }
17.     // 当数据库版本号增加时被自动调用
18.     @Override
19.     public void onUpgrade(SQLiteDatabase db, int oldVersion,
                              int newVersion) {
20.         db.execSQL(DBHelper.DROP_TABLE_STUDENT);
21.         onCreate(db);
22.     }
23. }
```

步骤6：实现Student实体类和DAO类

根据t_student表的字段结构，在entitiy包中创建Student类实现Serializable接口，这个类包含_id、name、classmate和age 4个属性，创建相应的getter/setter方法，再创建无参的构造方法和包含name、classmate和age属性的构造方法，代码如下：

```
1.  public class Student implements Serializable {
2.      private int _id;
3.      private String name;
4.      private String classmate;
5.      private int age;
6.      public Student() {
7.      }
8.      public Student(String name, String classmate, int age) {
9.          this.name = name;
10.         this.classmate = classmate;
11.         this.age = age;
12.     }
13.     // 省略所有的getter/setter方法
14. }
```

在dao包中创建StudentDao类，创建insert()、delete()、update()、selectAll()等4个方法实现增删改查功能。insert()和update()方法的参数是Student对象，delete()的参数是Student对象的_id属性。这些功能的代码前面已详细讲解过，此处给出完整代码。

```
1.  public class StudentDao {
2.      private DBHelper dbHelper;
3.      public StudentDao(Context context) {
4.          dbHelper = new DBHelper(context);
5.      }
6.      // 插入一条数据
7.      public void insert(String name, String classmate, int age) {
8.          // 打开数据库
9.          SQLiteDatabase db = dbHelper.getWritableDatabase();
10.         // 封装数据
11.         ContentValues values = new ContentValues();
12.         values.put("name", name);
13.         values.put("classmate", classmate);
14.         values.put("age", age);
```

```
15.          // 执行语句
16.          db.insert("t_student", null, values);
17.          // 关闭数据库
18.          db.close();
19.      }
20.      // 更新数据
21.      public void update(String name, String classmate, int age) {
22.          SQLiteDatabase db = dbHelper.getWritableDatabase();
23.          ContentValues values = new ContentValues();
24.          values.put("name", name);
25.          values.put("classmate", classmate);
26.          values.put("age", age);
27.          db.update("t_student", values, "_id=?", new String[]{"1"});
28.          db.close();
29.      }
30.      // 删除一条数据
31.      public void delete(int _id) {
32.          SQLiteDatabase db = dbHelper.getWritableDatabase();
33.          db.delete("t_student", "_id=?", new String[]{String.valueOf
                    (_id)});
34.          db.close();
35.      }
36.      // 查询所有数据
37.      public List<Student> selectAll() {
38.        List<Student> students = new ArrayList<>();
39.        // 打开数据库
40.        SQLiteDatabase db = dbHelper.getReadableDatabase();
41.        // 查询
42.        Cursor cursor = db.query("t_student", null, null, null, null,
                          null, null);
43.        // 将查询结果转为List
44.        while (cursor.moveToNext()) {
45.          Student student=new Student(cursor.getString(cursor.
                              getColumnIndex("name")),
46.                 cursor.getString(cursor.getColumnIndex("classmate")),
47.                 cursor.getInt(cursor.getColumnIndex("age")));
48.            student.set_id(cursor.getInt(cursor.getColumnIndex("_id")));
49.            students.add(student);
50.        }
51.        // 关闭数据库
52.        cursor.close();
53.        db.close();
54.        // 返回结果
55.        return students;
56.      }
57. }
```

步骤7：实现 RecyclerView 的适配器类

创建 RecyclerView 的适配器类 StudentAdapter 继承 RecyclerView.Adapter 类，重写 onCreateViewHolder()、onBindViewHolder()和 getItemCount()方法。通过构造方法获取数据库表的数据集合和 Context 对象，在 onCreateViewHolder()方法中注册 item_student 布局获取 View 对象，设置 item 监听器，构造自定义的 ViewHolder 对象，并设置显示其内容的视图。调用 onBindViewHolder()绑定相应的数据。创建内部类 MyViewHolder 继承 RecyclerView.ViewHolder 类完成 item 的布局控件的初始化。

创建 item 单击事件监听的接口，在其中声明 onItemClick()回调方法用于响应每条数据的单击事件，完整代码如下：

```java
1.  public class StudentAdapter extends RecyclerView.Adapter
        <StudentAdapter.ViewHolder> {
2.      private List<Student> datas;
3.      public StudentAdapter(List<Student> datas) {
4.          this.datas = datas;
5.      }
6.      @NonNull
7.      @Override
8.      public ViewHolder onCreateViewHolder(@NonNull ViewGroup
                                              parent, int viewType) {
9.          View view = LayoutInflater.from(parent.getContext())
10.                 .inflate(R.layout.item_student, parent, false);
11.         ViewHolder viewHolder = new ViewHolder(view);
12.         // 设置 view 的 onClick 事件监听
13.         view.setOnClickListener(new View.OnClickListener() {
14.             @Override
15.             public void onClick(View view) {
16.                 // 触发回调接口对象的单击事件
17.                 // 通过 view 的 setTag/getTag()方法传递 item 的 position 位置
18.                 if(mOnItemClickListener != null) {
19.                     mOnItemClickListener.onItemClick(view, (int)
                            view.getTag());
20.                 }
21.             }
22.         });
23.         return viewHolder;
24.     }
25.     @Override
26.     public void onBindViewHolder(ViewHolder holder, int position) {
27.         Student student = datas.get(position);
28.         holder.nameItem.setText(student.getName());
29.         holder.classmateItem.setText(student.getClassmate());
30.         holder.ageItem.setText(String.valueOf(student.getAge()));
31.         holder.itemView.setTag(position);
32.     }
33.     // 定义 item 单击的回调接口
34.     public interface OnItemClickListener {
35.         void onItemClick(View view, int position);
36.     }
37.     // 定义回调接口的对象及 set 方法
38.     private OnItemClickListener mOnItemClickListener = null;
39.     public void setOnItemClickListener(OnItemClickListener
                                           onItemClickListener) {
40.         mOnItemClickListener = onItemClickListener;
41.     }
42.     @Override
43.     public int getItemCount() {
44.         return datas.size();
45.     }
46.     // ViewHolder 类
47.     public static class ViewHolder extends RecyclerView.ViewHolder {
48.         TextView nameItem;
49.         TextView classmateItem;
50.         TextView ageItem;
51.         public ViewHolder(View itemView) {
52.             super(itemView);
53.             nameItem = itemView.findViewById(R.id.tv_name);
54.             classmateItem = itemView.findViewById(R.id.tv_classmate);
55.             ageItem = itemView.findViewById(R.id.tv_age);
56.         }
57.     }
58. }
```

步骤 8：完成 MainActivity 类的按钮的功能

MainActivity 类实现 OnClickListener 接口，重写 onClick()方法，在 onCreate()方法中完成数据初始化和控件对象初始化功能。在 initView()方法中编写按钮的事件监听设置、RecyclerView 的适配器创建和 ItemClick 的事件监听。

```java
1.  public class MainActivity extends AppCompatActivity
2.          implements View.OnClickListener {
3.      private List<Student> datas;
4.      private StudentDao dao;
5.      private Student currentStudent;
6.      private StudentAdapter adapter;
7.
8.      @Override
9.      protected void onCreate(Bundle savedInstanceState) {
10.         super.onCreate(savedInstanceState);
11.         setContentView(R.layout.activity_main);
12.         // 获取数据库的数据
13.         dao = new StudentDao(this);
14.         datas = dao.selectAll();
15.         // 初始化控件
16.         initView();
17.     }
18.     private void initView() {
19.         // 初始化控件
20.         Button btnAdd = findViewById(R.id.btn_add);
21.         Button btnUpdate = findViewById(R.id.btn_update);
22.         Button btnDelete = findViewById(R.id.btn_delete);
23.         // 设置按钮监听器
24.         btnAdd.setOnClickListener(this);
25.         btnUpdate.setOnClickListener(this);
26.         btnDelete.setOnClickListener(this);
27.         // RecyclerView 控件的初始化、设置布局管理器和动画
28.         RecyclerView recyclerView = findViewById(R.id.rv_students);
29.         recyclerView.setLayoutManager(new LinearLayoutManager(this));
30.         recyclerView.setItemAnimator(new DefaultItemAnimator());
31.         // 设置 RecyclerView 控件的 Adapter
32.         adapter = new StudentAdapter(datas);
33.         recyclerView.setAdapter(adapter);
34.         // adapter 添加 item 的单击事件的监听
35.         adapter.setOnItemClickListener(new StudentAdapter.
                                            OnItemClickListener() {
36.             @Override
37.             public void onItemClick(View view, int position) {
38.                 currentStudent = datas.get(position);
39.                 Toast.makeText(MainActivity.this, "第" + (position +
                                    1) + "条",
40.                         Toast.LENGTH_SHORT).show();
41.             }
42.         });
43.     }
44. }
```

下述代码中的第 5～15 行是单击"添加"和"修改"按钮的处理代码，通过打开 InsertActivity 界面输入 Student 对象的每个属性的值，然后返回 MainActivity 更新数据列表，因此使用 startActivityForResult()打开 InsertActivity；第 42～49 行调用 onActivityResult()回调方法处理 InsertActivity 返回的信息；第 16～37 行处理"删除"按钮的功能，使用提示对话框确定是否删除此条记录，确定删除后调用第 51～54 行的 changeData()方法更新数据列表

```java
1.  @Override
2.  public void onClick(View v) {
3.      Intent intent = new Intent(this, InsertActivity.class);
4.      switch (v.getId()) {
5.          case R.id.btn_add:
6.              startActivityForResult(intent, 100);
7.              break;
8.          case R.id.btn_update:
9.              // 将选中的 student 传递给 InsertActivity
10.             Bundle bundle = new Bundle();
11.             bundle.putSerializable("student", currentStudent);
                // Student 类需序列化
12.             intent.putExtra("flag", 1);
13.             intent.putExtras(bundle);
14.             startActivityForResult(intent, 101);
15.             break;
16.         case R.id.btn_delete:
17.             new AlertDialog.Builder(this).setTitle("删除").setMessage("确认删除？")
18.                     .setPositiveButton("确定",
19.                             new DialogInterface.OnClickListener() {
20.                                 @Override
21.                                 public void onClick(DialogInterface dialog,
                                            int which) {
22.                                     dao.delete(currentStudent.get_id());
                                        // 删除数据
23.                                     dialog.dismiss();
24.                                     changeData();
25.                                     adapter.notifyDataSetChanged();
                                        // 刷新 RecyclerView 列表
26.                                 }
27.                             })
28.                     .setNegativeButton("取消",
29.                             new DialogInterface.OnClickListener() {
30.                                 @Override
31.                                 public void onClick(DialogInterface dialog,
                                            int which) {
32.                                     dialog.dismiss();
33.                                 }
34.                             }).show();
35.             break;
36.     }
37. }
38.
39. @Override
40. protected void onActivityResult(int requestCode, int resultCode,
                    Intent data) {
41.     super.onActivityResult(requestCode, resultCode, data);
42.     if ((requestCode == 100 || requestCode == 101) && resultCode == RESULT_OK) {
43.         // 通过改变 adapter 刷新 RecyclerView 列表
44.         changeData();
45.         adapter.notifyDataSetChanged();
46.     }
47. }
48. // 重新装载数据
49. private void changeData() {
50.     datas.clear();
51.     datas.addAll(dao.selectAll());
52. }
```

步骤9：完成 InsertActivity 类的功能

InsertActivity 用于添加和更新数据，界面如图 5.9 所示。InsertActivity 通过 Bundler 对象从 MainActivity 获取需要更新的数据，当数据添加或修改完成后，调用 setResult()方法返回 MainActivity，更新数据，完整代码如下：

```java
1.  public class InsertActivity extends AppCompatActivity
2.          implements View.OnClickListener {
3.      private EditText etName;
4.      private EditText etAge;
5.      private Spinner spClassmate;
6.      private StudentDao studentDao = new StudentDao(this);
7.      private Student currentStudent;
8.      private boolean isUpdate = false; // 添加或更新的标识符
9.      @Override
10.     protected void onCreate(@Nullable Bundle savedInstanceState) {
11.         super.onCreate(savedInstanceState);
12.         setContentView(R.layout.activity_insert);
13.         // 初始化控件对象
14.         initView();
15.         // 判断是否有数据需要加载
16.         Intent intent = getIntent();
17.         Bundle bundle = intent.getExtras();
18.         if(bundle != null) {
19.             currentStudent = (Student) bundle.get("student");
20.         }
21.         // 控件加载数据
22.         if(currentStudent != null) {
23.             isUpdate = true;
24.             etName.setText(currentStudent.getName());
25.             etAge.setText(String.valueOf(currentStudent.getAge()));
26.             // 设置 Spinner 值
27.             SpinnerAdapter spinnerAdapter = spClassmate.getAdapter();
28.             for(int i = 0; i < spinnerAdapter.getCount(); i++) {
29.                 if(spinnerAdapter.getItem(i).toString()
30.                         .equals(currentStudent.getClassmate())) {
31.                     spClassmate.setSelection(i);
32.                     break;
33.                 }
34.             }
35.         }
36.     }
37.     private void initView() {
38.         etName = findViewById(R.id.et_name);
39.         spClassmate = findViewById(R.id.sp_classmate);
40.         etAge = findViewById(R.id.et_age);
41.         // 初始化控件对象
42.         Button btnConfirm = findViewById(R.id.btn_confirm);
43.         Button btnCancel = findViewById(R.id.btn_cancel);
44.         btnConfirm.setOnClickListener(this);
45.         btnCancel.setOnClickListener(this);
46.     }
47.     @Override
48.     public void onClick(View view) {
49.         switch (view.getId()) {
50.             case R.id.btn_confirm:
51.                 // 将输入的数据封装成 student 对象
52.                 Student student = new Student(etName.getText().toString(),
53.                         spClassmate.getSelectedItem().toString(),
54.                         Integer.parseInt(etAge.getText().toString()));
```

```
55.              if(isUpdate) {
56.                  student.set_id(currentStudent.get_id());
57.                  studentDao.update(student);      // 更新数据
58.              } else {
59.                  studentDao.insert(student);      // 插入数据
60.              }
61.              // 返回MainActivity,刷新RecyclerView
62.              setResult(RESULT_OK, new Intent());
63.              finish();
64.              break;
65.          case R.id.btn_cancel:
66.              finish();
67.              break;
68.          }
69.      }
70. }
```

步骤10：运行程序

编码完成后，在模拟器上运行程序，测试添加、修改和删除功能。

5.4 内容提供者

前面讲解的三种存储技术都只限于在应用程序的内部访问，跨应用程序之间的数据共享功能则由内容提供者（ContentProvider）实现，开发者可以将自己的数据开放给其他应用程序，屏蔽直接开放数据库带来的安全问题。

5.4.1 内容提供者简介

内容提供者是 Android 系统的四大组件之一，它提供了应用程序之间共享数据的机制和存储方式，允许一个应用程序访问另一个应用程序的数据，同时还能保证被访问数据的安全性。它有两种应用场景：一种是通过代码访问其他应用中已有的内容提供者，Android 系统提供了多种内容提供者，如联系人、浏览器、相机、短信等；另一种是在应用程序中创建 ContentProvider，与其他应用程序共享数据。需要注意的是，提供内容提供者的应用程序，无论它是否启动，其他应用程序都可以通过提供的接口操作它的内部数据。

内容提供者是 Android 实现在不同应用程序之间共享数据的标准 API。应用程序通过内容提供者以一个或多个表的形式提供数据访问的接口；其他应用程序则通过 ContentResolver 类访问这些数据。内容提供者以 Android 自带的联系人的 ContentProvider 为例说明工作原理，如图 5.11 所示。

在图 5.11 中，联系人应用程序通过 ContentProvider 暴露它的数据库数据，这些数据是以 URI 格式构成的唯一标志，A 应用程序通过 ContentResolver 操作联系人暴露的数据，联系人的 ContentProvider 将操作结果通过 ContentResolver 返回给 A 应用程序。

ContentResolver 提供类似数据库的增删改查操作，但它无法直接使用数据库表名，而使用 URI 标识数据，此 URI 也称为内容 URI。URI 全称是 Uniform Resource Identifier，中文翻译为统一资源标识符，用以标识数据的名称、位置等信息，它主要由协议 scheme、权限 authority 和路径 path 三部分组成，一个标准的内容 URI 格式如图 5.12 所示。

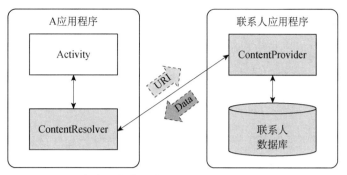

图 5.11　内容提供者、应用程序和存储空间之间的交互

图 5.12　URI 的组成结构

协议 scheme 的值为 Android 应用的标准前缀"content://"；权限 authority 用于区分不同的应用程序，为了避免冲突，一般采用应用程序的包名进行命名；路径 path 用于区分不同的表及被请求的特定记录的 id，示例中请求的是 student 表的第 1 条记录，如果请求表的所有记录，"/1"则需省略。另外需要在 AndroidManifest.xml 配置清单文件中注册 ContentProvider，并对权限属性进行声明，示例代码如下所示：

```
1.  <provider
2.      android:name=".MyContentProvider"
3.      android:authorities="com.example.provider"
4.      android:enabled="true"
5.      android:exported="true" />
```

下面列出一些 Android 系统自带的 ContentProvider 的 URI 类型。
- 联系人的 URI：content://com.android.contacts/contacts。
- 联系人的电话 URI：content://com.android.contacts/data/phones。
- SD 卡的音频文件 URI: content://media/external/audio/media。
- 日历的 URI：content://com.android.calendar/calendars。

5.4.2　创建 ContentProvider

实现跨应用程序的数据共享的功能，可以通过继承 ContentProvider 抽象类实现，需要实现 ContentProvider 类的 6 个抽象方法，示例代码如下：

```
1.  public class MyContentProvider extends ContentProvider {
2.      public MyContentProvider() {
3.      }
4.      @Override
5.      public boolean onCreate() {
6.          return false;
7.      }
8.      @Override
9.      public int delete(Uri uri, String selection, String[] selectionArgs) {
10.         return 0;
11.     }
12.     @Override
```

```
13.    public Uri insert(Uri uri, ContentValues values) {
14.        return null;
15.    }
16.    @Override
17.    public Cursor query(Uri uri, String[] projection, String selection,
18.                    String[] selectionArgs, String sortOrder) {
19.        return null;
20.    }
21.    @Override
22.    public int update(Uri uri, ContentValues values, String selection,
23.                    String[] selectionArgs) {
24.        return 0;
25.    }
26.    @Override
27.    public String getType(Uri uri) {
28.        return null;
29.    }
30. }
```

从上面的方法声明可以看出，操作数据的 4 个方法与 SQLite 数据库操作的方法类似，最重要的差别是将表名改成了 URI 标识符，而这个参数则由 ContentResolver 对象传递过来。下面简单介绍一下每个方法的含义。

① onCreate()：初始化 ContentProvider 时调用，完成数据库创建等初始化工作，返回 true 表示初始化成功，返回 false 则表示失败。尽管 ContentProvider 也是系统组件，但它不像 Activity 一样有完整的生命周期方法，它只有 onCreate()方法用于初始化。

② getType(Uri uri)：根据 URI 的值返回相应的 MIME 类型。

③ insert(Uri uri, ContentValues values)：添加一条值为 values 的 URI 对应的数据，添加成功则返回这条数据的 URI。

④ update(Uri uri, ContentValues values, String select, String[] selectionArgs)：更新与 select 条件匹配的 URI 对应的数据，返回受影响的数据的行数。

⑤ delete(Uri uri, String select, String[] selectionArgs)：删除与 select 条件匹配的 URI 对应的数据，返回受影响的数据的行数。

⑥ query(Uri uri, String[] projection, String selection, String[] selectionArgs, String sortOrder)：查询与 where 条件匹配的 URI 对应的数据。

访问不同的数据需要根据 Uri 参数进行区分，UriMatcher 类就是用来匹配这些不同的 Uri 的，该类提供一个 addURI()方法注册 URI 到 UriMatcher 对象中，通过调用 UriMatcher 的 match()方法匹配已注册的 URI，判断出访问哪种类型的数据。UriMatcher 相关的示例代码如下：

```
1.  public class StudentProvider extends ContentProvider {
2.      public static final int TABLE1_DIR = 0;
3.      public static final int TABLE1_ITEM = 1;
4.      public static final int TABLE2_DIR = 2;
5.      public static final int TABLE2_ITEM = 3;
6.      public static final String AUTHORITY = "com.example.provider";
7.      private static UriMatcher uriMatcher;
8.      static {
9.          uriMatcher = new UriMatcher(UriMatcher.NO_MATCH);
10.         uriMatcher.addURI(AUTHORITY, "table1", TABLE1_DIR);
11.         uriMatcher.addURI(AUTHORITY, "table1/#", TABLE1_ITEM);
12.         uriMatcher.addURI(AUTHORITY, "table2", TABLE2_DIR);
13.         uriMatcher.addURI(AUTHORITY, "table2/#", TABLE2_ITEM);
14.     }
15.     @Nullable
```

```
16.      @Override
17.      public Cursor query(@NonNull Uri uri, String[] projection,
                            String selection,
18.                         String[] selectionArgs, String sortOrder) {
19.          Cursor cursor = null;
20.          switch(uriMatcher.match(uri)) {
21.              case TABLE1_DIR:
22.                  // 查询table1所有数据
23.                  break;
24.              case TABLE1_ITEM:
25.                  // 查询table1单条数据
26.                  break;
27.              case TABLE2_DIR:
28.                  // 查询table2所有数据
29.                  break;
30.              case TABLE2_ITEM:
31.                  // 查询table2单条数据
32.                  break;
33.          }
34.          return cursor;
35.      }
36.      // 省略其他代码
37. }
```

上述代码的第 2~5 行代码定义的 4 个整型常量分别表示 table1 所有数据、table1 单条数据、table2 所有数据和 table2 单条数据。第 8~16 行代码创建 UriMatcher 对象，调用它的 addURI()方法注册 URI，这个方法需要 3 个参数，第一个参数是 Uri 的 authority 部分；第二个参数是 Uri 的 path 部分，"#"表示匹配任意长度的数字通配符；第三个参数是匹配成功的返回码。调用 query()方法时通过 UriMatcher 的 match()方法对传入的 Uri 参数进行匹配，根据返回码调用代码进行数据查询，insert()、update()和 delete()方法的实现与 query()类似。

最后讲解一下 getType()方法，它的作用是返回指定 Uri 的数据 MIME 类型，MIME 的全称是 Multipurpose Internet Mail Extensions，用来指定某种扩展名的文件的打开方式的类型，如：png 格式图片的 MIME 字符串是"image/png"。针对内容 URI 的 MIME 类型，对于单个数据应以"vnd.android.cursor.item"开头，多个数据应以"vnd.android.cursor.dir/"开头，然后加上"vnd.<authority>.<path>"，getType()的示例代码如下：

```
1.  public String getType(@NonNull Uri uri) {
2.      switch (uriMatcher.match(uri)) {
3.          case TABLE1_DIR:
4.              return "vnd.android.cursor.dir/vnd.com.example.
                                              provider.table1";
5.          case TABLE1_ITEM:
6.              return "vnd.android.cursor.item/vnd.com.example.
                                              provider.table1";
7.          case TABLE2_DIR:
8.              return "vnd.android.cursor.dir/vnd.com.example.
                                              provider.table2";
9.          case TABLE2_ITEM:
10.             return "vnd.android.cursor.item/vnd.com.example.provider.
                                              table2";
11.     }
12.     return null;
13. }
```

至此，我们已经详细讲解了创建 ContentProvider 的步骤，下面给出案例 5-4 的 student 数据库的 t_student 表的 ContentProvider 创建的完整代码，其中的 DBHelper 类直接使用上个案例中的类定义。

【案例 5-5】实现学生表的 ContentProvider。

步骤 1：创建项目

启动 Android Studio，创建名为 D0505_ContentProvider 的项目，选择 Empty Activity 项，将包名改为 com.example.provider，单击 Finish 按钮，等待项目构建完成。

步骤 2：导入 DBHelper 类

从案例 5-4 中将 DBHelper.java 文件拷贝到项目包下。

步骤 3：创建 StudentProvider

在包名上右击，选择 New->Other->Content Provider，进入创建 Provider 对话框，如图 5.13 所示，设置 Class Name 为 StudentProvider，URI Authorities 为 com.example.provider，单击 Finish 按钮完成创建。向导完成的 StudentProvider 还会在 AndroidManifest.xml 中完成注册。

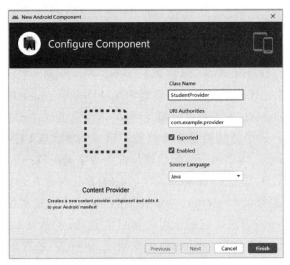

图 5.13　创建 Provider 对话框

步骤 4：编写增删改查的功能代码

参考讲解的知识点，编写 student 表的 ContentProvider 的 5 个抽象方法，代码如下：

```java
1.  public class StudentProvider extends ContentProvider {
2.      // SQLiteDatabase 必须声明为成员变量
3.      private SQLiteDatabase db;
4.      public static final int STUDENT_DIR = 0;
5.      public static final int STUDENT_ITEM = 1;
6.      public static final String AUTHORITY = "com.example.provider";
7.      private static UriMatcher uriMatcher;
8.      static {
9.          uriMatcher = new UriMatcher(UriMatcher.NO_MATCH);
10.         uriMatcher.addURI(AUTHORITY, "student", STUDENT_DIR);
11.         uriMatcher.addURI(AUTHORITY, "student/#", STUDENT_ITEM);
12.     }
13.     @Override
14.     public boolean onCreate() {
15.         DBHelper dbHelper = new DBHelper(getContext());
16.         db = dbHelper.getWritableDatabase();
17.         return true;
18.     }
```

```java
19.     @Nullable
20.     @Override
21.     public Cursor query(@NonNull Uri uri, String[] projection,
                    String selection,
22.                 String[] selectionArgs, String sortOrder) {
23.         Cursor cursor = null;
24.         switch(uriMatcher.match(uri)) {
25.             case STUDENT_DIR:
26.                 // 查询t_student表的所有数据
27.                 cursor = db.query("t_student", projection,
                            selection, selectionArgs,
28.                         null, null, sortOrder);
29.                 break;
30.             case STUDENT_ITEM:
31.                 // 查询t_student表的单条数据
32.                 String id = uri.getPathSegments().get(1);
33.                 cursor = db.query("t_student", projection, "_id=?",
                                new String[]{id},
34.                         null, null, sortOrder);
35.                 break;
36.         }
37.         return cursor;
38.     }
39.     @Nullable
40.     @Override
41.     public Uri insert(@NonNull Uri uri, ContentValues contentValues) {
42.         Uri newUri = null;
43.         long newId = 0;
44.         switch (uriMatcher.match(uri)) {
45.             case STUDENT_DIR:
46.             case STUDENT_ITEM:
47.                 newId = db.insert("t_student", null, contentValues);
48.                 newUri = Uri.parse("content://" + AUTHORITY +
                                "/student/" + newId);
49.                 break;
50.         }
51.         return newUri;
52.     }
53.     @Override
54.     public int delete(@NonNull Uri uri, String select, String[]
                    selectionArgs) {
55.         int count = 0;
56.         switch (uriMatcher.match(uri)) {
57.             case STUDENT_DIR:
58.                 count = db.delete("t_student", select, selectionArgs);
59.                 break;
60.             case STUDENT_ITEM:
61.                 String id = uri.getPathSegments().get(1);
62.                 count = db.delete("t_student", "_id=?", new
                                String[]{id});
63.                 break;
64.         }
65.         return count;
66.     }
67.     @Override
68.     public int update(@NonNull Uri uri, ContentValues contentValues,
69.                 String select, String[] selectionArgs) {
70.         int count = 0;
71.         switch (uriMatcher.match(uri)) {
72.             case STUDENT_DIR:
73.                 count = db.update("t_student", contentValues,
```

```
74.                break;
75.            case STUDENT_ITEM:
76.                String id = uri.getPathSegments().get(1);
77.                count =db.update("t_student",contentValues,"_id=?",
                            new String[]{id});
78.                break;
79.        }
80.        return count;
81.    }
82.    /**
83.     * 内容提供器必须提供的方法,用于获取 Uri 对象所对应的 MIME 类型
84.     * MIME 字符串的组成:
85.     *      以 vnd 开头,
86.     *      内容以路径结尾,则后接 android.cursor.dir/
87.     *      url 以 id 结尾,则后接 android.cursor.item/
88.     *      最后接上 vnd.<authority>.<path>
89.     */
90.    @Nullable
91.    @Override
92.    public String getType(@NonNull Uri uri) {
93.        switch (uriMatcher.match(uri)) {
94.            case STUDENT_DIR:
95.                return "vnd.android.cursor.dir/vnd.com.example.
                                    provider.student";
96.            case STUDENT_ITEM:
97.                return "vnd.android.cursor.item/vnd.com.example.
                                    provider.student";
98.        }
99.        return null;
100.    }
101. }
```

上述第 21~38 行代码段是查询方法 query() 的实现,根据 STUDENT_DIR 和 STUDENT_ITEM 标识符调用 SQLiteDatabase 类的 query() 方法查询所有数据和单条数据,insert()、update() 和 delete() 方法也相应完成添加、修改和删除的功能。

至此 ContentProvider 创建完成,它所提供的增删改查的方法必须使用 ContentResolver 对象进行访问,接下来讲解 ContentResolver 的使用方法。

5.4.3 访问其他应用程序的数据

Android 提供 ContentResolver 对象来获取 ContentProvider 提供的数据,此对象称为内容解析器,Context 类的 getContentResolver() 可以直接获取此对象,它也提供了类似 ContentProvider 的增删改查方法访问这些数据。

【案例 5-5(续)】通过改造上小节案例 5-4 的 SQLite 案例,学会使用 StudentProvider 来完成 t_student 表的增删改查。由于我们对上个案例进行了良好的分层设计,因此只需要修改 StudentDao 类的方法即可实现。

步骤 1:创建项目

启动 Android Studio,创建名为 D0505_ContentResolver 的项目,选择 Empty Activity 项,将包名改为 com.example.resolver,单击 Finish 按钮,等待项目构建完成。

步骤 2:导入 D0504_SQLite 项目

可以将 D0504_SQLite 项目的布局文件、java 目录的文件复制到相应目录,并导入

RecyclerView 的依赖库中,也可以使用 File 菜单下的 New->Import Module 将整个模块导入。

步骤 3:修改 StudentDao 类

修改 StudentDao 类,定义 ContentResolver 对象和内容 URI 字符串常量,将打开数据库的语句改为解析 Uri,使用 ContentResolver 对象完成增删改查功能,完整代码如下:

```
1.  public class StudentDao {
2.      private final Context context;
3.      private final static String CONTENT_URI = "content://com.example.
                                                    provider/student";
4.      public StudentDao(Context context) {
5.          this.context = context;
6.      }
7.      // 插入一条数据
8.      public void insert(Student student) {
9.          Uri uri = Uri.parse(StudentDao.CONTENT_URI);    // 解析 Uri
10.         // 封装数据
11.         ContentValues values = new ContentValues();
12.         values.put("name", student.getName());
13.         values.put("classmate", student.getClassmate());
14.         values.put("age", student.getAge());
15.         context.getContentResolver().insert(uri, values);   // 执行语句
16.     }
17.     // 更新数据
18.     public void update(Student student) {
19.         // 解析 Uri
20.         Uri uri = Uri.parse(StudentDao.CONTENT_URI + "/" +
                                student.get_id());
21.         // 封装数据
22.         ContentValues values = new ContentValues();
23.         values.put("name", student.getName());
24.         values.put("classmate", student.getClassmate());
25.         values.put("age", student.getAge());
26.         context.getContentResolver().update(uri, values, null,
                                                null);   // 执行语句
27.     }
28.     // 删除一条数据
29.     public void delete(int _id) {
30.         Uri uri = Uri.parse(StudentDao.CONTENT_URI + "/" + _id);
            // 解析 Uri
31.         context.getContentResolver().delete(uri, null, null);
            // 执行语句
32.     }
33.     // 查询所有数据
34.     public List<Student> selectAll() {
35.         List<Student> students = new ArrayList<>();
36.         Uri uri = Uri.parse(StudentDao.CONTENT_URI);    // 解析 Uri
37.         // 查询
38.         Cursor cursor = context.getContentResolver().query
                            (uri, null, null,null,null);
39.         // 将查询结果转为 List
40.         assert cursor != null;
41.         while (cursor.moveToNext()) {
42.             Student student = new Student(cursor.getString(
43.                     cursor.getColumnIndex("name")),
44.                     cursor.getString(cursor.getColumnIndex
                                ("classmate")),
45.                     cursor.getInt(cursor.getColumnIndex("age")));
46.             student.set_id(cursor.getInt(cursor.getColumnIndex("_id")));
```

```
47.                students.add(student);
48.            }
49.            cursor.close();    // 关闭数据库
50.            return students;   // 返回结果
51.       }
52.       // 查询一条数据
53.       public Student select(int _id) {
54.            Student student = null;
55.            Uri uri = Uri.parse(StudentDao.CONTENT_URI + "/" + _
                            id);   // 解析Uri
56.            // 查询
57.            Cursor cursor = context.getContentResolver().query(uri,
                            null,null,null,null);
58.            // 获取查询结果
59.            assert cursor != null;
60.            if (cursor.moveToNext()) {
61.                student = new Student(cursor.getString(cursor.
                                    getColumnIndex("name")),
62.                        cursor.getString(cursor.getColumnIndex("classmate")),
63.                        cursor.getInt(cursor.getColumnIndex("age")));
64.            }
65.            cursor.close();    // 关闭数据库
66.            return student;    // 返回结果
67.       }
68. }
```

步骤4：运行项目

将本项目安装到模拟器后运行，测试内容提供者项目提供的数据库表的增删改查功能。

5.5 数据库框架 Room

SQLite 数据库的使用与 Java 的 JDBC 技术类似，也存在诸多缺点，如编写大量的重复代码；为每个查询实现对象映射；难以实施数据库迁移；测试困难。

为了解决这些问题，Google 在 2018 年发布了 Room 数据库持久层框架，它是在 SQLite 的基础上提供抽象层的持久存储库，基于对象关系映射（Object Relationship Mapping，ORM）模型，使得数据库访问更健壮、高效和简洁。

针对以上 SQLite 数据库访问的缺点，Room 框架具有以下特点：

- 使用动态代理，减少了重复代码量。
- 使用编译时注解，在编译过程中就完成对 SQL 语句的语法检验。
- 数据库迁移相对方便。
- 保证数据库的可测试性。
- 保证数据库操作远离主线程。

Room 库主要包括三个组件：Database、Entity 和 Dao。Database 类是数据库拥有者，作为与持久层数据的底层连接的访问节点，运行时通过 Room.databaseBuilder()获取 Database 实例对象，并且创建一个数据库，Database 类必须满足以下条件：

- 必须为继承自 RoomDatabase 的抽象类。
- 在类的头部添加与实体类关联的注解。
- 包含一个无参的抽象方法，返回使用@Dao 注解的 Dao 类。

Entity 实体类代表数据库表的实体类,默认情况下 Room 框架将实体类的所有属性与表的字段一一对应;实体类的属性提供 getter/setter 方法或声明为 public 访问类型供外部访问,在类的声明中使用@Entity 注解标识该实体类与表的对应关系,使用@Primarykey、@ColumnInfo 等注解表示表字段的属性。

Dao 类(Data Access Objects 数据访问对象的缩写)是 Room 框架的主要组件,负责定义访问数据库的方法,Room 框架会在编译时创建每个 DAO 实例,它可以定义为一个接口或抽象类,使用@Dao 注解标识 Dao 类,使用@Insert、@Delete、@Update 和@Query 4 种注解标识表数据的增删改查操作。

这三个组件之间的关系如图 5.14 所示。

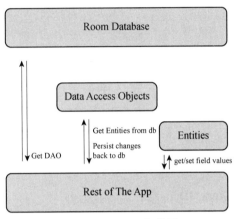

图 5.14　Room 架构图

使用 Room 库进行数据库操作一般包括以下几个步骤:
① 添加 Room 框架的依赖库。
② 创建数据库的实体 Entity。
③ 创建数据库访问的 DAO。
④ 创建数据库 Database。
⑤ 封装数据库与业务逻辑交互的 Repository。

接下来通过重构 5.3 节的 SQLite 数据库案例讲解 Room 框架的使用方法。

【案例 5-6】使用 Room 框架重构案例 5-4 项目的学生信息表的增删改查功能。

步骤 1:创建项目

启动 Android Studio,创建名为 D0506_Room 的项目,选择 Empty Activity 项,将包名改为 com.example.room,单击 Finish 按钮,等待项目构建完成。

步骤 2:导入 D0504_SQLite 项目

可以将 D0504_SQLite 项目的布局文件、java 目录的文件复制到相应目录中,并导入 RecyclerView 的依赖库,也可以使用 File 菜单下的 New->Import Module 将整个模块导入。

步骤 3:添加 Room 的依赖库

在 app/build.gradle 文件中添加 Room 的依赖项,代码如下:

```
1.  dependencies {
2.      def room_version = "2.2.5"
3.      implementation "androidx.room:room-runtime:$room_version"
4.      annotationProcessor "androidx.room:room-compiler:$room_version"
5.  }
```

步骤 4：修改 entity 包的 Student 类

打开 Student 类文件，添加@Entity 注解与数据库表关联，给 id 属性添加自增主键的注解@PrimaryKey 和字段名的@ColumnInfo 注解，给其他属性添加@ColumnInfo 注解，如果属性名称与表的字段名称相同，就无须添加 name 属性，详细代码如下：

```java
1.  @Entity(tableName = "t_student")
2.  public class Student implements Serializable {
3.      @PrimaryKey(autoGenerate = true)
4.      @ColumnInfo(name = "_id")
5.      private int id;
6.      @ColumnInfo
7.      private String name;
8.      @ColumnInfo
9.      private String classmate;
10.     @ColumnInfo
11.     private int age;
12.     public Student() {
13.     }
14.     public Student(String name, String classmate, int age) {
15.         this.name = name;
16.         this.classmate = classmate;
17.         this.age = age;
18.     }
19.     // 省略 getter/setter 方法
20. }
```

步骤 5：修改 dao 包的 StudentDao 类

打开 StudentDao 类文件，将此类改为接口，删除构造方法和增删改查方法的实现代码，添加对应的 DAO 注解，示例代码如下：

```java
1.  @Dao
2.  public interface StudentDao {
3.      // 插入一条数据
4.      @Insert(onConflict = OnConflictStrategy.IGNORE)
5.      void insert(Student student);
6.      // 更新数据
7.      @Update
8.      void update(Student student);
9.      // 删除一条数据
10.     @Delete
11.     void delete(Student student);
12.     // 查询所有数据
13.     @Query("SELECT * FROM t_student")
14.     List<Student> selectAll();
15.     // 查询一条数据
16.     @Query("SELECT * FROM t_student WHERE _id=:id")
17.     Student select(int id);
18. }
```

第 1 行代码使用@Dao 注解标识 StudentDao 是一个 Dao 接口，第 4、7、10、13 和 16 行代码标识各个方法上的@Insert、@Update、@Delete 和@Query 注解来完成表的增删改查功能。

@Query 注解用于声明需要查询的 SQL 语句，查询参数的占位符用冒号（:）表示，冒号后加上变量的名称，在编译处理此查询时，Room 会将:id 绑定参数与 id 参数进行匹配，Room 框架通过参数名称进行匹配，如果不匹配则会在编译时出错；@Query 还支持 update、delete 两种 SQL 语句。

@Insert 和@Update 注解可设置参数 onConflict 表示处理数据冲突时采取的策略,有效的策略包括 REPLACE、ABORT、IGNORE 三种策略,REPLACE 表示用新数据替代旧数据,ABORT 表示回滚冲突的事务,IGNORE 表示保持已有的数据。

步骤 6:创建 Database 类

首先删除 util 包下的 DBHelp 类,创建名为 InfoRoomDatabase 的抽象类继承自 androidx.room.RoomDatabase 类,使用@Database 注解,其中的 entities 参数包含@Entity 注解修饰的实体类的集合,version 是数据库版本号,exportSchema 是将数据库 schema 输出到指定的目录中,默认值为 true。

由于数据库创建是非常消耗资源的工作,所以推荐将数据库类设计为单例模式,避免创建多个数据库对象消耗资源。另外,需要特别注意的是,Room 框架对数据库的操作不能在主线程执行,否则会报异常,第 18 行调用 allowMainThreadQueries()允许在主线程中查询,在后续章节讲解 Adnroid 的多线程之后再做完善,第 9、10 行设置了写操作的线程池用于数据库表的写操作,以解决不能在主线程中执行的问题,代码如下:

```java
1.  @Database(entities = {Student.class}, version = 1, exportSchema =
            false)
2.  public abstract class InfoRoomDatabase extends RoomDatabase {
3.      public static String DB_NAME = "info.db";
4.      private static volatile InfoRoomDatabase INSTANCE;
5.      private static final int NUMBER_OF_THREADS = 4;
6.      // 获取Dao的抽象方法
7.      public abstract StudentDao getStudentDao();
8.      // 数据库写操作的线程池
9.      public static final ExecutorService writeExecutor =
10.             Executors.newFixedThreadPool(NUMBER_OF_THREADS);
11.     // 单例模式
12.     public static InfoRoomDatabase getInstance(final Context context) {
13.         if (INSTANCE == null) {
14.             synchronized (InfoRoomDatabase.class) {
15.                 if (INSTANCE == null) {
16.                     INSTANCE = Room.databaseBuilder(context.
                                getApplicationContext(),
17.                             InfoRoomDatabase.class, InfoRoomDatabase.
                                DB_NAME)
18.                             .allowMainThreadQueries()
19.                             .build();
20.                 }
21.             }
22.         }
23.         return INSTANCE;
24.     }
25.     // 清除Database实例
26.     public void cleanUp() {
27.         INSTANCE = null;
28.     }
29. }
```

步骤 7:封装 Repository

在 com.example.room 包下创建 repository 包,然后创建 StudentReposity 类封装业务逻辑代码,创建 insert()、update()、delete()方法实现增删改的功能。需要注意的是,它们需要在子线程中完成,selectAll()和 select()方法分别实现查询所有和查询单个的功能,代码如下:

```java
1.  public class StudentReposity {
2.      private final StudentDao studentDao;
```

```
3.   public StudentReposity(Application application) {
4.       InfoRoomDatabase db = InfoRoomDatabase.getInstance(application);
5.       studentDao = db.getStudentDao();
6.   }
7.   // 在子线程中执行修改操作
8.   public void insert(final Student student) {
9.       InfoRoomDatabase.writeExecutor.execute(new Runnable() {
10.          @Override
11.          public void run() {
12.              studentDao.insert(student);
13.          }
14.      });
15.  }
16.  // 在子线程中执行更新操作
17.  public void update(final Student student) {
18.      InfoRoomDatabase.writeExecutor.execute(new Runnable() {
19.          @Override
20.          public void run() {
21.              studentDao.update(student);
22.          }
23.      });
24.  }
25.  // 在子线程中执行删除操作
26.  public void delete(final Student student) {
27.      InfoRoomDatabase.writeExecutor.execute(new Runnable() {
28.          @Override
29.          public void run() {
30.              studentDao.delete(student);
31.          }
32.      });
33.  }
34.  // 查询所有
35.  public List<Student> selectAll() {
36.      return studentDao.selectAll();
37.  }
38.  // 查询单个
39.  public Student select(int id) {
40.      return studentDao.select(id);
41.  }
42. }
```

步骤 8：修改 MainActivity 类

打开 MainActivity 类文件，将原有的 StudentDao 对象替换为 StudentReposity 对象，在 onCreate()方法中完成它的创建，代码如下：

```
1.  @Override
2.  protected void onCreate(Bundle savedInstanceState) {
3.      super.onCreate(savedInstanceState);
4.      setContentView(R.layout.activity_main);
5.      // 获取数据库的数据
6.      studentReposity = new StudentReposity(this.getApplication());
7.      datas = studentReposity.selectAll();
8.      // 初始化控件
9.      initView();
10. }
```

以上案例只是简单地介绍了 Room 框架的使用，它的数据类型转换、数据库迁移等高级应用都未涉及，可以参考官方文档进行实践。Room 数据库框架支持多种类型的注解，最常用的注解如表 5.6 所示。

表 5.6　Room 框架常用注解及含义说明

注解类型	注解名称	含义描述
Entity 相关	@Entity	声明类为表对应的实体类，参数：表名 tableName、表的索引 indices、主键 primaryKeys、外键 foreignKeys、忽略字段 ignoredColumns 和集成父类索引 inheritSuperIndices
	@PrimaryKey	声明为主键，参数：自动创建 autoGenerate
	@ColumnInfo	声明为表的字段名，参数：字段名称 name
	@ForeignKey	声明为外键，参数：表名 entity、主类字段 parentColumns、辅助类字段 childColumns
	@Ignore	Room 编译忽略该属性
	@Index	声明建立索引，参数：添加索引的字段 value、索引名称 name 和唯一键 unique
	@Embedded	声明为嵌套字段，被修饰的属性中的所有字段都在表中
Dao 相关	@Dao	声明类为 Dao 类
	@Insert	声明 Dao 类的插入方法，参数：冲突策略 onConflict
	@Update	声明 Dao 类的更新方法，参数：冲突策略 onConflict
	@Delete	声明 Dao 类的删除方法，参数：冲突策略 onConflict
	@Query	声明 Dao 类的查询方法，参数：SQL 语句
	@Transaction	声明 Dao 类的事务方法
	@Relation	声明 Dao 类的表关系的方法
Database 相关	@Database	声明类为数据库类，参数：实体类集合 entities、数据库版本 version、数据库 schema 输出 exportSchema，一般为单例对象

在使用 Room 库时需要注意以下几点：

● @Insert、@Delete、@Update 注解标识的方法在事务中执行。

● Room 框架仅允许一个事务运行，其他事务排队运行。

● @Tranaction 注解标识的方法不能为 final、private 或 abstract，但如果该方法也包含 @Query 则可以为 abstract。

● 如果查询的同时还包含@Relation 注解的查询存在多个查询，使用@Transaction 可以在一个事务中进行多个查询，避免因为其他的事务导致查询报错。

5.6　本章小结

本章知识点较多，主要讲解了 Android 的 4 种数据存储技术，首先介绍了存储简单键值对数据的 SharedPreferences 类；然后讲解了 Android 的存储空间的基本概念，在此基础上讲解了文件存储的使用方法，同时讲解了 Android 6.0 版本及以上的动态权限申请的概念和实现；然后讲解了数据库存储技术，Android 内置 SQLite 数据库，它的使用与 Java 的 JDBC 技术类似；接着讲解了跨应用程序的数据共享技术 ContentProvider，最后讲解了 Google 的 Room 数据库框架的应用。通过本章的学习，读者可以在实际项目中根据场景选择不同的数据存储技术。

习　题

一、选择题

1. 下列选项中，不属于 Android 的数据存储方式的是（　　）。

A. SQLite 数据库 B. ContentProvider
C. Map D. 文件存储

2. 下面关于文件存储的描述中，正确的是（　　）。
A. 内部存储的存储路径通常为 mnt/sdcard 目录
B. 内部存储可以将数据存储到 SD 卡上
C. 外部存储文件是不安全的
D. 外部存储的文件可以被其他应用程序所共享

3. 下列选项中，属于获取 SharedPreferences 的实例对象的方法是（　　）。
A. SharedPreferences.Editor B. getPreferences()
C. getSharedPreferences() D. 以上方法都不对

4. 下列选项中，属于 Environment 类中获得 SD 根目录的方法的是（　　）。
A. getDataDirectory() B. getExternalStoragePublicDirectory()
C. getExternalStorageState() D. getDownloadCacheDirectory()

5. 下面关于数据存储方式的描述中，正确的是（　　）。
A. SharedPreferences 是四大组件之一
B. ContentProvider 通过 openFileInput()和 openFileOutput()方法读取设备的文件
C. SQLite 是 Android 自带的一个轻量级的数据库
D. SQLite 数据库运算速度比较慢，占用资源较多

6. 下列选项中，属于指定文件只能被当前程序读写的操作模式的是（　　）。
A. MODE_PRIVATE B. MODE_APPEND
C. MODE_WORLD_READABLE D. MODE_WORLD_WRITEABLE

7. 下列选项中，属于内容提供者的是（　　）。
A. Activity B. ContentProvider C. ContentResolver D. ContentObserver

8. 下列选项中，属于获取 ContentResolver 实例对象的是（　　）。
A. new ContentResolver() B. ContentProvider.newInstance()
C. getContentResolver() D. ContentProvider.getContentResolver()

9. 下面关于 ContentProvider 的 URI Authorities 描述中，正确的是（　　）。
A. 类名 B. 唯一标识 C. URI 名称 D. 包名

二、简答题

1. Android 的数据存储方式有哪些？分别有何特点？
2. 简述数据添加到 SQLite 数据库的编程步骤。
3. 简述 SharedPreferences 保存数据和读取数据的方法。
4. 简述内容提供者的工作原理。
5. 简述 Room 数据库框架的架构原理。

三、编程题

1. 设计班级课程表的展示界面，使用 SQLite 数据库存储课程信息，完成增删改查功能。
2. 设计个人中心界面，将个人中心的设置信息存储在 SharedPreference 中。
3. 使用 Room 框架重构第 1 题。

第 6 章 服务与广播

通过前几章的学习，大家基本掌握了界面的使用及数据存储的相关知识，但对于完成一个应用来说，这还远远不够，在一个完整的开发过程中，除了要设计并实现相应的界面功能，与用户进行必要的交互外，开发者还需要在"暗地"里完成一些操作，这类操作因为不需要与用户有直接的交互动作，所以免去了界面设计。

除此之外，还有一类功能，它就像校园里、商场里的广播一样，通过此种方式将某个信息广而告之，当听众从这条广播获取到它所需的信息后，会做出相应的动作。这两类功能就是本章将要讨论的服务（Service）和广播接收者（Broadcast Receiver），接下来逐一讨论它们的基本概念和实现方式。

本章学习目标：

- 了解 Service 的概念
- 了解 Broadcast Receiver 的概念
- 理解 Service 的生命周期
- 掌握 Service 的基本应用
- 掌握 Broadcast Receiver 的基本应用

 ## 6.1 服务

我们常说 Android 有四大组件，服务 Service 就是其中之一。Service 是一个可以在后台执行长时间运行操作而不提供用户界面的应用组件。Service 可由其他应用组件启动，即使用户切换到其他应用也会在后台持续运行。此外，其他组件可以绑定到 Service，与之交互，也可以执行进程间的通信（InterProcess Communication，IPC）。例如，网络请求、播放音乐、执行文件 I/O 或与内容提供者 ContentProvider 交互等，均可以在后台进行。

Service 不是一个单独的进程，也不是一个单独的线程，它作为 Android 组件运行在主线程上，如果想在 Service 中处理耗时的操作，则必须在 Service 中启动多线程，以降低 Activity

无响应（Application Not Responding，ANR）的风险。总而言之，需要明确的是，Service 虽然可以在后台执行，但不代表 Service 会执行在多线程中，多线程的概念将在下一章节详细介绍，此处可以形象地理解成工厂中的多条生产线。

6.1.1 服务的基本概念

如果某个程序组件需要在运行时向用户呈现界面，或者程序需要多次与用户交互，就需要使用 Activity，否则就应该考虑使用 Service。归纳起来，Service 有以下的应用场景。

- 解决不需要界面的问题。诸如开发一个推送功能，需要在后台维持，在等待推送消息时并不需要界面展示任何数据，这时候就可以使用 Service 实现。
- 在 Activity 界面启动后台操作。例如，用户在界面执行下载文件的操作，此界面需要等待下载结束，如果此时启动一个新的 Activity，必然会遮挡当前界面而造成下载中断，此场景可以使用 Service 实现。
- 通过 Service 实现远程调用。通过定义 AIDL（Android Interface Definition Language）服务，跨应用程序调用 Service。先定义一个远程调用接口，然后为该接口提供一个 IBinder 实现类，客户端获取了远程的 Service 的 IBinder 对象的代理后，通过该 IBinder 对象去回调远程 Service 的属性或方法。

Service 按照运行类型可分成前台服务和后台服务，前台服务会在通知栏显示通知，告诉用户正在执行的操作，当服务被中止时通知也会消失，常见的如音乐播放，其后台服务是默认的运行方式，当服务被终止时用户是看不到效果的，典型的应用还有天气更新、日期同步等。为了减少后台服务过度消耗手机资源造成的系统性能降低问题，自 Android 8.0 及以后，Google 对 Service 进行了诸多改进，限制开发者滥用后台服务，推荐使用计划作业 JobScheduler。

Service 按照启动方式分为 startService()启动方式和 bindSerivce()绑定方式，启动方式用于启动一个无须与 Activity 等组件进行通信的后台服务，一旦启动，服务即可在后台无限期运行，即使应用程序退出服务仍在运行，停止服务调用 stopSelf()或 stopService()；bindService() 方式用于启动需要进行通信的后台服务，它提供一个客户端-服务器接口，允许组件与服务进行交互，该服务与所有组件之间的绑定全部取消后便会销毁，调用 unbindService()解绑服务。这两种方式也可组合，这种情况下必须调用 stopSelf()或 stopService()停止服务。

6.1.2 服务的生命周期

和 Activity 一样，Service 也有生命周期，通过调用一系列的生命周期回调方法监测 Service 的状态变化，从而在合适的时候执行相关的操作。Service 有两种不同的启动方式，其生命周期方法及执行流程都有所区别，如图 6.1 所示。

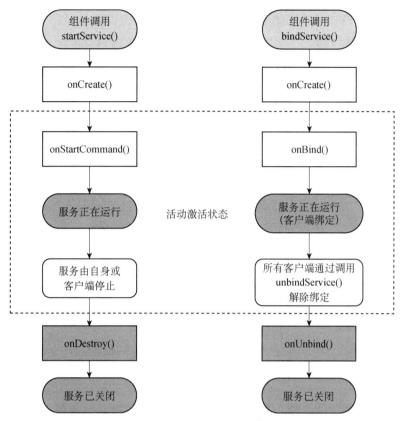

图 6.1 Service 的生命周期

根据生命周期流程描述,两种启动模式共提供了 5 个回调方法,详细说明如表 6.1 所示。

表 6.1 Service 的回调方法描述

方法名称	功能描述
onCreate()	当 Service 首次被创建时调用此方法完成初始化任务,该方法在整个生命周期中只被调用一次,如果 Service 已在运行则不会调用此方法
onStartCommand()	当 Activity 等组件通过调用 startService()启动 Service 时调用,Service 启动后可在后台无限期运行。此方式启动的服务需要调用 stopSelf()或 stopService()停止
onBind()	当 Activity 等组件通过调用 bindService()方法绑定 Service 时调用,这种方式实现的 Service 类必须通过返回 IBinder 提供的接口与 Service 通信
onUnbind()	当最后一个绑定的组件解绑或调用 unbindService()方法时执行,该方法的返回值默认为 false,表示不允许重复绑定。返回 false 之后再次执行 bindService()依然能够成功,但是 unbindService()无效,因此允许重复绑定务必返回 true
onDestroy()	当 Service 不再使用且将被销毁时调用。此方法用于清理所有资源,如线程、注册侦听器、接收器等,此方法也只会被调用一次

这 5 个回调方法中只有 onStartCommand()具有返回值,它必须是以下三种常量值之一:

● START_NOT_STICKY,如果服务在 onStartCommand()返回后被系统终止,除非有待传递的挂起 Intent,否则系统不会重建服务。这是最安全的选项。

● START_STICKY,如果服务在 onStartCommand()返回后被系统终止,则会尝试重新创建并调用 onStartCommand(),但不会重新传递最后一个 Intent。此返回值适用于不执行命令但可无限期运行并等待任务的媒体播放器等场景。

● START_REDELIVER_INTENT，如果服务在 onStartCommand()返回后被系统终止，则会重建服务，并通过传递给服务的最后一个 Intent 调用 onStartCommand()。所有挂起的 Intent 均依次传递。此返回值适用于主动执行应立即恢复的任务（如下载文件）的服务。

接下来通过一个案例理解 Service 的创建和生命周期回调方法的调用。

【案例 6-1】采用两种启动 Service 的方式展示生命周期方法的变化过程，界面如图 6.2 所示。

步骤 1：创建项目

启动 Android Studio，创建名为 D0601_ServiceLifeCycle 的项目，选择 Empty Activity 项，将包名改为 com.example.servicelifecycle，单击 Finish 按钮，等待项目构建完成。

步骤 2：构建界面布局

打开 activity_main.xml，根布局采用 ConstraintLayout，由于布局较为简单，代码不再提供。

步骤 3：创建 Service

图 6.2　主界面布局

右击包名，选择 New->Service->Service，打开创建对话框，填写 Class Name，创建两个名为 MyService 的 Service，如图 6.3 所示。

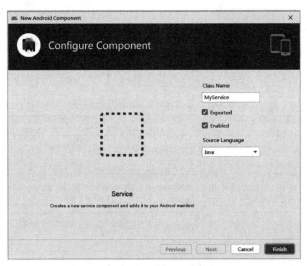

图 6.3　创建 Service 对话框

单击 Finish 按钮之后，系统会创建 MyService 类，并在 AndroidManifest.xml 中注册该 Service，代码如下：

```
1.  <service
2.      android:name=".MyService"
3.      android:enabled="true"
4.      android:exported="true"/>
```

其中，android:enabled 属性用于设置是否由系统实例化服务，如果值为"false"，则该服务被禁用，默认值为 true。android:exported 属性用于设置其他应用的组件是否能调用服务或与之交互，如果值为"false"，则表示同一个应用的组件可以启动或绑定服务，默认值为"true"。

在生成的 MyService 代码框架中加入生命周期的回调方法，代码如下：

```
1.  public class MyService extends Service {
2.      private static final String TAG = MyService.class.getSimpleName();
3.      public MyService() {
4.      }
5.      @Override
6.      public void onCreate() {
7.          super.onCreate();
8.          Log.i(TAG, "onCreate: 服务被创建");
9.      }
10.     @Override
11.     public int onStartCommand(Intent intent, int flags, int startId) {
12.         Log.i(TAG, "onStartCommand: 服务启动");
13.         return super.onStartCommand(intent, flags, startId);
14.     }
15.     @Override
16.     public IBinder onBind(Intent intent) {
17.         Log.i(TAG, "onBind: 服务被绑定");
18.         return null;
19.     }
20.     @Override
21.     public void onRebind(Intent intent) {
22.         super.onRebind(intent);
23.         Log.i(TAG, "onRebind: 服务被再次绑定");
24.     }
25.     @Override
26.     public boolean onUnbind(Intent intent) {
27.         Log.i(TAG, "onUnbind: 服务解绑");
28.         return true;
29.     }
30.     @Override
31.     public void onDestroy() {
32.         Log.i(TAG, "onDestroy: 服务被销毁");
33.         super.onDestroy();
34.     }
35. }
```

上述代码通过日志输出 Service 的生命周期回调方法的调用过程。

步骤 4：启动 Service

打开 MainActivity 类文件，编写 4 个按钮的单击事件监听器，初始化 TextView 控件，然后分别编写启动服务、绑定服务两种方式的代码。

（1）启动服务

在 onClick()方法中实现"启动服务"和"停止服务"按钮的事件处理，代码如下：

```
1.  public void onClick(View view) {
2.      Intent intent;
3.      switch (view.getId()) {
4.          case R.id.btn_start:
5.              intent = new Intent(this, MyService.class);
6.              startService(intent);
7.              break;
8.          case R.id.btn_stop:
9.              intent = new Intent(this, MyService.class);
10.             stopService(intent);
11.             break;
12.     }
13. }
```

运行项目，单击 3 次"启动服务"按钮，Logcat 窗口看到如图 6.4 所示的日志输出。

图 6.4　startService 的日志输出

单击"停止服务"按钮，Logcat 的日志输出如图 6.5 所示。

图 6.5　stopService 的日志输出

由日志输出可以看出，onstartCommand()方法在每次单击"启动服务"按钮都被调用，而 onCreate()只被调用一次，单击"停止服务"按钮多次也只调用一次 onDestroy()方法。

（2）绑定服务

在 onClick()方法中实现"绑定服务"和"解绑服务"按钮的事件处理，代码如下：

```
1.   public void onClick(View view) {
2.       Intent intent;
3.       switch (view.getId()) {
4.           case R.id.btn_bind:
5.               intent = new Intent(this, MyService.class);
6.               bindService(intent, connection, BIND_AUTO_CREATE);
7.               break;
8.           case R.id.btn_unbind:
9.               unbindService(connection);
10.              break;
11.      }
12.  }
```

上述第 6 行代码的 binderService()方法用于绑定服务，它有 3 个参数，第一个参数 intent 用来指定服务，第二个参数 connection 是 ServiceConnection 接口的实现类对象，提供 Activity 与 Service 交互的桥梁，具体应用在下节详细讲解，它在 MainActivity 类中的定义如下代码所示；第三个参数是服务的常量标志。

```
1.   ServiceConnection connection = new ServiceConnection() {
2.       @Override
3.       public void onServiceConnected(ComponentName name, IBinder service) {
4.           // 服务绑定后的操作
5.           Log.i("MainActivity", "onServiceConnected()被调用");
6.       }
7.       @Override
8.       public void onServiceDisconnected(ComponentName name) {
9.           // 服务被意外销毁后的操作
```

```
10.            Log.i("MainActivity", "onServiceDisconnected()意外销毁被调用");
11.        }
12. };
```

运行项目，单击 3 次"绑定服务"按钮，Logcat 窗口看到如图 6.6 所示的日志输出结果。

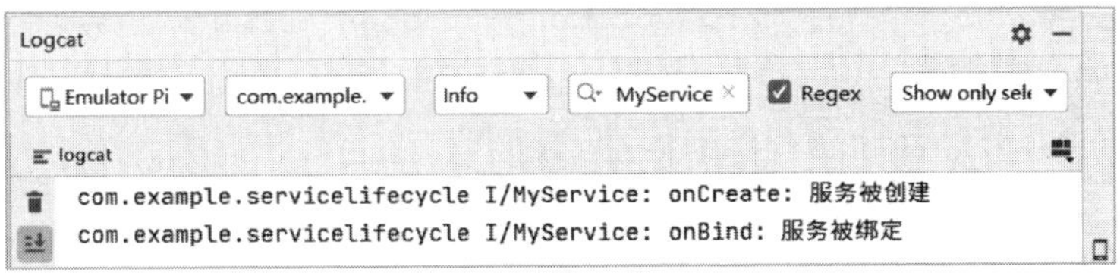

图 6.6 bindService 的日志输出

单击"解绑服务"按钮，Logcat 的日志输出如图 6.7 所示。

图 6.7 unbindService 的日志输出

由日志输出可以看出，单击 3 次"绑定服务"按钮 onBind()方法不会被重复调用，由于 onBind()的返回值为 null，MainActivity 类定义的 ServiceConnection 接口实现类的方法也不会被调用，单击"解绑服务"按钮会调用 onUnbind()和 onDestroy()方法。

本案例通过日志输出的方式帮助理解 Service 的生命周期，接下来讲解 Activity 如何与 Service 进行数据交互。

6.1.3　Activity 和 Service 的交互

当 Service 启动后，我们会将部分不需要界面的任务交由 Service 执行，但是有些时候用户也需要和 Service 进行交互，比如，后台执行的音乐播放，用户应能通过 Activity 界面的控制按钮对播放过程进行控制；又比如，后台执行的下载任务，用户也应能通过前台的 Activity 暂停、继续下载任务等。这些例子表明，虽然后台任务都在后台默默执行，但也需要与用户进行交互。这可以通过前台的 Activity 与后台的 Service 建立通信连接，将用户的意图传达给 Service。

【案例 6-2】以模拟下载为例实现 Activity 和 Service 之间的通信，界面如图 6.8 所示。

步骤 1：创建项目

启动 Android Studio，创建名为 D0602_DownloadService 的项目，选择 Empty Activity 项，将包名改为 com.example.service，单击 Finish 按钮，等待项目构建完成。

步骤 2：构建界面布局

打开 activity_main.xml，根布局采用 ConstraintLayout，添加 3

图 6.8 主界面布局

个按钮和一个进度条 ProgressBar，ProgressBar 设置样式 style 为横向、最大值 max 为 100，初始界面不显示进度条，因此设置它的 visibility 为 gone，代码如下：

```
1.  <ProgressBar
2.      android:id="@+id/progressBar"
3.      style="@style/Widget.AppCompat.ProgressBar.Horizontal"
4.      android:layout_width="0dp"
5.      android:layout_height="wrap_content"
6.      android:layout_marginStart="8dp"
7.      android:layout_marginTop="32dp"
8.      android:layout_marginEnd="8dp"
9.      android:visibility="gone"
10.     android:max="100"
11.     app:layout_constraintEnd_toEndOf="parent"
12.     app:layout_constraintStart_toStartOf="parent"
13.     app:layout_constraintTop_toBottomOf="@+id/btn_download" />
```

步骤 3：创建 DownloadService 类

使用向导创建 DownloadService 类，绑定方式启动 Service 需要实现 onBind()回调方法的 IBinder 接口，该接口的作用是将 DownloadService 实例传递给 Activity，实现两者之间的交互。

首先添加日志的标签的字符串变量和进度条变量，代码如下：

```
1.  public class DownloadService extends Service {
2.      public static final String TAG = DownloadService.class.
                                            getSimpleName();;
3.      private int progress = 0;
4.  }
```

（1）创建 MyBind 类

在 DownloadService 类中定义一个内部类 MyBinder 继承自 Binder 类，由于 Binder 实现了 IBinder 接口，相当于 MyBinder 是 IBinder 实现类的子类；然后在其中定义 getService()方法用于创建 DownloadService 类的实例对象，代码如下：

```
1.  class MyBinder extends Binder {
2.      public DownloadService getService() {
3.          return DownloadService.this;
4.      }
5.  }
```

定义 MyBind 类之后，则需要将 onBind()方法的返回值改为 MyBind 的实例对象，代码如下：

```
1.  @Override
2.  public IBinder onBind(Intent intent) {
3.      Log.d(TAG, "onBind()被调用");
4.      return new MyBinder();
5.  }
```

（2）创建进度变化的监听接口

接下来，创建监听进度条数值发生改变的接口。在 3.6.3 小节中讲解了 Fragment 通过创建监听回调接口与 Activity 之间进行数据传递，Service 也可采用这种方式实现进度条的监听，代码如下：

```
1.  // 注册回调接口的方法
2.  private OnProgressListener onProgressListener;
3.  public void setOnProgressListener(OnProgressListener
                                      onProgressListener) {
4.      this.onProgressListener = onProgressListener;
```

```
5.    }
6.    // 更新进度的回调接口
7.    public interface OnProgressListener {
8.        void onProgress(int progress);
9.    }
```

(3) 实现下载功能

在类中定义 startDownload()方法实现下载功能，此处通过 Thread 类的 sleep()方法模拟下载，并调用进度变化的监听方法，传递进度条的数值，代码如下：

```
1.  // 模拟下载任务，每秒钟更新一次
2.  public void startDownload() {
3.      Log.d(TAG, "startDownload()被调用");
4.      progress = 0;
5.      new Thread(new Runnable() {
6.          @Override
7.          public void run() {
8.              while (progress < 100) {
9.                  progress += 5;
10.                 // 进度发生变化时通知 MainActivity
11.                 if(onProgressListener != null){
12.                     onProgressListener.onProgress(progress);
13.                 }
14.                 try {
15.                     Thread.sleep(1000);
16.                 } catch (InterruptedException e) {
17.                     e.printStackTrace();
18.                 }
19.             }
20.         }
21.     }).start();
22. }
```

步骤 4：实现 MainActivity 的功能

绑定服务需要 ServiceConnection 接口的实现类对象，并实现它的两个方法，分别用于处理服务连接之后和失去连接需要完成的任务。onServiceConnected()方法的第二个参数是从 Service 传入的 IBinder 对象，通过该对象就可调用 Service 类的方法，实现相互间的通信，如调用它的监听器方法触发设置进度条的变化事件，代码如下：

```
1.  private ProgressBar progressBar;
2.  private DownloadService downloadService;
3.  // 服务连接接口
4.  private ServiceConnection connection = new ServiceConnection() {
5.      @Override
6.      public void onServiceDisconnected(ComponentName name) {
7.          Log.i(TAG, "onServiceDisconnected()意外销毁被调用");
8.      }
9.      @Override
10.     public void onServiceConnected(ComponentName name, IBinder iBinder) {
11.         Log.i(TAG, "onServiceConnected()");
12.         downloadService = ((DownloadService.MyBinder) iBinder).getService();
13.         downloadService.setOnProgressListener(new DownloadService.
                                                  OnProgressListener(){
14.             @Override
15.             public void onProgress(int progress) {
16.                 progressBar.setProgress(progress);
17.                 // 下载完成则移除进度条
18.                 If (progress == 100) {
```

```
19.                    Toast.makeText(MainActivity.this, "下载完成",
20.                            Toast.LENGTH_SHORT).show();
21.                    progressBar.setVisibility(View.GONE);
22.                }
23.            }
24.        });
25.    }
26. };
```

接下来实现界面组件的初始化,以及监听事件的处理,代码如下:

```
1.  public class MainActivity extends AppCompatActivity
2.          implements View.OnClickListener {
3.      ...
4.      @Override
5.      protected void onCreate(Bundle savedInstanceState) {
6.          super.onCreate(savedInstanceState);
7.          setContentView(R.layout.activity_main);
8.          // 初始化控件
9.          progressBar = findViewById(R.id.progressBar);
10.         Button btnBind = findViewById(R.id.btn_bind);
11.         Button btnUnbind = findViewById(R.id.btn_unbind);
12.         Button btnDownload = findViewById(R.id.btn_download);
13.         // 设置按钮的单击事件监听器
14.         btnBind.setOnClickListener(this);
15.         btnUnbind.setOnClickListener(this);
16.         btnDownload.setOnClickListener(this);
17.     }
18.     @Override
19.     public void onClick(View v) {
20.         switch (v.getId()) {
21.             case R.id.btn_bind:
22.                 // 绑定服务
23.                 Intent bindIntent = new Intent(this, DownloadService.
                                                    class);
24.                 bindService(bindIntent, connection, BIND_AUTO_CREATE);
25.                 break;
26.             case R.id.btn_download:
27.                 // 下载
28.                 if (downloadService != null) {
29.                     downloadService.startDownload();
30.                     progressBar.setVisibility(View.VISIBLE);
31.                 }
32.                 break;
33.             case R.id.btn_unbind:
34.                 // 解绑服务
35.                 unbindService(connection);
36.                 break;
37.         }
38.     }
39. }
```

第 28~31 行当 downloadService 对象不为空时,调用它的 startDownload()方法模拟下载功能。

6.1.4 前台服务

由于后台服务优先级相对较低,当系统出现内存不足等异常情况时,就可能被回收,另外由于后台服务应用不当会影响系统性能，因此，从 Android 8.0 版本开始，系统限制了后台服务，推荐使用前台服务或计划作业。例如，天气类 App 在状态栏中的天气预报，音乐类 App 放在前台状态栏的播放控制器，词典类 App 放在前台的快捷方式等。

简而言之，前台服务相对于后台服务优先级更高、不易被回收，并且前台服务必须在状态栏显示通知，即使用户停止与应用的交互，前台服务仍会继续运行。在【案例 6-2】中添加一个"启动前台服务"的按钮，单击后启动前台服务，具体步骤如下：

步骤 1：创建 ForegroundService

在包名下使用向导创建 ForegroundService 类，重写 onCreate()、onStartCommand() 和 onDestroy() 方法，代码结构如下：

```java
1.  public class ForegroundService extends Service {
2.      private static final String TAG = ForegroundService.class.
                                            getSimpleName();
3.      public ForegroundService() {
4.      }
5.      @Override
6.      public void onCreate() {
7.          super.onCreate();
8.          Log.i(TAG, "onCreate()被调用");
9.      }
10.     @Override
11.     public IBinder onBind(Intent intent) {
12.         Log.i(TAG, "onBind()被调用");
13.         return null;
14.     }
15.     @Override
16.     public int onStartCommand(Intent intent, int flags, int startId) {
17.         Log.i(TAG, "onStartCommand()被调用");
18.         return START_STICKY_COMPATIBILITY;
19.     }
20.
21.     @Override
22.     public void onDestroy() {
23.         Log.i(TAG, "onDestroy()被调用");
24.         super.onDestroy();
25.     }
26. }
```

步骤 2：添加前台服务的通知 Notification

从 Android 8.0 版本开始，创建通知需要设置通知通道，使用 NotificationManager 对象将通知通道注册到应用之后才能发送通知，代码如下：

```java
1.  private static final String CHANNEL_ID = "服务消息";  // 通知通道的id
2.  private static final int NOTIFICATION_ID = 100;     // 通知 ID,不能设为0
3.  // Android 8.0创建通知通道
4.  @RequiresApi(api = Build.VERSION_CODES.O)
5.  private void createChannel() {
6.      NotificationManager notificationManager = (NotificationManager)
7.              getSystemService(Context.NOTIFICATION_SERVICE);
8.      // 创建通知通道需要三个参数：通道 ID、通道 name 和重要性
9.      NotificationChannel channel = new NotificationChannel
                                    (CHANNEL_ID, "下载服务",
10.             NotificationManager.IMPORTANCE_HIGH);
11.     // 设置通知参数
12.     channel.setDescription("下载");      // 设置描述
```

```
13.     channel.enableLights(true);          // 设置允许提示灯
14.     channel.setLightColor(Color.RED);    // 设置提示灯颜色
15.     channel.enableVibration(true);       // 设置允许震动
16.     channel.setVibrationPattern(new long[]{0, 1000, 500, 1000});
        // 设置震动
17.     // 创建通知
18.     if (notificationManager != null) {
19.         notificationManager.createNotificationChannel(channel);
20.     }
21. }
```

然后使用 NotificationCompat.Builder 类创建通知，设置通知图标、标题、内容、进度条及单击展开的操作等属性，并根据当前 Android 的版本是否为 8.0 及以上版本加载通知通道，代码如下：

```
1.  private Notification notification;
2.  private NotificationCompat.Builder builder;
3.  // 创建通知
4.  private void createNotification() {
5.      // Android8.0 及以上版本创建通知通道
6.      if (Build.VERSION.SDK_INT >= Build.VERSION_CODES.O) {
7.          createChannel();
8.      }
9.      // 设置通知的单击操作
10.     Intent intent = new Intent(this, MainActivity.class);
11.     PendingIntent pendingIntent = PendingIntent.getActivity
                                     (this, 0, intent, 0);
12.     // 创建通知
13.     builder = new NotificationCompat.Builder(this, CHANNEL_ID);
14.     notification = builder.setSmallIcon(R.mipmap.ic_launcher)
        // 通知小图标
15.             .setContentTitle("图片下载")         // 通知标题
16.             .setContentText("下载中...")         // 通知内容
17.             .setWhen(System.currentTimeMillis()) // 设定通知显示的时间
18.             .setProgress(100, 0, false)          // 设置进度条
19.             .setContentIntent(pendingIntent)
20.             .build();                            //创建通知
21. }
```

步骤 3：模拟下载任务

修改上个案例的 startDownload()方法，添加更新通知栏进度条的代码，以及下载结束移除前台服务的代码如下：

```
1.  // 模拟下载任务，每秒钟更新一次
2.  public void startDownload() {
3.      Log.i(TAG, "startDownload()被调用");
4.      progress = 0;
5.      NotificationManagerCompat notificationManager=
        NotificationManagerCompat.from(this);
6.      new Thread(new Runnable() {
7.          @Override
8.          public void run() {
9.              while (progress < 100) {
10.                 progress += 5;
11.                 // 更新通知栏的进度条
12.                 builder.setProgress(100, progress, false)
13.                         .setContentText("下载" + progress + "%");
```

```
14.                    notificationManager.notify(NOTIFICATION_ID, builder.
                                                 build());
15.                    try {
16.                        Thread.sleep(1000);
17.                    } catch (InterruptedException e) {
18.                        e.printStackTrace();
19.                    }
20.                }
21.                // 下载完成将进度条归零
22.                builder.setContentText("下载完成").setProgress(0, 0, false);
23.                notificationManager.notify(NOTIFICATION_ID, builder.build());
24.                //移除前台服务
25.                stopForeground(true);
26.            }
27.        }).start();
28. }
```

步骤4：设置前台服务

由于每次下载完成都会调用 stopForeground()停止前台服务，因此需要在 onStartCommand()方法中调用创建通知的 createNotification()方法、设置前台服务的 startForeground()方法和 startDownload()方法执行下载任务，在 onDestroy()方法中调用 stopForeground()停止前台服务。代码如下：

```
1.  @Override
2.  public int onStartCommand(Intent intent, int flags, int startId) {
3.      Log.i(TAG, "onStartCommand()被调用");
4.      createNotification();   // 创建服务通知
5.      startForeground(NOTIFICATION_ID, notification);   // 设置为前台服务
6.      startDownload();
7.      return START_STICKY_COMPATIBILITY;
8.  }
9.  @Override
10. public void onDestroy() {
11.     Log.i(TAG, "onDestroy()被调用");
12.     super.onDestroy();
13.     try {
14.         stopForeground(true);
15.     } catch (Exception e) {
16.         e.printStackTrace();
17.     }
18. }
```

步骤5：MainActivity 类实现按钮功能

打开 MainActivity 类文件，在 onCreate()方法中添加"启动前台服务"按钮设置监听器的代码，在 onClick()方法中添加前台服务启动的代码。

```
1.  @Override
2.  public void onClick(View v) {
3.      switch (v.getId()) {
4.          ...
5.          case R.id.btn_start_foreground:
6.              intent = new Intent(this, ForegroundService.class);
7.              // Android 8.0 以上版本使用 startForegroundService 启动前台服务
8.              if (Build.VERSION.SDK_INT >= Build.VERSION_CODES.O) {
9.                  startForegroundService(intent);
10.             } else {
11.                 startService(intent);
12.             }
```

```
13.            break;
14.        }
15. }
```

步骤6：运行项目

在运行项目之前，需要在 AndroidManifest.xml 文件中为<uses-permission>标签添加前台服务的权限。

```
1. <manifest xmlns:android="http://schemas.android.com/apk/res/android"
2.     package="com.example.service">
3.     <uses-permission android:name="android.permission.FOREGROUND_
                                      SERVICE" />
4.     <application ...>
5. </manifest>
```

单击"启动前台服务"按钮，便会在状态栏出现通知的标志，单击后可以看到进度条的变化，下载完成则通知消失，运行界面如图 6.9、图 6.10 所示。

图 6.9　前台服务的通知界面

图 6.10　单击通知后的界面

6.2　广播机制

在商场、学校等公共场所，为了方便地让所有人知晓一个消息，常常会采用广播的形式。Android 系统也有类似的广播机制进行系统级别的消息通知，比如当手机电池电量低的时候会弹出"提示电量不足"的消息窗口、当网络发生变化时也会弹出"正在切换移动网络"的提示消息，还有很多的类似场景都是 Android 的广播机制在发挥作用，系统传递出一些信息，使得应用程序能够响应这些信息，进而完成相应的操作。

Android 系统内置了数量众多的广播，只要涉及手机的基本操作，基本都会发出相应的系统广播，但应用程序不会全盘接收，应用程序可以通过注册感兴趣的广播进行过滤，这种机制使得 Android 的广播机制更为灵活。Android 提供了相关的 API，允许应用程序与 Android 系统、其他 Android 应用之间相互发送和接收广播。发送广播要借助第 4 章学过的 Intent，而接收广播需要借助于广播接收器 BroadcastReceiver，在学习它之前先了解一下广播的基本概念。

6.2.1 广播机制简介

Android 系统的广播机制是基于消息发布（Publish）-订阅（Subscribe）的事件驱动模型构建的，广播发送者负责发布消息，广播接收者需要先订阅消息，然后才能接收消息。广播发送者无须知道广播接收者是否存在或者何时收到，发送和接收实现完全的解耦。广播机制的具体实现流程如图 6.11 所示。

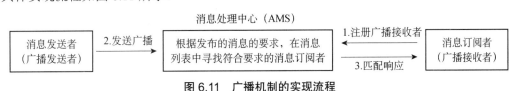

图 6.11 广播机制的实现流程

① 广播接收者 BroadcastReceiver 通过 Binder 机制在 AMS（Activity Manager Service）消息处理中心进行注册。

② 广播发送者通过 Binder 机制向 AMS 消息处理中心发送广播。

③ AMS 根据广播发送者的要求，在已注册列表中查找符合要求的广播接收者，将广播发送到消息循环队列。

④ AMS 将广播发送到广播接收者相应的消息循环队列中。

⑤ 广播接收者通过消息循环执行此广播，回调其 onReceive()方法进行相关的处理。

以上的实现流程解决了广播如何发送和广播接收者如何接收这两个问题，由此得出 Android 广播机制包含的三个要素：广播、注册和接收者。接下来对广播发送者和广播接收者进行讲解。

6.2.2 广播接收器

Android 系统提供 BroadcastReceiver 类定义广播接收器，它提供了处理广播的抽象方法 onReceive()。自定义广播接收器必须作为 BroadcastReceiver 的子类，重写 onReceive()方法。广播接收器的注册分为动态注册和静态注册两种方式，动态注册是在 Activity 类中通过代码进行注册的，并在 Activity 销毁时注销；静态注册是在配置清单文件 AndroidManifest.xml 中进行注册的，接下来通过接收系统广播讲解广播接收器的创建及两种注册方式。

Android 系统内置了多个系统广播，当特定的条件发生时由系统自动发出，如网络状态发生变化、电池的电量发生变化、时间发生变化等。完整的系统广播清单可以查看 Android SDK 路径"platforms/<任意 android api 版本>/data"目录下的 broadcast_actions.txt，在实际项目中可以根据场景合理使用。

【案例 6-3】使用两种注册方式创建广播接收器接收系统的锁屏、解锁广播，使用默认界面即可。

步骤 1：创建项目

启动 Android Studio，创建名称为 D0603_BroadcastReceiver 的项目，选择 Empty Activity 项，将包名改为 com.example.broadcastreceiver，单击 Finish 按钮，等待项目构建完成。

步骤 2：创建 BroadcastReceiver

（1）静态注册

在包名上右击，选择 New->Other->Broadcast Receiver，打开创建 BroadcastReceiver 的对话框，在 Class Name 输入框中填写类名称为 ScreenReceiver，默认勾选 Exported 和 Enabled，它们的含义与创建 Service 组件相同，如图 6.12 所示。

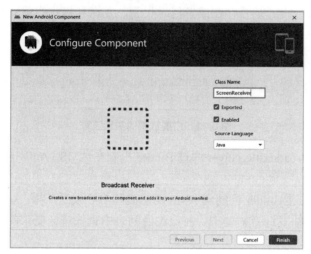

图 6.12　创建静态注册的 BroadcastRecevier 的对话框

单击 Finish 按钮就会创建继承自 BroadcastReceiver 的 ScreenReceiver 类，并在配置清单文件中添加了<receiver>标签，添加<intent-filter>子标签，增加锁屏、解锁的<action>，代码如下：

```
1.   <receiver
2.       android:name=".ScreenReceiver"
3.       android:enabled="true"
4.       android:exported="true">
5.       <intent-filter>
6.           <action android:name="android.intent.action.SCREEN_ON" />
7.           <action android:name="android.intent.action.SCREEN_OFF" />
8.       </intent-filter>
```

然后重写 onReceive()方法，使用日志输出解锁和锁屏的广播信息，代码如下：

```
1.   public class ScreenReceiver extends BroadcastReceiver {
2.       private static final String TAG = ScreenReceiver.class.
                                              getSimpleName();
3.       @Override
4.       public void onReceive(Context context, Intent intent) {
5.           final String action = intent.getAction();
6.           if (Intent.ACTION_SCREEN_OFF.equals(action)) {
7.               Log.i(TAG, "屏幕锁屏了");
8.           } else if(Intent.ACTION_SCREEN_ON.equals(action)) {
9.               Log.i(TAG, "屏幕解锁了");
10.          }
11.      }
12.  }
```

运行项目，切换锁屏和解锁，查看 Logcat 的日志输出没有结果，这是因为受 Android 8.0 及以上版本的后台执行的限制，系统广播中不明确指定启动组件的广播（称为隐式广播）的静态注册已失效，需要改为动态注册方式进行接收。

（2）动态注册

首先，删除 AndroidManifest.xml 配置 ScreenReceiver 类的<receiver>标签，然后打开 MainActivity 类文件，在 onCreate()方法中注册 ScreenReceiver 对象，在 onDestroy()方法中注销广播接收器，代码如下：

```
1.  public class MainActivity extends AppCompatActivity {
2.      private ScreenReceiver screenReceiver;
3.      @Override
4.      protected void onCreate(Bundle savedInstanceState) {
5.          super.onCreate(savedInstanceState);
6.          setContentView(R.layout.activity_main);
7.          screenReceiver = new ScreenReceiver();  //实例化广播接收器
8.          IntentFilter intentFilter = new IntentFilter();
            // 实例化 IntentFilter
9.          // 添加 action
10.         intentFilter.addAction(Intent.ACTION_SCREEN_OFF);
11.         intentFilter.addAction(Intent.ACTION_SCREEN_ON);
12.         registerReceiver(screenReceiver, intentFilter); // 注册广播接收器
13.     }
14.     @Override
15.     protected void onDestroy() {
16.         super.onDestroy();
17.         unregisterReceiver(screenReceiver);    // 注销广播接收器
18.     }
19. }
```

再一次运行项目，切换锁屏和解锁，查看 Logcat 的日志输出，结果如图 6.13 所示。

图 6.13　广播接收器接收系统广播

静态注册由于常驻内存，每次发送广播，广播接收器都会消耗资源，如果多个应用静态注册了基于系统事件的隐式广播，就会导致快速、连续的资源消耗，严重影响用户体验。为了限制对系统进行广播，特别是隐式广播的滥用，从 Android 8.0 版本开始禁止静态注册隐式广播，但隐式广播可以使用动态注册进行接收；系统显式广播或自定义广播依旧可以使用静态注册。静态注册与动态注册的特点及应用场景如表 6.2 所示。

表 6.2　静态注册与动态注册的区别

注册方式	特点	应用场景
静态注册	常驻内存，不受应用程序是否启动的约束，随时可以使用。缺点：耗电、占内存、无法取消	需要时刻监听广播
动态注册	非常驻内存，灵活，可以控制注册和注销；跟随组件的生命周期变化，在组件结束前必须注销。缺点：只有注册后才能使用	需要特定时刻监听广播

6.2.3　自定义广播

接收系统广播只需要创建广播接收器即可。当系统广播不能满足实际需求时，Android 还

可以创建自定义广播，编写对应的广播接收器。广播传递的消息内容的载体实质上就是 Intent 对象，调用 Intent 对象的 setAction()方法设置意图动作，调用 putExtra()方法传递自定义信息，然后调用 sendBroadcast()和 sendOrderedBroadcast()方法发送广播。在编码实现之前，先了解一下自定义广播的分类。

1. 从执行方式角度分类

Android 广播从执行方式的角度分为两种类型。

（1）普通广播（Normal Broadcast）

普通广播相对于有序广播，也被称为无序广播，它是一种完全异步执行的广播，在广播发出后所有的接收器几乎在同一时刻接收到广播消息，所有的接收者都被同等对待，没有先后顺序，如图 6.14 所示。标准广播的执行效率高，但接收者无法将处理结果交给下一个接收者，也无法阻止其他接收器收到此广播，换句话说，标准广播不能被截断。

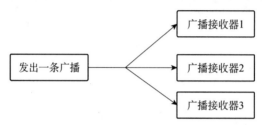

图 6.14　普通广播的发送与接收

（2）有序广播（Ordered Broadcast）

有序广播是一种按照顺序执行的广播，广播发出后，同一时刻只有一个广播接收器能够收到广播，当这个广播接收器执行完毕后才会继续传播。有序广播是按照优先级的顺序进行接收的，而且每个接收者可以修改操作，下一个就能得到上一个修改后的结果，除此之外，接收到广播的接收者有权截断正在传递的广播，后续的广播接收器就无法接收到这个广播，如图 6.15 所示。

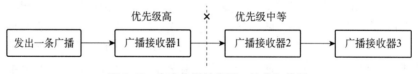

图 6.15　有序广播的发送、接收和截断

2. 从作用域角度分类

Android 广播从作用域角度也可以分为两种类型。

（1）全局广播（Global Broadcast）

全局广播指可以被其他应用程序都能接收到的广播，比如系统发出的广播，它存在数据泄露、恶意攻击等一些安全风险。

（2）本地广播（Local Broadcast）

本地广播指只能在应用程序内部传递的广播，广播接收器也只能接收来自本应用程序的广播，不仅提高了安全性，传递效率也更高。

接下来通过一个案例了解自定义广播的发送和接收，界面如图 6.16 所示。

【案例 6-4】自定义广播的创建、发送和接收。

步骤 1：创建项目

启动 Android Studio，创建名为 D0604_CustomBroadcast 的项目，选择 Empty Activity，将包名改为 com.example.custombroadcast，单击 Finish 按钮，等待项目构建完成。

步骤 2：设计界面布局，初始化界面控件对象

打开 main_activity.xml 文件，根布局采用 ConstraintLayout，输入框 EditText 使用了 TextInputLayout 布局，界面布局较为简单，代码不再给出。

在 MainActivity 类文件的 onCreate()方法中完成控件的初始化、按钮的事件监听，代码如下：

图 6.16　主界面布局

```
1.  private TextView tvReceive;
2.  private EditText etContent;
3.  @Override
4.  protected void onCreate(Bundle savedInstanceState) {
5.      super.onCreate(savedInstanceState);
6.      setContentView(R.layout.activity_main);
7.      // 初始化控件对象
8.      etContent = findViewById(R.id.et_content);
9.      tvReceive = findViewById(R.id.tv_receive);
10.     Button btnNormal = findViewById(R.id.btn_normal);
11.     Button btnOrdered = findViewById(R.id.btn_ordered);
12.     Button btnLocal = findViewById(R.id.btn_local);
13.     // 设置监听器
14.     btnNormal.setOnClickListener(this);
15.     btnOrdered.setOnClickListener(this);
16.     btnLocal.setOnClickListener(this);
17. }
```

步骤 3：创建普通广播接收器类

本案例使用动态注册，不再利用广播接收器的创建向导，通过创建类的方式创建广播接收器类。在包名下右击，选择 New->Java Class，创建 MyNormalReceiver 类继承自 BroadcastReceiver 抽象类，并重写 onReceive()方法，通过 onReceive()方法的第 2 个参数 Intent 接收广播数据，代码如下：

```
1.  public class MyNormalReceiver extends BroadcastReceiver {
2.      private final Context context;
3.      public MyNormalReceiver(Context context) {
4.          this.context = context;
5.      }
6.      @Override
7.      public void onReceive(Context context, Intent intent) {
8.          final String action = intent.getAction();
9.          if (MainActivity.NORMAL_ACTION.equals(action)) {
10.             // 接收数据，并将数据显示在界面上
11.             String data = intent.getStringExtra(MainActivity.ACTION_
                                                   EXTRA_KEY);
12.             if (data != null) {
13.                 data = "普通广播收到的内容：" + data;
14.                 ((MainActivity) this.context).setContent(data);
15.             }
16.         }
17.     }
18. }
```

上述代码的第 2 行定义了 MainActivity 类的 Context 上下文对象用于将收到的信息传递给 MainActivity 显示，onReceive()方法在第 11~15 行代码段获取 intent 传递的数据，并调用 setContent()方法将数据显示到主界面的文本框中。

步骤 4：MainActivity 注册普通广播并发送

普通广播的注册与上个案例的编码相同，定义 registerReceivers()方法用于注册广播，然后在 onCreate()方法中调用；第 9~13 行代码段为"全局发送"按钮的单击事件的处理，输入框的内容调用 intent 的 putExtra()方法获取，将 intent 作为 sendBroadcast()方法的参数发送广播，最后在 onDestroy()方法中注销广播接收器，代码如下：

```
1.  private MyNormalReceiver myNormalReceiver;
2.  public static final String NORMAL_ACTION="com.example.custombroadcast.
                                NORMAL_ACTION";
3.  public static final String ACTION_EXTRA_KEY = "data";
4.  @Override
5.  public void onClick(View view) {
6.      Intent intent = new Intent();
7.      String content = etContent.getText().toString();
8.      switch (view.getId()) {
9.          case R.id.btn_normal:
10.             Intent.setAction(MainActivity.NORMAL_ACTION);
11.             intent.putExtra(MainActivity.ACTION_EXTRA_KEY, content);
12.             sendBroadcast(intent);
13.             break;
14.     }
15. }
16. // 文本框加载广播的内容
17. public void setContent(String data) {
18.     tvReceive.setText(data);
19. }
20. // 广播注册
21. private void registerReceivers() {
22.     myNormalReceiver = new MyNormalReceiver(this);
23.     IntentFilter intentFilter = new IntentFilter(MainActivity.
                                NORMAL_ACTION);
24.     registerReceiver(myNormalReceiver, intentFilter);    // 注册普通广播
25. }
26. @Override
27. protected void onDestroy() {
28.     super.onDestroy();
29.     unregisterReceiver(myNormalReceiver); // 注销广播接收器
30. }
```

步骤 5：创建 3 个有序广播接收器类

参照步骤 3 的方法创建 3 个有序广播类，分别为 MyOrderReceiver、MyOrderReceiver1 和 MyOrderReceiver2，其中 MyOrderReceiver 为最终的有序广播。MyOrderReceiver1 类的示例代码如下：

```
1.  public class MyOrderReceiver1 extends BroadcastReceiver {
2.      private final static String TAG = MyOrderReceiver1.class.
                                getSimpleName();
3.      private final Context context;
4.      public MyOrderReceiver1(Context context) {
5.          this.context = context;
6.      }
7.      @Override
8.      public void onReceive(Context context, Intent intent) {
```

```
9.            final String action = intent.getAction();
10.           if (MainActivity.ORDER_ACTION.equals(action)) {
11.               // 接收数据,并将数据显示在界面上
12.               String data = getResultData();
13.               if (data != null) {
14.                   data = "有序广播1收到的内容: " + data;
15.                   Log.i(TAG, data);
16.                   ((MainActivity) this.context).setContent(data);
17.                   setResultData("10000");
18.                   // abortBroadcast();
19.               }
20.           }
21.       }
22. }
```

上述第 12 行的 getResultData()方法用于接收上一个有序广播发送的数据,第 17 行的 setResultData()方法将字符串数据传递给下一个有序广播,实现数据的传递。第 18 行的 abortBroadcast()方法用于中止优先级比它低的有序广播。

步骤 6:MainActivity 注册有序广播并发送

在 MainActivity 类的 registerReceivers()方法中增加注册这 3 个有序广播的代码,并调用 setPriority()方法设置优先级,默认优先级为 0,数值越大优先级越高。

```
1. private MyOrderReceiver myOrderReceiver;
2. private MyOrderReceiver1 myOrderReceiver1;
3. private MyOrderReceiver2 myOrderReceiver2;
4. // 广播注册
5. private void registerReceivers() {
6.     ...
7.     // 注册有序广播 1
8.     myOrderReceiver1 = new MyOrderReceiver1(this);
9.     intentFilter = new IntentFilter(MainActivity.ORDER_ACTION);
10.    intentFilter.setPriority(1000);
11.    registerReceiver(myOrderReceiver1, intentFilter);
12.    // 注册有序广播 2
13.    myOrderReceiver2 = new MyOrderReceiver2(this);
14.    intentFilter = new IntentFilter(MainActivity.ORDER_ACTION);
15.    intentFilter.setPriority(200);
16.    registerReceiver(myOrderReceiver2, intentFilter);
17.    // 注册有序广播 3
18.    myOrderReceiver = new MyOrderReceiver(this);
19.    intentFilter = new IntentFilter(MainActivity.ORDER_ACTION);
20.    intentFilter.setPriority(0);
21.    registerReceiver(myOrderReceiver, intentFilter);
22. }
```

在 onClick()方法中添加"有序发送"按钮的处理代码,具体如下:

```
1. public static final String ORDER_ACTION = "com.example.custombroadcast.
                                              ORDER_ACTION";
2. @Override
3. public void onClick(View view) {
4.     Intent intent = new Intent();
5.     String content = etContent.getText().toString();
6.     switch (view.getId()) {
7.         ...
8.         case R.id.btn_ordered:
9.             intent.setAction(MainActivity.ORDER_ACTION);
10.            //有序广播,可被拦截,可终止,可修改数据
11.            sendOrderedBroadcast(intent, null, new MyOrderReceiver
```

```
                              (this), null,
                Activity.RESULT_OK, content, null);
13.         break;
14.     }
15. }
```

第 11 行 sendOrderedBroadcast()是发送有序广播的方法，涉及的参数比较多，分别说明如下。
① intent：传递数据的 Intent 对象。
② receiverPermission：指定广播接收器的权限，没有权限限制则设为 null。
③ resultReceiver：指定最终广播接收器对象，不管广播是否被终止都会执行。
④ scheduler：Handler 类型的对象，此概念将在第 7 章中进行说明，此处设为 null。
⑤ initialCode：传递初始化的整型数，此处设为 Activity.RESULT_OK。
⑥ initialData：传递初始化的 result 字符串数据，广播接收器调用 getResultData()方法接收，此处为输入框的内容；与 getResultData()方法对应的是 setResultData()方法，它可以将数据传递到下一个广播接收器。
⑦ initialExtras：初始化的 Bundle 类型的 result 数据，广播接收器调用 getResultExtras()方法接收 Bundle 数据，此处设为 null。

最后在 onDestroy()方法中注销这 3 个有序广播。

```
1. @Override
2. protected void onDestroy() {
3.     super.onDestroy();
4.     // 注销广播接收器
5.     unregisterReceiver(myNormalReceiver);
6.     unregisterReceiver(myOrderReceiver1);
7.     unregisterReceiver(myOrderReceiver2);
8.     unregisterReceiver(myOrderReceiver);
9. }
```

步骤 7：创建本地广播接收器，在 MainActivity 注册并发送

参考步骤 3 创建本地广播接收器 MyLocalReceiver 类，代码与 MyNormalReceiver 基本雷同。

注册和发送本地广播需要在 app/build.gradle 中添加 LocalBroadcastManager 组件的依赖，代码如下：

```
1. dependencies {
2.     implementation 'androidx.localbroadcastmanager:
       localbroadcastmanager:1.0.0'
3. }
```

创建本地广播与普通广播的最大区别在于，本地广播需要使用 LocalBroadcastManager 类的实例注册、发送、注销广播，相关的代码如下：

```
1.  public static final String LOCAL_ACTION = "com.example.custombroadcast.
                                               LOCAL_ACTION";
2.  private MyLocalReceiver myLocalReceiver;
3.  private LocalBroadcastManager instance;
4.  // 广播注册
5.  private void registerReceivers() {
6.      ...
7.      // 注册本地广播
8.      myLocalReceiver = new MyLocalReceiver(this);
9.      instance = LocalBroadcastManager.getInstance(this);
10.     intentFilter.addAction(MainActivity.LOCAL_ACTION);
11.     instance.registerReceiver(myLocalReceiver, intentFilter);
```

```
12. }
13. @Override
14. public void onClick(View view) {
15.     Intent intent = new Intent();
16.     String content = etContent.getText().toString();
17.     switch (view.getId()) {
18.         ...
19.         case R.id.btn_local:
20.             intent.setAction(MainActivity.LOCAL_ACTION);
21.             intent.putExtra(MainActivity.ACTION_EXTRA_KEY, content);
22.             instance.sendBroadcast(intent);
23.             break;
24.     }
25. }
26. @Override
27. protected void onDestroy() {
28.     super.onDestroy();
29.     ...
30.     // 注销广播接收器
31.     instance.unregisterReceiver(myLocalReceiver);
32. }
```

步骤 8：运行项目

启动模拟器运行项目，输入框输入 5000，分别单击三个按钮，文本框显示的结果如图 6.17 所示。

图 6.17 三种类型广播的运行结果

3 个有序广播的数据传递的日志在 Logcat 中显示的结果如图 6.18 所示。

图 6.18 有序广播数据传递的结果

在 MyOrderReceiver1 中调用 abortBroadcast()方法终止广播，文本框显示的结果如图 6.19

所示，数据传递的日志如图 6.20 所示。

图 6.19　有序广播被截断的结果

图 6.20　有序广播被截断的数据传递

从图 6.18、图 6.20 可以看出，有序广播无论是否截断，最终广播接收器都会接收到广播，这就是 sendOrderedBroadcast()方法中 resultReceiver 对象的含义。

为了方便，本案例只在同一个应用程序中演示了这 3 种广播的使用，大家可以自行创建一个新的应用程序，创建广播接收器，接收普通广播、有序广播发出的消息内容，而本地广播发出的消息内容则不能接收到。由于官方不推荐使用静态注册，所以本案例也未实现，大家可自行完善该案例。

最后提醒一点：当系统或应用程序发出广播时，Android 系统的包管理对象就会检查所有已安装的包配置文件有无匹配的 action，如果有且满足接收条件的 action，那么就调用这个 BroadcastReceiver，获取 BroadcastReceiver 对象并执行 onReceive()方法，BroadcastReceiver 是在 intent 匹配后再实例化的，而且每次都重新实例化。由此可见，广播的生命周期是非常短暂的，不要在广播接收器中执行耗时操作，以免引发系统的 ANR。

6.2.4　最佳实践

● 优先使用本地广播，如果不需要向其他应用的组件发送广播，使用本地广播的效率更高，既无须进行进程间通信，也不会产生任何系统级的广播开销。

● 优先使用动态注册而非静态注册，静态注册的广播可能导致系统启动大量的应用程序，从而对设备性能和用户体验造成严重影响。Android 8.0 及以上版本也对静态注册做出了限制。

● 通过权限限制广播，当注册接收器时，任何应用都可以向你应用的接收器发送潜在的恶意广播。可以在发送广播时指定权限或者发送本地广播。

● onReceive()方法调用返回后就可能被系统回收,如需做耗时操作,应尽量使用 goAsync() 开启后台任务或使用 JobScheduler。

6.3 本章小结

本章主要讲解了 Android 四大组件中的 Service 和 BroadcastReceiver,这两部分的知识点在实际开发中使用频率很高,尤其是处理一些后台任务及全局通知的场景。本章首先讲解了 Service 的基本概念和生命周期,以及如何实现与 Activity 的通信,针对高版本 Android 对后台服务的限制,讲解了前台服务的使用;然后讲解了 Android 的广播机制、对广播接收器 BroadcastReceiver 的注册,广播分类、广播发送和接收也进行了详细讲解,并给出了广播开发的最佳实践。

习 题

一、填空题

1. 广播的发送有两种方式,分别为_____和_____。
2. 代码注册广播需要使用_____方法,解除广播需要使用_____方法。
3. 继承 BroadcastReceiver 会重写_____方法。
4. 广播接收者优先级是在_____属性中声明的。
5. 创建服务时,必须要继承_____类,绑定服务时,必须要实现服务的_____方法。
6. 服务的开启方式有两种,分别是_____和_____。

二、选择题

1. 关于广播的作用,下列说法中正确的是(   )。
   A. 它主要用来接收系统发布的一些消息
   B. 它可以进行耗时的操作
   C. 它可以启动一个 Activity
   D. 它可以帮助 Activity 修改用户界面
2. 下列关于 BroadcastReceiver 的说明中错误的是(   )。
   A. 用来接收广播的 Intent
   B. 一个广播 Intent 只能被一个订阅了此广播的 BroadcastReceiver 所接收
   C. 对有序广播,系统会根据接收者声明的优先级按顺序逐个执行接收者
   D. 广播接收者优先级数值越大优先级别越高
3. 在清单文件中,注册广播使用的节点是(   )。
   A. <activity>   B. <broadcast>
   C. <receiver>   D. <broadcastreceiver>
4. 第一次启动服务会调用(   )方法。
   A. onCreat()   B. onStart()
   C. onResume()   D. onStartCommand()

5. 下列方法中，不属于 Service 生命周期的是（　　）。
   A. onStop()　　　　B. onStart()　　　　C. onResume()　　　　D. onDestory()
6. 下列选项中，属于绑定服务特点的是（　　）（多选题）。
   A. 以 BindService()方法开启　　　　B. 调用者关闭后服务关闭
   C. 必须实现 ServiceConnection()　　D. 使用 stopService()方法关闭服务
7. Service 与 Acitvity 的共同点是（　　）（多选题）。
   A. 都是四大组件之一　　　　　　　B. 都有 onResume()方法
   C. 都可以被远程调用　　　　　　　D. 都可以自定义美观界面

三、简答题

1. 简述 Android 的广播机制及应用场景。
2. 简述 Android 广播的收发方法。
3. 简述调用 startService()或 bindService()方法启动服务的区别。

四、编程题

结合所学知识完成发送短信功能，实现模拟器之间的短信收发。

第 7 章 网络编程

通过前面章节的讲解，学习了如何在程序中使用服务和广播执行一些特定的任务，这类任务一般不体现在与用户的界面交互上，而是在后台完成任务，这也给我们带来了一些全新的认识。本章节将学习 Android 如何与网络进行通信，包括 Android 的多线程、HTTP 通信、数据封装与解析、数据异步刷新等。通过本章节的讲解，应用将可以和网络服务进行数据交互，不再仅限于本地应用。

本章学习目标：

- 了解 Android 的多线程
- 了解 HTTP 协议
- 理解网络通信的方式
- 掌握 Android 的网络通信的实现
- 掌握常用的网络框架的应用

 ## 7.1　Android 的多线程

当 Android 应用程序启动时，系统会启动一个 Linux 进程，该进程包含一个主线程运行所有的组件，此线程也被称为 UI 线程。该线程负责 Android 系统事件、用户输入事件、UI 绘制、服务等任务的串行执行，需要保持较高的响应速度，所以主线程不允许进行耗时的操作，如长时间的网络请求、I/O 操作等，容易导致用户界面卡顿、运行不流畅等问题，引起 ANR（Application Not Responding，应用无响应，卡顿时间超过 5 秒）异常。由于 Android 系统采用的是单线程模型，从 Android 4.0 开始，Android 规定：所有耗时的任务都必须启动子线程（也称为工作线程）实现，以保证 UI 线程的运行不受影响。

7.1.1　多线程的概念

Android 中的多线程本质上就是 Java 的多线程，创建 Java 多线程有 3 种方式。

1. 继承 Thread 类

继承 Thread 类创建线程可以创建自定义类继承自 Thread 类，也可以使用匿名类直接创建示例对象，自定义类可以被复用，而匿名类对象使用简洁但不能被复用，可以根据具体场景进行选择，示例代码如下：

```
1.  public static final String TAG = MainActivity.class.getSimpleName();
2.  @Override
3.  protected void onCreate(Bundle savedInstanceState) {
4.      super.onCreate(savedInstanceState);
5.      setContentView(R.layout.activity_main);
6.      // 方式1：匿名类启动线程
7.      new Thread("Thread匿名类线程") {
8.          @Override
9.          public void run() {
10.             Log.i(TAG, Thread.currentThread().getName());
11.         }
12.     }.start();
13.     // 方式2 启动线程
14.     new MyThread("MyThead").start();
15. }
16. // 方式2：继承Thread类
17. private static class MyThread extends Thread {
18.     public MyThread(@NonNull String name) {
19.         super(name);
20.     }
21.     @Override
22.     public void run() {
23.         Log.i(TAG, Thread.currentThread().getName());
24.     }
25. }
```

2. 实现 Runnable 接口

Java 创建线程只能采用 Thread 类对象，实现 Runnable 接口的类需要通过 Thread 类对象启动线程，实现 Runnable 接口也可以通过自定义类或匿名类两种方式，示例代码如下：

```
1.  public static final String TAG = MainActivity.class.getSimpleName();
2.  @Override
3.  protected void onCreate(Bundle savedInstanceState) {
4.      super.onCreate(savedInstanceState);
5.      setContentView(R.layout.activity_main);
6.      // 方式1：匿名类实现Runnable接口
7.      new Thread(new Runnable() {
8.          @Override
9.          public void run() {
10.             Log.i(TAG, Thread.currentThread().getName());
11.         }
12.     }).start();
13.     // 方式2 启动线程
14.     new Thread(new MyRunnable());
15. }
16. // 方式2：实现Runnable接口
17. static class MyRunnable implements Runnable {
18.     @Override
19.     public void run() {
20.         Log.i(TAG, Thread.currentThread().getName());
21.     }
22. }
```

3. 使用 Callable 接口和 Future 接口

Callable 类似于 Runnable，提供一个 call()方法作为线程执行体，并有返回值，Java 提供

future 接口，它定义了一些公共方法来控制关联的 Callable 任务，获取 call()方法的返回值。Java 提供了实现 Future、Runnable 接口的 FutureTask 类，FutureTask 类的实例可以作为 Thread 的目标对象，这是唯一能有简单返回值的创建线程的方法。示例代码如下：

```
1.  public static final String TAG = MainActivity.class.getSimpleName();
2.  @Override
3.  protected void onCreate(Bundle savedInstanceState) {
4.      super.onCreate(savedInstanceState);
5.      setContentView(R.layout.activity_main);
6.      // 使用 Callback 和 Future 接口
7.      final FutureTask<Integer> task = new FutureTask<>(new
                                          Callable<Integer>() {
8.          @Override
9.          public Integer call() throws Exception {
10.             int I = 10;
11.             Log.i(TAG, Thread.currentThread().getName() + i);
12.             return I;
13.         }
14.     });
15.     // 启动线程
16.     new Thread(task, "有返回值的线程").start();
17.     try {
18.         Log.i(TAG, "task 返回值为" + task.get());
19.     } catch (ExecutionException | InterruptedException e) {
20.         e.printStackTrace();
21.     }
22. }
```

特别注意的是，线程创建时需要调用 start()才能启动，并非调用 run()方法。线程启动后会经历新建、就绪、运行、阻塞和死亡 5 种状态，这些状态之间的转换如图 7.1 所示。

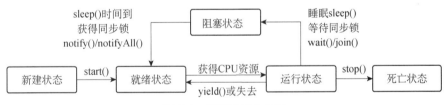

图 7.1　线程的状态转换

由于 Android 的 UI 控件是线程不安全的，多线程的并发访问会给 UI 控件带来不可预期的状态，所以 Android 系统不允许子线程更新 UI 界面控件，所有的界面更新必须在 UI 线程进行。Android 的多线程开发必须遵守以下规则：

- 禁止阻塞 UI 线程。
- 禁止在 UI 线程之外访问 Android UI 组件。

为了实现子线程与 UI 主线程之间的线程间通信，Android 提供了 4 种常用的方法，即 Handler 与 Thread 的组合、ThreadPoolExecutor、AsyncTask、IntentService。

AsyncTask 是 Android 提供的轻量级的异步处理辅助类，用以替代 Handler 与 Thread 的组合，简化多线程编程，将子线程和主线程由一些具有特定职责的方法分别实现，包括预处理方法 onPreExecute()、后台执行任务的方法 onInBackground()、更新进度的方法 publishProgress()和处理返回结果的方法 onPostExecute()，但由于存在 Context 泄露、回调泄露导致崩溃等问题，在 Android 11 版本中 AsyncTask 被正式弃用，Android 给出了替代方案，即使用 java.util.concurrent 包的相关类，如 ThreadPoolExecutor、FutureTask。

IntentService 是一种特殊的 Service，用来接收并处理通过 Intent 传递的异步请求，客户端通过调用 startService（Intent）启动一个 IntentService，利用子线程处理请求，处理完成后自动结束 Service，相对比较简单。

Handler 是 Android 系统提供的一套异步消息传递机制，ThreadPoolExecutor 提供一组线程池，管理多个线程并行执行，提高多线程的效率，接下来重点讲解这两种多线程实现方式。

7.1.2 Handler 消息传递机制

Handler 是 Android 系统提供的异步消息处理机制，用于多线程场景下将子线程中需要更新 UI 的信息传递给 UI 主线程，实现子线程对 UI 的更新操作，保证线程安全。这套机制包括了 Message、ThreadLocal、MessageQueue、Handler 及 Looper 等多线程通信的相关类，这些类的功能描述如表 7.1 所示。

表 7.1　Handler 相关类的功能描述

类名	功能描述
Handler	负责发送和处理消息，实现子线程与 UI 线程之间的消息通信
Looper	每个线程只能有一个消息循环体，负责创建 MessageQueue，循环读取消息并交给 Handler 处理
Message	Handler 接收和处理的消息对象，线程间通信的载体
MessageQueue	单链表结构的消息队列，先进先出管理 Message，初始化 Looper 并创建与之相关的 MessageQueue
ThreadLocal	Handler 内部的数据存储类，保证每个线程拥有同一个类的不同对象，在线程的生命周期内有效

图 7.2 解释了这几个类之间的关系及执行过程，首先，第（1）～（3）步在主线程中创建 Looper 对象、Handler 对象和 MessageQueue 对象，MessageQueue 对象创建之后 Looper 就自动进行消息循环，Handler 对象会自动绑定到主线程的 MessageQueue 和 Looper 对象上；然后第（4）～（5）步子线程使用 Handler 对象发送消息到消息队列中，消息内容就是子线程对 UI 的操作信息；第（6）步当消息循环读取到此消息后就会取出后进行分发，第（7）步 Handler 对象接收到 Looper 发送的消息并进行处理，完成整个消息传递过程。总结一下，Handler 机制的工作流程主要包括 4 个步骤：创建异步通信对象->子线程发送消息->消息循环->主线程消息处理。

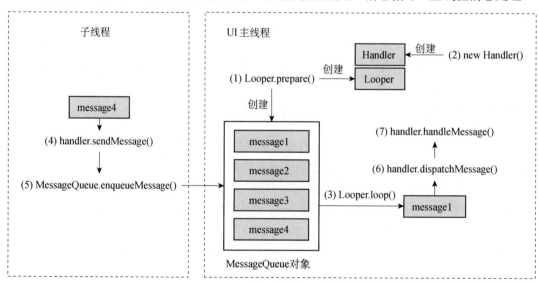

图 7.2　Android 消息机制的执行流程

接下来通过一个案例实现 Handler 的消息传递。

【案例 7-1】计算 1+2+…+100 的值，每隔 100 毫秒计算一次，每累加 5 个数使用 Handler 更新进度条的值，运行效果如图 7.3、图 7.4 所示。

图 7.3 进度条的更新界面

图 7.4 最终的计算结果的界面

步骤 1：创建项目

启动 Android Studio，创建名为 D0701_Multithread 的项目，选择 Empty Activity，将包名改为 com.example.multithread，单击 Finish 按钮，等待项目构建完成。

步骤 2：设计界面布局，初始化界面控件对象

打开 activity_main.xml 文件，增加一个按钮、进度条 ProgressBar 和 TextView，起始界面不显示进度条，TextView 先显示进度条的更新值，计算完成后显示最终计算结果。初始化界面控件的代码如下：

```
1.   private TextView tvMsg;
2.   private ProgressBar progressBar;
3.   @Override
4.   protected void onCreate(Bundle savedInstanceState) {
5.       super.onCreate(savedInstanceState);
6.       setContentView(R.layout.activity_main);
7.       // 初始化控件对象
8.       tvMsg = findViewById(R.id.tv_msg);
9.       progressBar = findViewById(R.id.progress_bar);
10.      Button btnImage = findViewById(R.id.btn_image);
11.      // 设置按钮的单击事件监听器
12.      btnImage.setOnClickListener(new View.OnClickListener() {
13.          @Override
14.          public void onClick(View v) {
15.              // 添加计算的逻辑代码
16.          }
17.      });
18.  }
```

步骤 3：创建 Handler 对象处理消息

将整个运算逻辑分为开始、进行和结果三个部分，分别发送消息给 Handler 进行处理，开始阶段显示进度条，进行阶段每累加 5 个值发送进度条更新消息，运算结束后发送计算结果

并隐藏进度条,代码如下:

```java
1.  // 定义3种消息类型
2.  private static final int START_NUM = 1;
3.  private static final int ADDING_NUM = 2;
4.  private static final int ENDING_NUM = 3;
5.  private final Handler numHandler = new Handler(Looper.getMainLooper()) {
6.      @Override
7.      public void handleMessage(@NonNull Message msg) {
8.          switch (msg.what) {
9.              case START_NUM:
10.                 // 计算开始
11.                 progressBar.setVisibility(View.VISIBLE);
12.                 break;
13.             case ADDING_NUM:
14.                 // 计算进行中
15.                 progressBar.setProgress(msg.arg1);
16.                 tvMsg.setText("已完成" + msg.arg1 + "%");
17.                 break;
18.             case ENDING_NUM:
19.                 // 计算结束
20.                 progressBar.setVisibility(View.GONE);
21.                 tvMsg.setText("计算结果:" + msg.arg1);
22.                 numHandler.removeCallbacks(numThread);
23.                 break;
24.         }
25.     }
26. };
```

Handler 对象直接从 Message 对象获取信息,Message 类包含了消息必要的描述和属性数据,属性字段包括 what、arg1、arg2、obj、replyTo 等,其中 arg1 和 arg2 用来存储整型数据,what 用来保存消息标志,obj 是 Object 类型的任意对象,replyTo 是消息管理器,会关联到一个 Handler。Message 对象通过调用 Handler 的 obtainMessage()或 Message 类的静态方法 obtain()直接获取,不建议使用 new 的方式新建 Message 对象。

步骤 4:创建子线程发送消息

打开 MainActivity 文件,在按钮的 onClick()事件处理方法中创建子线程发送消息,代码如下:

```java
1.  btnImage.setOnClickListener(new View.OnClickListener() {
2.      @Override
3.      public void onClick(View v) {
4.          numThread = new Thread(new Runnable() {
5.              @Override
6.              public void run() {
7.                  int result = 0;
8.                  // 发送空消息
9.                  numHandler.sendEmptyMessage(START_NUM);
10.                 for (int i = 0; i < 101; i++) {
11.                     try {
12.                         Thread.sleep(100);
13.                         result += i;
14.                     } catch (InterruptedException e) {
15.                         e.printStackTrace();
16.                         return;
17.                     }
18.                     if (i % 5 == 0) {
19.                         // 设置 Message 的属性值,并发送消息
```

```
20.                        Message msg = Message.obtain();
21.                        msg.what = ADDING_NUM;
22.                        msg.arg1 = i;
23.                        numHandler.sendMessage(msg);
24.                    }
25.                }
26.                Message msg = numHandler.obtainMessage();
27.                msg.what = ENDING_NUM;
28.                msg.arg1 = result;
29.                numHandler.sendMessage(msg);
30.            }
31.        });
32.        numThread.start();
33.    }
34.});
```

第 20～23 行是组装消息数据和发送的代码，也可以通过调用 Handler 的 sendToTarget() 方法进行简化，如 numHandler.obtainMessage(ADDING_NUM, i).sendToTarget()，其中 i 值是作为 Object 对象发送的，获取消息时要进行类型转换，代码为 progressBar.setProgress((Integer) msg.obj)。

Handler 发送消息的方法大致有以下几种：post（Runnable）、postAtTime（Runnable，long）、postDelayed（Runnable，Object，long）、sendEmptyMessage（int）、sendMessage（Message）、sendMessageAtTime（Message，long）、sendMessageDelayed（Message，long）。

以 post 开头的方法用于指定子线程对象，以匿名内部类的形式发送 Runnable 对象，重写 run() 方法直接对 UI 进行更新；以 send 开头的方法发送 Message 信息，用 Handler 的 handleMessage() 方法接收消息并对 UI 进行更新。Android 切换主线程更新 UI 还提供了一些简化的方法，包括 view.post（Runnable）、view.postDelayed（Runnable，Object，long）、view.postAtTime（Runnable，long）、activity.runOnUiThread（Runnable）。

通过以上的 Handler 工作机制分析及开发实践，对 Android 的 Handler 的线程编程有了较为清晰的理解，最后将线程 Thread、消息循环器 Looper、消息处理器 Handler 之间的关系总结如下：

- 一个线程只能绑定一个循环器，但一个 Thread 可以有多个处理器。
- 一个循环器可绑定多个处理器。
- 一个处理器只能绑定一个循环器。

7.1.3 ThreadPoolExecutor 线程池技术

在 Android 的多线程开发中，当遇到多线程任务众多的情况时，如果为每个任务创建一个线程，就会出现频繁创建、销毁的问题，不仅会占用大量的资源、增加时间，还会因为资源竞争出现问题，降低系统性能，造成界面卡顿。最好的解决方案就是使用线程池。线程池为多个任务重用线程，使得资源开销被分摊到了多个任务上。

使用线程池的好处：

- 降低资源消耗。通过复用线程，减少线程创建、销毁带来的消耗。
- 提高响应速度。当任务到达时能立即执行，无须等待线程的创建。
- 控制线程并发量。
- 提供简单管理。循环执行，统一分配、调优和监控。

Android 的线程池的概念来源于 Java 的 Executor 接口，具体实现类为 ThreadPoolExecutor。一般在创建线程池时，使用 Executor 的子类 ThreadPoolExecutor 创建，通过它的构造方法可以创建不同类型的线程池。线程池的线程分为两种：核心线程和普通线程，核心线程是线程池长期存活的线程，即便闲置也不会被销毁；普通线程有一定的寿命，如果闲置时间超过寿命则会被销毁。

ThreadPoolExecutor 类的构造方法的定义如下：

```
1.  public ThreadPoolExecutor(int corePoolSize,
2.              int maximumPoolSize,
3.              long keepAliveTime,
4.              TimeUnit unit,
5.              BlockingQueue<Runnable> workQueue,
6.              ThreadFactory threadFactory,
7.              RejectedExecutionHandler handler)
```

其中，这 7 个参数的含义介绍如下。

- corePoolSize：线程池中核心线程的数量。
- maximumPoolSize：线程池的最大线程数量。
- keepAliveTime：线程的存活时间。
- unit：枚举类型，表示 keepAliveTime 的单位，常用的取值有：TimeUnit.SECONDS（秒）、TimeUnit.MILLISECONDS（毫秒）等。
- workQueue：暂存任务的阻塞队列，用来保存已经提交但未执行的任务，不同线程池的排队策略不同，类型可以为数组、链表等。
- threadFactory：线程工厂接口，用于创建线程池中的线程，通常使用默认值。
- handler：任务的拒绝策略，线程池关闭或已达最大线程数被拒绝时，会抛出一个 RejectedExecutionException 异常，默认使用 ThreadPoolExecutor.AbortPolicy。

由以上定义可以看出，使用 ThreadPoolExecutor 创建线程池的参数众多，所以官方推荐使用 Executors 类的工厂方法创建，它封装了众多不同功能的线程池，使得创建线程池简单化，常用的线程池类型包括以下几个。

- newFixedThreadPool：固定容量的线程池，可控制线程最大并发数。
- newCachedThreadPool：可缓存的线程池，线程数量无限制。
- newSingleThreadExecutor：单线程的线程池，有且仅有一个子线程，遵循先进先出的规则。
- newScheduledThreadPool：可调度的线程池，支持定时以指定周期循环执行。

线程池的运行主要分成两部分：任务管理和线程管理，任务管理充当生产者的角色，当任务提交后，线程池会根据以下情况进行任务分配；线程管理是消费者的角色，它们被统一分配在线程池内，根据任务请求分配线程，具体流程如图 7.5 所示。

（1）如果运行的线程数少于 corePoolSize，则创建子线程执行任务。

（2）如果运行的线程数大于等于 corePoolSize，且任务队列未满时，将任务加入 BlockingQueue 队列。

（3）如果任务加入 BlockingQueue 队列已满，则创建新的线程处理任务。

（4）如果运行的线程数大于 maximumPoolSize，任务会被拒绝执行，拒绝策略由 RejectedExecutionHandler.rejectedExecution()方法执行。

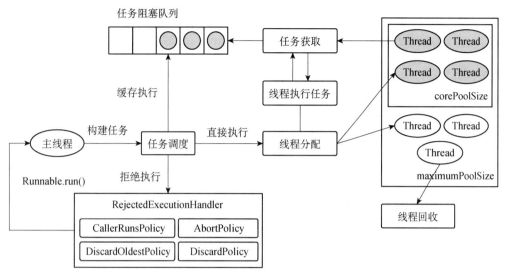

图 7.5 线程池的工作流程

使用 ThreadPoolExecutor 的基本步骤包括以下几步。
（1）创建任务类，重写 run()方法。
（2）使用 Executors 的工厂方法创建线程池。
（3）向线程池提交任务。
（4）关闭线程池。

使用线程池的示例代码如下：

```
1.  @Override
2.  protected void onCreate(Bundle savedInstanceState) {
3.      super.onCreate(savedInstanceState);
4.      setContentView(R.layout.activity_main);
5.      // 使用 Executors 的工厂方法创建固定大小的线程池
6.      final ExecutorService executor = Executors.newFixedThreadPool(2);
7.      try {
8.          for (int i = 1; i <= 3; i++) {
9.              // 分配任务，并将其加入到线程池
10.             String taskName = "任务-" + i;
11.             executor.execute(new Task(taskName));
12.             Log.i(TAG, "put " + taskName);
13.         }
14.         executor.shutdown();
15.         executor.awaitTermination(1, TimeUnit.HOURS);
16.         Log.i(TAG, "线程池中启动的任务已全部完成");
17.     } catch (Exception e) {
18.         e.printStackTrace();
19.     }
20. }
21. //任务类
22. public static class Task implements Runnable, Serializable {
23.     //保存任务所需要的数据
24.     private Object data;
25.     public Task(Object tasks) {
26.         this.data = tasks;
27.     }
28.     public void run() {
29.         // 处理一个任务，输出日志
```

```
30.         Log.i(TAG, Thread.currentThread().getName());
31.         Log.i(TAG, "start " + data);
32.         try {
33.             //便于观察，等待一段时间
34.             Thread.sleep(2000);
35.         } catch (Exception e) {
36.             e.printStackTrace();
37.         }
38.         data = null;
39.     }
40. }
```

以上案例创建了一个任务类 Task 实现 Runnable 接口，在第 28～39 行的 run()方法中日志输出了线程和任务信息，然后使用 Executors.newFixedThreadPool()工厂方法创建了数量为 2 的固定大小线程池，第 8～13 行代码执行了 3 个任务，程序结束时使用 shutdown()方法关闭线程池。关闭线程池有三种方法，包括 shutdown()、shutdownNow()、awaitTermination()。

● shutdown()：将线程池的状态置为 SHUTDOWN，但不会立即停止，需要停止接收外部提交的任务、等待队列中的所有任务执行完之后才会真正停止。

● shutdownNow()：将线程池状态置为 STOP，企图立即停止，但不一定会停止，需要先停止接收外部提交的任务、忽略队列中等待的任务、尝试中断正在运行的任务、返回未执行的任务队列之后会立即停止。

● awaitTermination()：阻塞当前线程，直到所有已提交的任务执行完成、超时时间到或者线程被终端，然后返回 true 或 false。

代码运行后的日志输出的结果如图 7.6 所示，从结果可以看出任务-2 和任务-3 都使用了线程池中的第 2 个线程。

图 7.6　线程池执行任务的输出结果

7.2　WebView 控件

随着 HTML5 的普及，很多 App 都会内嵌 WebView 加载 HTML5 页面，即原生与 HTML5 共存，这就是目前比较流行的"混合式开发"的一种实现方式。WebView 是 Android 提供的原生控件，实现与前端页面信息响应交互，使得 Android 应用能显示网页内容，相当于内置的浏览器。从 Android 4.4 开始 WebView 改为基于 Chromium 浏览器内核，不仅提升了性能，对 HTML5、CSS3 和 JavaScript 也有良好的支持。它除了具有常用 View 的属性之外，还提供了 URL 请求、页面加载、页面渲染及与 JavaScript 交互等功能，可以加载网页、HTML 片段

及与原生 App 交互。

在使用之前，需要在 AndroidManifest.xml 配置文件中增加 INTERNET 网络访问的权限。WebView 的用法非常简单，首先在布局文件中增加 WebView 控件，代码如下：

```
1.  <WebView
2.      android:id="@+id/web_view"
3.      android:layout_width="match_parent"
4.      android:layout_height="match_parent" >
5.  </WebView>
```

然后在 Activity 类文件中添加 WebView 加载网页的代码，如下所示：

```
1.  @Override
2.  protected void onCreate(Bundle savedInstanceState) {
3.      super.onCreate(savedInstanceState);
4.      setContentView(R.layout.activity_main);
5.      // 加载页面
6.      WebView webView = findViewById(R.id.web_view);
7.      webView.loadUrl("https://www.baidu.com/");
8.      // 当前页面打开网页
9.      webView.setWebViewClient(new WebViewClient() {
10.         //重定向 url 请求，返回 true 表示拦截此 url，返回 false 表示不拦截此 url
11.         @Override
12.         public boolean shouldOverrideUrlLoading(WebView view,
                                                    String url) {
13.             // 重定向 url
14.             if (url.startsWith("content://")) {
15.                 url = url.replace("content://", "https://");
16.                 webView.loadUrl(url);
17.             }
18.             // 在本页面的 WebView 中打开，防止外部浏
                   览器打开此链接
19.             view.loadUrl(url);
20.             return false;
21.         }
22.     });
23. }
```

图 7.7　WebView 打开百度首页的界面

加载页面的代码也非常简单，使用 loadUrl()方法加载网址即可显示网页。第 7 行的 setWebViewClient()方法用于加载 WebViewClient 对象，重写它的 shouldOverrideUrlLoading()方法实现在 App 当前界面加载网页，而不是跳转到浏览器打开网页，还可以进行 url 的重定向。运行后就能呈现百度首页的页面，如图 7.7 所示。

WebView 类的常用 API 的定义及功能描述如表 7.2 所示。

表 7.2　WebView 的常用 API 的定义及功能描述

方法类型	方法名称	功能描述
加载类	LoadUrl(String url)	加载 url 地址的方法，url 的取值如下。 网址：https://www.baidu.com； 手机本地 HTML：file:///asset/test.html； asset 目录的 HTML 文件：content://com.example.provider/ test.html
	loadData(String data, String mimeType, String encoding)	加载 HTML 片段，data 为需要截取的内容，mimeType 为内容类型，encoding 为内容的编码方式

续表

方法类型	方法名称	功能描述
状态类	onResume()	生命周期方法，WebView 为活跃状态时调用
	onPause()	生命周期方法，WebView 切换到后台暂停时调用
	destroy()	生命周期方法，WebView 销毁时调用
	pauseTimers()	当应用程序被切换到后台时回调
	resumeTimers()	恢复正常状态时调用
操作网页类	canGoBack()/canGoForward()	是否可以后退/是否可以前进
	goBack()/goForward()	回退/前进
	goBackOrForward(intsteps)	前进或者后退指定的位置，正数为前进/负数为后退
缓存类	clearCache(true)	清除 WebView 产生的所有缓存
	clearFormData()	清除表单数据

WebViewClient 类主要用于辅助 WebView 处理通知、请求等事件，常用方法如表 7.3 所示。如果打开网页时不希望调用系统浏览器，而是在本 WebView 中显示，则需调用 shouldOverrideUrlLoading()方法，如第 12～21 行代码段所示，返回值为 false 表示将 url 交给当前的 WebView 加载，即正常的加载状态，true 则表示 url 已经被处理，WebView 不会再对这个 url 进行加载，可以用于屏蔽一些网址，实现黑名单机制。

表 7.3　WebViewClient 类的常用方法及功能描述

方法名称	功能描述
onPageStarted()	页面开始加载时调用，可以显示加载进度条
onPageFinished()	页面完成加载时调用，隐藏加载进度条，表示页面已经完成加载
onLoadResource()	页面每次加载资源时调用
shouldOverrideUrlLoading()	实现在当前应用内完成页面加载
onReceivedError()	页面加载发生错误时调用，可以跳转到自定义的错误提醒页面
onReceivedHttpError()	页面加载请求时发生错误时调用
onReceivedSslError()	页面加载资源时发生错误时调用
shouldOverrideKeyEvent()	覆盖按键默认的响应事件，根据自身的需求在单击按键时加入相应的逻辑
onScaleChanged()	页面的缩放比例发生变化时调用，可以根据当前的缩放比例重新调整 WebView 中显示的内容，如修改字体大小、图片大小等
shouldInterceptRequest()	可以根据请求携带的内容判断是否需要拦截请求

以上是 WebView 的简单应用，WebView 和 JavaScript 交互的高级应用就不在此处赘述。

7.3　基于 HTTP 的网络访问

7.3.1　HTTP 协议简介

HTTP 协议（HyperText Transfer Protocol，超文本传输协议）是 Internet 应用最为广泛的应用层网络协议,它规定了浏览器和服务器之间的通信规则,基于 TCP/IP 通信协议传递 HTML

文件、网络图片等数据。HTTP 协议是一种请求/响应式的协议,浏览器作为 HTTP 客户端通过 URL(Uniform Resource Location,统一资源定位符)向 HTTP 服务器端发送请求,服务器端接收到请求后做出响应,向客户端发送响应信息,如图 7.8 所示。

图 7.8　HTTP 的传输模式

HTTP 协议的特点包括简单性、无连接性、无状态性和灵活性。

● HTTP 简单性:客户端向服务器端请求服务时,只需传送请求方法和路径,常用的请求方法有 GET、HEAD、POST 等,每种方法规定了请求的类型不同。

● HTTP 无连接性:限制每次连接只处理一个请求,服务器端处理完客户端的请求即断开连接,可以有效节省传输时间。

● HTTP 无状态性:HTTP 协议是无状态的,它不保存请求和响应之间的通信状态,任何两次请求之间没有依赖关系,可以更快速处理大量事务,确保协议的可伸缩性。

● HTTP 灵活性:HTTP 协议允许传输任意类型的数据,传输的内容类型由 Content-Type 标记。

7.3.2　使用 HttpURLConnection

Android 系统发送 HTTP 请求访问网络有两种方式:java.net 提供的 HttpURLConnection 和 org.apache.http 提供的 HttpClient。HttpURLConnection 是一个 Java JDK 自带的、轻量级的 HTTP 客户端,它的 API 简单且体积小,更容易使用和扩展。由于 HttpURLConnection 具备更高的效率,能有效减少响应缓存并较低功耗,因此 Google 官方在 Android 6.0 及以上版本中弃用了 HttpClient。

使用 HttpURLConnection 的流程大致如下。

① 获取 HttpURLConnection 对象:通过调用 URL 对象的 openCollection()方法来实现。

② 调用 HttpURLConnection 对象的方法设置 HTTP 请求方法、连接超时及请求响应头等参数。

③ 调用 getInputStream()方法获得服务器端返回的输入流,利用 Java 的 I/O 流读取信息。

④ 调用 disconnect()方法关闭 HTTP 连接,释放资源。

接下来通过一个案例实现 HttpURLConnection 的用法。

【案例 7-3】使用 HttpURLConnection 获取百度首页的 HTML 文本。

步骤 1:创建项目

图 7.9　HttpURLConnection 请求网络数据

启动 Android Studio,创建名为 D0703_Network 的项目,选择 Empty Activity,将包名改为 com.example.network,单击

Finish 按钮,等待项目构建完成。

步骤 2:设计界面布局,初始化界面控件对象

打开 activity_main.xml 文件,增加一个按钮和带滚动的文本框,展示的内容是百度首页的 HTML 文本,通过浏览器的解析变成网页展示给用户。

由于 Android 对网络请求的限制,需要开启单独的线程发送网络请求。打开 MainActivity 类文件,在 onCreate()方法中初始化控件对象,设置按钮的单击事件监听器,在 onClick()方法中开启子线程,调用自定义方法 doGet()发送网络请求,代码如下:

```java
1.  @Override
2.  protected void onCreate(Bundle savedInstanceState) {
3.      super.onCreate(savedInstanceState);
4.      setContentView(R.layout.activity_main);
5.      // 初始化控件对象,设置事件监听器
6.      tvResponse = findViewById(R.id.tv_response);
7.      Button btnSend = findViewById(R.id.btn_request);
8.      btnSend.setOnClickListener(new View.OnClickListener() {
9.          @Override
10.         public void onClick(View v) {
11.             new Thread(new Runnable() {
12.                 @Override
13.                 public void run() {
14.                     String result = doGet(BAIDU_URL, "");  // 获取网页文本
15.                     if (!TextUtils.isEmpty(result)) {
16.                         showResponse(result);   // 显示在文本框中
17.                     }
18.                 }
19.             }).start();
20.         }
21.     });
22. }
```

步骤 3:编写网络请求方法

按照之前描述的流程,创建 doGet()方法,传递网址和请求参数,使用 HttpURLConnection 发送网络请求,获取请求数据,代码如下:

```java
1.  /**
2.   * GET 方式的网络请求
3.   * @param httpUrl 网址
4.   * @param queryParam 请求字符串,格式为 key1=value1&key2=value2
5.   * @return 服务器端返回的字符串
6.   */
7.  private String doGet(String httpUrl, String queryParam) {
8.      String result = null; // 返回字符串
9.      try {
10.         if (!TextUtils.isEmpty(queryParam)) {
11.             httpUrl += "?" + queryParam;
12.         }
13.         URL url = new URL(httpUrl);
14.         HttpURLConnection conn = (HttpURLConnection) url.openConnection();
15.         // 设置请求方法、连接超时和读超时时间
16.         conn.setRequestMethod("GET");
17.         conn.setConnectTimeout(5000);
18.         conn.setReadTimeout(5000);
19.         // 获取输入流
20.         InputStream is = conn.getInputStream();
21.         // 使用输入流读取网页脚本
```

```
22.          BufferedReader reader = new BufferedReader(new
                                     InputStreamReader(is));
23.          StringBuilder builder = new StringBuilder();
24.          String line = null;
25.          while ((line = reader.readLine()) != null) {
26.              builder.append(line).append(System.getProperty
                                            ("line.separator"));
27.          }
28.          result = builder.toString();
29.          // 关闭资源、断开连接
30.          reader.close();
31.          is.close();
32.          conn.disconnect();
33.      } catch (IOException e) {
34.          e.printStackTrace();
35.      }
36.      return result;
37. }
```

HttpConnection 在调用 getInputStream()时才会建立连接，因此无须调用它的 connect()方法。第 22～27 行调用 BufferedReader 对象读取服务器端返回的数据流，使用 StringBuilder 对象进行存放，数据读完后将数据返回。需要注意的是，doGet()在子线程中被调用，由于 Android 不允许在子线程中进行 UI 操作，因此此处不能将数据直接写入 TextView，而调用 runOnUiThread()方法来实现，代码如下：

```
1. private void showResponse(String result) {
2.     runOnUiThread(new Runnable() {
3.         @Override
4.         public void run() {
5.             tvResponse.setText(result);
6.         }
7.     });
8. }
```

步骤 4：运行项目

由于网络请求涉及用户隐私，在运行之前需要在 AndroidManifest.xml 文件中声明访问权限，否则会发生"Permission denied"的异常，如图 7.10 所示。

```
1. <manifest xmlns:android="http://schemas.android.com/apk/res/android"
2.     package="com.example.network">
3.     <uses-permission android:name="android.permission.INTERNET" />
4. </manifest>
```

图 7.10　网络权限被阻止的异常信息

以上是 GET 请求方式获取服务器端的数据，发送数据给服务器端可使用 POST 请求方式，可以通过调用 HttpURLConnection 的 setReqeustMethod()方法进行设置，获取 OutputStream 对象，然后使用 OutputStream 发送请求数据，示例代码如下：

```
1. URL url = new URL(httpUrl);
2. HttpURLConnection conn = (HttpURLConnection) url.openConnection();
3. // 设置请求方法、连接超时和读超时时间
```

```
4.  conn.setRequestMethod("POST");
5.  conn.setConnectTimeout(5000);
6.  conn.setReadTimeout(5000);
7.  conn.setDoOutput(true);
8.  conn.setDoInput(true);
9.  // post 方式发送参数
10. OutputStream os = conn.getOutputStream();
11. BufferedWriter writer = new BufferedWriter(
12.         new OutputStreamWriter(os, StandardCharsets.UTF_8));
13. writer.write(queryParam);
14. writer.flush();
15. writer.close();
16. os.close();
```

7.3.3 解析 JSON 数据

在网络传输过程中，客户端和服务器端交互需要定义数据交换的格式，才能理解传递的数据的含义，目前最常用的是 JSON 格式，它的全称是 JavaScript Object Natation，含义是 JavaScript 对象表示法，是一种基于文本、独立于语言的轻量级数据交换格式，具备格式简单、体积小、独立于语言和平台及自描述性等优点，网络传输更省带宽和流量，但它的可读性较差，没有 XML 直观。

JSON 数据有对象和数组两种形式，其中，对象数据是一对大括号{}封装的键值对"key:value"，其中关键字 key 为字符串，值 value 可以为字符串、数值、boolean、null、对象或数组。数组数据则使用一对中括号[]进行封装，其中的属性定义与对象相同。使用浏览器的 JSON 格式化工具，层次清晰的结构示例如图 7.11 所示。

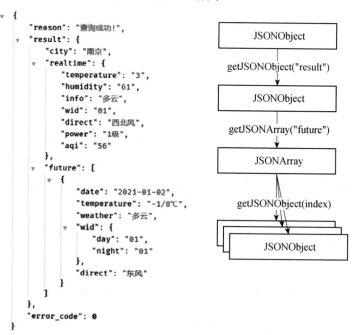

图 7.11 格式良好的 JSON 字符串及 JSONObject 解析

JSON 格式的数据层次清晰，键值对的映射也一目了然，图 7.11 所示的数据的最外层可以解析为 reason、result 和 error_code 三个对象，result 又包含 realtime、city 对象和 future 数组对象，以此类推就可以层层解析 JSON 数据。

解析和生成 JSON 字符串的过程就是数据的反序列化和序列化的过程，反序列化是将字节序列成 Java 对象供项目使用，序列化是将 Java 对象转为字节序列进行网络传输。JSON 序列化和反序列化的方法有很多，官方提供了 JSONObject，Google 提供了 GSON 开源库，还有很多第三方的开源库，如阿里提供的 FastJSON、FasterXML 公司的 Jackson 等。JSONObject 的常用方法如表 7.4 所示，但它的代码较为繁杂，解析数据量大容易出错，解析速度和效率较差，只适合处理层级简单的数据，此处不再详细讲解。GSON、FastJSON 等开源库简化了 JSON 数据的解析，解析方法简单，且性能更高，更适合解析大型、结构复杂的 JSON 数据。

表 7.4 JSONObject 类的常用方法说明

方法名	功能描述
JSONObject()	构造一个空的 JSONObject 对象
JSONObject(str)	根据 JSON 字符串的键值对构造一个 JSONObject 对象
JSONArray(str)	根据 JSON 字符串的键值对构造一个 JSONArray 数组
getJSONArray(str)	返回 str 映射的 JSONArray 数组对象
getJSONObject(str)	返回 str 映射的 JSONObject 对象
getXxx(str)	返回 str 映射的值，Xxx 类型：boolean、double、int、long、String
length()	返回此对象中的名称/值映射的数量
put(str, value)	设置关键字为 String 类型的值，类型可以为基本类型、对象，返回 JSONObject 对象
optXxx(str)	与 getXxx()含义相同，如果不存在则返回默认值，不报异常，推荐使用

GSON 是 Google 提供的快速 JSON 解析和转化工具，Android 官方也推荐使用 GSON，它简化了通过 JSONObject 和 JSONArray 对 JSON 字段逐个解析的烦琐，可以便捷快速解析 JSON 数据，减少出错的概率，提升了代码质量。

GSON 是 Google 推出的支持 JSON 与 Java 对象转换的开源库，它能将字符串直接解析为对象，解析速度更快，项目网站为 https://github.com/google/gson。GSON 提供 toJson()方法来序列化 Java 对象，它可以将 Java 基本类型、对象、集合和 JsonElement 对象等转为 JSON 格式的字符串；fromJson()方法用于反序列化 Java 对象，将 JSON 字符串转为以上类型的 Java 对象，其中，JsonElement 是 GSON 提供的 JSON 对象，它包含 JsonObject、JsonArray、JsonPrimitive 和 JsonNull 等子类。常用的方法定义如表 7.5 所示。

表 7.5 GSON 常用方法的功能描述

类型	方法名称	功能描述
创建实例	Gson()	构造方法创建
	GsonBuilder().create()	使用 GsonBuilder 对象的 create()方法创建
反序列化	fromJson(String json, Class<T> clazz)	将指定的 JSON 反序列化为指定类对象
	fromJson(String json, Type type)	将指定的 JSON 反序列化为指定泛型对象
	fromJson(JsonElement json, Class<T> clazz)	将指定的 JSON 反序列化为指定类对象
	fromJson(Reader json, Class<T> clazz)	从输入流读取 JSON 序列化为指定类对象
序列化	toJson(Object src)	将 Java 对象序列化为 JSON 字符串
	toJson(Object src, Type type)	将 Java 泛型对象序列化为 JSON 字符串
	toJson(JsonElement jsonElement)	将 JsonElement 转换为 JSON 字符串

续表

类型	方法名称	功能描述
JsonElement	getAsBoolean()	元素作为布尔值获取,其他 Java 类型类似
	isJsonNull()	验证此元素是否为 null 值
JsonObject	addProperty(String property, String value)	添加键值对的成员,名称必须是字符串
	getAsJsonObject(String property)	元素作为 JsonObject 获取
	remove(String property)	从 JsonObject 中删除指定属性

在使用 GSON 之前,还需要在项目中导入依赖库:

```
1.  dependencies {
2.      implementation 'com.google.code.gson:gson:2.8.6'
3.  }
```

接下来通过一个案例来讲解使用 GSON 解析 JSON 字符串。

图 7.12 获取天气信息的界面

【案例 7-3(续)】使用 GSON 解析聚合数据网站的天气预报数据。

步骤 1:设计界面布局,添加获取天气数据的控件

打开 D0703_Network 项目,修改 activity_main 的布局,添加一个用于填写城市名称的 EditText 和一个请求按钮。界面布局如图 7.12 所示。

步骤 2:了解聚合数据的天气预报 API

打开聚合数据 API 的 https://www.juhe.cn/docs,注册用户成功后进行登录,登录成功后进入个人中心,申请使用"天气预报"免费数据,获取 API 请求的 key 值;然后打开天气预报的网址 https://www.juhe.cn/docs/api/id/73,查阅提供的测试数据和示例代码。

步骤 3:创建数据类

图 7.4 所示的 JSON 字符串的反序列化,可以对照 JSON 数据自行创建 Java 类,也可以通过 Android Studio 的 GsonFormat 插件快速生成。首先在包名下创建 Weather 类,在类的编辑区域内右击,选择 generate->GsonFormat,打开如图 7.13 所示的对话框,将聚合数据网站的天气预报数据复制到编辑框,单击 OK 按钮得到图 7.14 所示的 Java 类的属性界面,勾选相关属性选项后单击 OK 按钮即可生成相应的 Java 类。

图 7.13 GsonFormat 对话框

图 7.14 生成 Java 类的属性定义界面

生成的代码如下所示：

```java
1.  public class Weather {
2.      private String reason;
3.      private Result result;
4.      @SerializedName("error_code")
5.      private int errorCode;
6.      // 省略 getter/setter 方法
7.      public static class Result {
8.          private String city;
9.          private Realtime realtime;
10.         private List<Future> future;
11.         // 省略 getter/setter 方法
12.         public static class Realtime {
13.             private String temperature;
14.             private String humidity;
15.             private String info;
16.             private String wid;
17.             private String direct;
18.             private String power;
19.             private String aqi;
20.             // 省略 getter/setter 方法
21.         public static class Future {
22.             private String date;
23.             private String temperature;
24.             private String weather;
25.             private Wid wid;
26.             private String direct;
27.             // 省略 getter/setter 方法
28.             public static class Wid {
29.                 private String day;
30.                 private String night;
31.                 // 省略 getter/setter 方法
32.             }
33.         }
34.     }
35. }
```

步骤 4：获取 Weather 数据

根据 API 文档的要求，天气预报 API 的请求参数为城市和申请的 key，而且需要将城市名称的编码转为 UTF-8。在 MainActivity 类中创建 urlencode()方法，它的参数为 Map 类型的键值对，使用 URLEncode.encode()方法将中文转换为 UTF-8 编码，并拼接为网络请求字符串，类似 key=xxx& city=%E8%8B%8F%E5%B7%9E：

```java
1.  private static String urlencode(Map<String, ?> data) {
2.      StringBuilder builder = new StringBuilder();
3.      for (Map.Entry<String, ?> entry : data.entrySet()) {
4.          try {
5.              builder.append(entry.getKey()).append("=")
6.                      .append(URLEncoder.encode(entry.getValue() + "",
                                "UTF-8"))
7.                      .append("&");
8.          } catch (UnsupportedEncodingException e) {
9.              e.printStackTrace();
10.         }
11.     }
12.     String result = builder.toString();
13.     return result.substring(0, result.lastIndexOf("&"));
14. }
```

接着创建 getWeather() 方法获取请求的天气数据，再利用 GSON 将其转为 Weather 对象，然后使用 Handler 更新 UI 控件的值，代码如下：

```
1.  // 天气预报的消息 id、接口地址和请求 key
2.  public final static int WEATHER_ID = 1;
3.  public final static String WEATHER_URL = "http://apis.juhe.cn/
                                              simpleWeather/query";
4.  public static String API_KEY = "换成你申请的key";
5.  private void getWeather(String city) {
6.      new Thread(new Runnable() {
7.          @Override
8.          public void run() {
9.              // 组合请求参数
10.             Map<String, Object> params = new HashMap<>();
11.             params.put("city", city);
12.             params.put("key", API_KEY);
13.             // 获取天气数据
14.             String result = doGet(WEATHERI_URL, urlencode(params));
15.             if (!TextUtils.isEmpty(result)) {
16.                 // 将 JSON 数据转为 Java 对象
17.                 Gson gson = new Gson();
18.                 Weather weather = gson.fromJson(result, Weather.class);
19.                 // handler 发送消息
20.                 Message msg = Message.obtain();
21.                 msg.what = WEATHER_ID;
22.                 msg.obj = weather;
23.                 handler.sendMessage(msg);
24.             }
25.         }
26.     }).start();
27. }
```

第 10～12 行代码段将请求参数使用 HashMap 存储，调用 doGet() 方法发送请求获取 JSON 字符串，第 17～18 行代码段利用 GSON 将 JSON 反序列化为 Weather 对象，GSON 反序列过程非常简单，重点在于 Java 类的设计要符合 JSON 字符串的结构。第 20～23 行代码段使用 Handler 的 message 将天气数据传递给 UI 控件显示，其中 msg.what 为消息 id，msg.obj 为 Weather 对象。

步骤 5：展示天气信息

创建 Handler 对象接收 getWeather() 传递的数据，将 Weather 对象的数据组装成 TextView 要显示的字符串。

```
1.  private final Handler handler = new Handler(Looper.getMainLooper()) {
2.      @Override
3.      public void handleMessage(@NonNull Message msg) {
4.          if (msg.what == WEATHER_ID) {
5.              Weather weather = (Weather) msg.obj;
6.              if (weather.getErrorCode() == 0) {
7.                  String result =
8.                      "\n 当前温度: " + weather.getResult().getRealtime().
                            getTemperature() +
9.                      "\n 当前天气: " + weather.getResult().getRealtime().
                            getInfo() +
10.                     "\n 空气质量: " + weather.getResult().getRealtime().
                            getAqi() +
11.                     "\n 明天天气: " + weather.getResult().getFuture().get(0).
                            getWeather() +
```

```
12.             "\n 明天温度: " + weather.getResult().getFuture().get(0).
                    getTemperature()+
13.             "\n 后天天气: " + weather.getResult().getFuture().get(1).
                    getWeather() +
14.             "\n 后天温度: " + weather.getResult().getFuture().get(1).
                    getTemperature();
15.         tvResponse.setText(result);
16.     }
17.   }
18. }
19. };
```

步骤 6：实现按钮的单击事件

最后，在 onCreate()方法中添加按钮的单击事件的实现代码，代码如下：

```
1.  @Override
2.  protected void onCreate(Bundle savedInstanceState) {
3.      ...
4.      EditText etCity = findViewById(R.id.et_city);
5.      Button btnWeather = findViewById(R.id.btn_weather);
6.      btnWeather.setOnClickListener(new View.OnClickListener() {
7.          @Override
8.          public void onClick(View v) {
9.              String city = etCity.getText().toString();
10.             if (!TextUtils.isEmpty(city)) {
11.                 getWeather(city);
12.             }
13.         }
14.     });
15. }
```

最后需要注意的是，从 Android 9.0 开始强制使用 https 请求，所有的 http 请求都会被阻塞，有两种常用的解决方案，第一种方案最简单，在 AndroidManifest.xml 清单文件中为 application 节点添加属性 android:usesCleartextTraffic="true"，同时需要将 build.gradle 文件中的 SDK 的最低版本设为 23。

```
1.  <application
2.      ...
3.      android:usesCleartextTraffic="true"
4.      ... />
```

第二种方案是配置网络安全性，在 res 目录上右击，选择 New->Android Resource File，打开创建资源文件的对话框，创建资源类型为 XML 的资源文件，文件名为 network_security_config.xml，设置如下属性：

```
1.  <?xml version="1.0" encoding="utf-8"?>
2.  <network-security-config>
3.      <!-- 默认配置，明文通信，使用系统证书 -->
4.      <base-config cleartextTrafficPermitted="true">
5.          <trust-anchors>
6.              <!-- trust system while release only -->
7.              <certificates src="system" />
8.          </trust-anchors>
9.      </base-config>
10. </network-security-config>
```

然后打开 AndroidManifest.xml 文件，给 application 节点添加属性 android:networkSecurityConfig="@xml/network_security_config"。

至此，使用 HttpURLConnection 和 GSON 获取、解析和展示网络数据的功能就完成了，接下来学习网络通信框架的应用。

7.4 网络访问框架

7.4.1 OkHttp 框架

OkHttp 是由 Square 公司提供的 HTTP 网络请求的开源客户端，是 Android 开发公认的最好网络请求框架，从 Android 4.4 开始 HttpURLConnection 的底层也采用 OkHttp 框架实现，它的项目地址为 https://square.github.io/okhttp。它具备以下良好的特性：
- 支持 HTTPS 和 HTTP 请求，以及对 HTTP/2 的支持。
- 使用连接池降低请求网络延迟。
- 透明的 GZIP 压缩减少响应数据的大小，节省流量。
- 缓存响应内容，避免重复的网络请求。
- 支持同步阻塞调用和异步回调调用。

OkHttp 网络请求包含的重要的类的描述如表 7.6 所示。

表 7.6　OkHttp 库的类和接口的说明

类或接口	功能描述
OkHttpClient	用于创建 Call 对象
OkHttpClient.Builder	封装 OkhttpClient 对象的请求配置参数，如超时时间、拦截器等
Request	封装了请求报文信息，包括 URL、请求方法、请求头等
Request.Builder	Request 的构造器类
RequestBody	封装了请求数据，作为 POST 请求方式的请求体
FormBody	封装了表单数据，作为 POST 请求方式的请求体
Call 接口	发送 HTTP 请求，是连接 Request 和 Response 的桥梁，同步发送使用 execute()方法，异步发送使用 enqueue()方法。enqueue()方法由于不阻塞线程，所以无须启动子线程发送网络请求，而同步方法则必须启动子线程
Response	封装了响应报文信息，包括状态码、响应头和响应体等

在使用 OkHttp 之前需要在 AndroidManifest.xml 文件中添加网络请求权限，如果使用了 OkHttp 中的 DiskLruCache 缓存，还需要声明存储权限，然后在项目中引入 OkHttp 的依赖库：

```
1.  dependencies {
2.      implementation 'com.squareup.okhttp3:okhttp:4.9.0'
3.  }
```

OkHttp 的使用过程比较简单，使用步骤如下：
① 创建 OkHttpClient 对象。
② 创建携带请求信息的 Request 对象。
③ 创建 Call 对象。
④ Call 发起同步或异步请求。

首先，创建一个 OkHttpClient 对象和 Request 对象，调用 OkHttpClient 的 newCall()方法

创建 Call 对象，调用 Call 的 enqueue()方法提交异步 GET 请求，请求被加入 Dispatcher 的 runningAsyncCalls 双端队列中，通过线程池执行，服务器返回的数据通过 Callback 接口的两个回调方法获取，onFailure()方法处理异常情况，onResponse()方法处理响应数据，通过它的参数 Response 对象的 body()用于获取响应主体的数据，再调用 string()方法转为字符串。发送 GET 请求的示例代码如下：

```
1.  // 创建 OkHttpClient 对象
2.  OkHttpClient client = new OkHttpClient().Builder().build();
3.  // 构建 Request 对象
4.  Request request = new Request.Builder().url(httpUrl).build();
5.  // 构建 Call 对象，发送异步请求
6.  client.newCall(request).enqueue(new Callback() {
7.      @Override
8.      public void onFailure(@NotNull Call call, @NotNull IOException e) {
9.          // 处理异常信息
10.     }
11.     @Override
12.     public void onResponse(@NotNull Call call, @NotNull Response response)
13.         throws IOException {
14.         // 处理服务器返回的响应数据
15.         if (response.isSuccessful()) {
16.             String body = response.body().string(); // 获取响应数据
17.         }
18.     }
19. });
```

OkHttp 的 POST 请求可以使用 RequestBody 封装 JSON 字符串、表单数据使用 FormBody 封装，上传文件可通过 MultipartBody 类实现。首先使用 RequestBody 的 create()方法来生成请求体，然后在构建 Request 对象时调用 post()方法来加载 RequestBody 对象，示例代码如下：

```
1.  // 设置传输数据的媒体类型
2.  MediaType mediaType = MediaType.Companion.parse("application/
                         json;charset=utf-8");
3.  // 创建请求体数据
4.  RequestBody body = RequestBody.Companion.create(json, mediaType);
5.  // P 创建 POST 请求
6.  Request request = new Request.Builder().post(body).url(httpUrl).build();
```

通过以上的代码示例，OkHttp 极大简化了网络请求的编码，但在使用过程中需要注意以下几点。

● 使用 OkHttpClient.Builder 来创建 okHttpClient 对象，使得该对象保持单例，使用同一个 OkHttpClient 对象执行所有请求，这是因为每个 OkHttpClient 实例都拥有自己的连接池和线程池，重用这些资源可以减少延时和节省资源。

● 每个 Call 只能执行一次，不能重复利用，否则会报异常。

● 在 onFailure()和 onResponse()回调方法中不能更新 UI 控件。

7.4.2 Glide 图片加载框架

Glide 图片加载框架是用于 Android 应用程序的快速高效的图片加载和缓存库，提供了高性能、可扩展的图片解码管道、资源池技术及易用的 API 库，具备以下特点。

● 加载类型多样化：Glide 支持图片、gif 动画、webp 图片等。

- 图片来源丰富，支持网络、文件、应用资源、二进制流和 Uri 对象等。
- 生命周期绑定：图片请求与页面生命周期绑定，实现动态管理。
- 使用链式调用方式，代码简洁。
- 提供丰富、灵活的 API，可以自定义图片。
- 高效的缓存和本地存储效率策略，支持多种缓存策略，内存开销小。

使用 Glide 之前，首先在 app/build.gradle 文件中添加依赖库，然后在 AndroidMainfest.xml 中添加网络访问的权限：

```
1. dependencies {
2.     implementation "com.github.bumptech.glide:glide:4.11.0"
3.     annotationProcessor 'com.github.bumptech.glide:compiler:4.11.0'
4. }
```

相比 Glide 3，Glide 4 新增了 compiler 注解依赖库，这个库用于生成 Generated API，它是 Glide 4 新引入的功能，使用注解处理器生成一个 API，在 Application 模块中可使用该 API 一次性地调用到 RequestBuilder、RequestOptions 和集成库所有的选项，也就是提供一套与 Glide 3 一致的流式 API 接口。

首先为应用运用程序创建一个带有@GlideModule 注解的、继承自 AppGlideModule 的类，代码如下所示。单击菜单 Builde->Rebuild project，编译生成与 GlideAppMoudle 在同一个包下的 GlideApp 类，通过 GlideApp.with()方式使用 Glide 4 的 Generated API。

```
1. @GlideModule
2. public class GlideAppModule extends AppGlideModule {
3.
4. }
```

在 Activity 中使用 GlideApp 加载图片资源，示例代码如下：

```
1. ImageView ivPic = findViewById(R.id.iv_pic);
2. GlideApp.with(this)
3.     .load("https://cdn.pixabay.com/photo/2020/12/25/09/46/
           dog-5858985__340.jpg")
4.     .into(ivPic);
```

上述代码中的 with()方法用于创建一个加载图片的对象，并将图片加载与 Activity 的生命周期进行绑定，当 Activity 的生命周期结束销毁时，图片加载也会停止，避免内存泄露。load()方法用于指定待加载的图片。into()方法用于将图片加载到哪个 ImageView 控件上。

除了这关键的三步骤之外，Glide 还提供了设置占位符、缓存策略、指定图片格式和大小、自定义裁剪等 API 功能，这些功能的方法描述如表 7.7 所示。

表 7.7 Glide 常用方法的功能描述

类型	方法名	功能描述
占位符	placeholder()	图片的加载过程中临时显示的图片，加载成功后替换掉
	error()	图片加载失败显示的图片
图片格式	asBitmap()	加载 Bitmap 图片
	asGif()	加载 Gif 动画
图片大小	override()	指定图片的尺寸
图片变换	transforms()	内置转换：circleCrop、fitCenter、centerCrop
缓存策略	diskCacheStrategy()	设置磁盘缓存策略，取值：NONE、DATA、RESOURCE、ALL、AUTOMATIC

续表

类型	方法名	功能描述
	skipMemoryCache(true)	取消内存缓存功能，默认已开启
动画效果	crossFade()	淡入淡出动画效果
	dontAnimation()	取消任何淡入淡出效果

使用这些方法的实例代码如下：

```
1.   GlideApp.with(this)
2.           .load("https://cdn.pixabay.com/photo/2020/03/31/19/20/
                dog-4988985__340.jpg")
3.           .centerCrop()    // 居中
4.           .circleCrop()    // 圆形
5.           .skipMemoryCache(true)    // 关闭内存缓存
6.           .diskCacheStrategy(DiskCacheStrategy.NONE)  // 关闭磁盘缓存
7.           .override(Target.SIZE_ORIGINAL)   // 指定尺寸
8.           .placeholder(R.drawable.pic)      // 临时占位符
9.           .error(R.drawable.ic_error)       // 错误占位符
10.          .into(ivPic);
```

 ## 7.5 本章小结

本章主要讲解了 Android 的多线程和网络编程的相关知识，由于 Android 对网络访问的线程要求，需要使用 Handler 传递异步数据，以及使用 ThreadPoolExecutor 类对多线程进行管理。Android 采用 HttpURLConnection 类进行网络访问及对 JSON 数据的解析，同时也讲解了主流的网络框架 OkHttp 和 Glide 的使用方法。通过本章的学习，希望读者能开发一些基于网络数据的较为复杂的应用程序，为今后开发商用的应用打下基础。

习　题

一、选择题

1. 下列关于多线程下载的说法中错误的是（　　）。
A. 多线程下载一般建议开 3～5 个线程，线程不要过多
B. 多线程下载受真实服务器带宽的影响
C. 多线程加速下载原理是单位时间内抢到服务器的资源越多得到的 CPU 资源越多
D. 多线程下载开的线程越多下载越快

2. 下面关于 HttpURLConnection 访问网络的描述中，正确的是（　　）。
A. 以 GET 方式访问网络 URL 的内容一般要大于 1KB
B. 以 GET 的方式提交的数据要比 POST 的方式相对安全
C. 使用 HttpURLConnection 访问网络时需要设置超时时间，防止连接被阻塞时无响应
D. 使用 GET 方式提交数据时，用户通过浏览器无法看到发送的请求数据

3. 下列选项中，属于 HttpURLConnection 提交数据后请求成功的状态码的是（　　）。
A. 100　　　　　　B. 200　　　　　　C. 404　　　　　　D. 500

4. 下列选项中，不属于 JSON 数据的是（　　）。

A. {"Provice":"Jiangsu","City":"Nanjing"}

B. ["abc", 12345, false, null]

C. [{"name":"LiYan","age":20}]

D. {"abc", 12345, false, null}

5. 下列对 Handler 的描述中正确的是（　　）。

A. 它可以在内部携带少量的信息，用于在不同线程之间交换数据

B. 它主要用于发送消息和处理消息

C. 它主要用来存放通过 Handler 发送的消息

D. 每个线程中的 MessageQueue 的管家

6. 关于 Android 中消息机制的说法中正确的是（　　）。

A. Handler 只能用来发送消息

B. Handler 是用来发送消息和处理消息的

C. MessageQueue 是用来收集消息并主动发送消息的

D. Looper 是主消息的循环器，Looper 是由 handler 创建的

二、简答题

1. 简述三种实现多线程方法的区别。

2. 简述使用 HttpURLConnection 访问网络的步骤。

3. 简述 Handler 消息传递机制的原理。

4. 简述 OkHttp 网络框架的基本原理。

三、编程题

1. 使用两种多线程的实现方法，完成 2 个窗口同时卖火车票；每个窗口卖 100 张，卖票速度都是 1s/张的功能。

2. 使用 OkHttp 框架重构【案例 7-3】的天气预报。

第 8 章　多媒体开发

经过前面章节的学习我们已经能开发一个较为完善的 Android 应用程序，本章将讲解 Android 多媒体开发的知识，包括音视频的播放、动画技术等，为应用程序提供更为丰富、炫酷的用户体验。

本章学习目标：

- 了解多媒体的基本概念
- 掌握 Android 音视频的播放方法
- 掌握 Android 三种动画的使用方法
- 理解各种动画的应用场景

 ## 8.1　多媒体简介

多媒体应用是智能手机主流的应用类型，听音乐、看电影、短视频等众多的娱乐方式都少不了多媒体功能的支持，Android 提供了常见的音频、视频的编解码机制，可以很方便地播放音视频。

多媒体的英文为 Multimedia，是文本、声音、图片、动画和影片等多种资源的综合。计算机系统中的多媒体是指两种或两种以上资源组合的一种人机交互式信息交流和传播媒体。多媒体系统是 Android 最为庞大的系统之一，涉及硬件抽象层、编解码、底层多媒体框架、Android 多媒体框架和应用层接口等多方面的内容，本节只涉及 Android 应用层的开发，包括音频 Audio 和视频 Video 的播放、动画过渡。

Android 应用层的音视频的接口都在 android.media 包内，主要包括以下几个。

- MediaPlayer 类：播放声音和视频的主要 API，提供获取、解码和播放等功能。
- AudioManager 类：管理设备上的音频源和音频输出。
- MediaRecorder：提供音频、视频录制等。
- AudioRecord：提供音频的录音接口，默认编码为 PCM_16_BIT。
- SoundPool：提供几种声音同时播放的功能，适用于游戏场景。
- Ringtone：提供系统自带的声音文件的播放功能，适用于系统提示音。

8.2 音频播放

MediaPlayer 类是 Android 多媒体框架最重要的组成部分，常用方法如表 8.1 所示，只需要极少的配置就能获取、解码及播放音频和视频，支持多种不同的媒体源，如本地资源、内部 URI 资源、外部或网络资源。

表 8.1 MediaPlayer 类的常用方法

方法名称	功能描述
create()	播放本地 res/raw/目录下的资源
setDataSource()	设置音频源，本地、网络资源均可
setAudioStreamType()	设置音乐格式，如 AudioManager.STREAM_MUSIC
prepare()	播放前的准备工作
start()	开始或恢复播放
pause()	暂停播放
reset()	重置为初始状态
seekTo()	拖曳进度
stop()	停止播放
release()	释放资源
isPlaying()	判断当前是否正在播放
getDuration()	获取音乐的最大长度（单位：毫秒）
getCurrentPosition()	获取当前的播放进度
setOnCompletionListener(onCompletionListener)	监听播放完成的事件
setOnErrorListener(OnErrorListener)	监听播放发生错误时的事件
setOnPreparedListener(OnPreparedListener)	监听视频装载完成的事件

使用 MediaPlayer 可以简单概括为获取 MediaPlayer 的实例、加载播放源、播放音频、控制播放、释放资源，实现步骤如下。

（1）实例化 MediaPlayer 类，加载要播放的媒体

获取 MediaPlayer 对象有两种方式：使用 new 新建 MediaPlayer 的实例和调用 MediaPlayer 的静态方法 create()，两者的区别在于使用 create()方法会直接加载播放文件，使用 new 创建 MediaPlayer 对象，需要设置播放文件。

```
1.  // 创建 MediaPlayer 对象
2.  MediaPlayer player = new MediaPlayer();
3.  // 设置音频类型
4.  player.setAudioStreamType(AudioManager.STREAM_MUSIC);
```

setAudioStreamType()方法用于设置音频类型，如 AudioManager.STREAM_MUSIC 为音乐类型、AudioManager.STREAM_ALARM 为闹钟类型等。

（2）设置播放文件

音频文件的来源可以为本地、SD 卡或网络，分别使用 MediaPlayer.create()静态方法加载 res/raw 目录下的音频文件，使用 setDataSource()实例方法加载本地文件或网络文件，示例代码如下：

```
1.  // 设置 res/raw 目录的本地播放源
2.  player = MediaPlayer.create(getApplicationContext(), R.raw.music);
3.  // 设置本地文件
4.  player.setDataSource("/sdcard/one.mp3");
5.  // 设置网络 URL 文件
6.  player.setDataSource("https://www.music.com/one.mp3");
```

需要注意的是，播放 SD 卡或网络的音频文件，需要在 AndroidManifest.xml 配置清单文件中添加 SD 读写、网络访问的权限。

（3）播放或暂停音频文件

使用 setDataSource()初始化音频文件之后，在播放之前必须调用 prepare()或 prepareAsync()将音频文件解析到内存，进入准备状态之后调用 start()方法进行播放。通过调用 start()、pause()和 seekTo()等方法在 start、pause 和 playbackCompleted 状态之间切换。但是，当调用 stop()后再重新准备 MediaPlayer 之前，不能再次调用 start()。

（4）播放完成

MediaPlayer 对象会消耗系统资源，播放完成后必须调用 release()方法释放资源。

接下来通过使用 MediaPlayer 播放音频的案例了解它的使用。

【案例 8-1】使用 MediaPlayer 创建一个简单的音乐播放器，可以播放、暂停、继续，初始界面如图 8.1 所示，使用 SeekBar 控件显示播放进度，单击播放按钮，将当前播放的进度和歌曲时长使用 TextView 控件显示，播放按钮也变为暂停按钮，播放运行界面如图 8.2 所示。

图 8.1 音乐播放器的初始界面

图 8.2 音乐播放器的播放界面

步骤 1：创建项目

启动 Android Studio，创建名为 D0801_MusicPlayer 的项目，选择 Empty Activity，将包名改为 com.example.musicplayer，单击 Finish 按钮，等待项目构建完成。

步骤 2：设计界面布局，初始化界面控件对象

首先，在 res 目录下创建 raw 目录，将一个 mp3 音频文件复制到此目录。

然后，在 MainActivity 类中创建 initView()方法初始化界面的控件，使用 Glide 框架加载图片，使用 circleCrop()方法设置图片为圆形，并设置三个按钮和进度条的监听事件，然后在 onCreate()中调用此方法，代码如下：

```
1.  // 初始化控件对象，设置单击事件监听器
2.  private TextView tvCurrent;
3.  private TextView tvTotal;
4.  private SeekBar seekBar;
5.  private ImageButton btnPlay;
6.  private MediaPlayer player;
7.  private void initView() {
8.      // 初始化控件
9.      tvCurrent = findViewById(R.id.tv_current);
10.     tvTotal = findViewById(R.id.tv_total);
11.     seekBar = findViewById(R.id.seek_bar);
12.     ImageView ivBackground = findViewById(R.id.iv_bg);
13.     // 设置图片为圆形，并居中显示
14.     GlideApp.with(this)
15.             .load(ResourcesCompat.getDrawable(getResources(),
16.                     R.drawable.background, this.getTheme()))
17.             .centerCrop()
18.             .circleCrop()
19.             .into(ivBackground);
20.     // 设置按钮的单击事件监听器
21.     ImageButton btnPrev = findViewById(R.id.btn_start);
22.     ImageButton btnEnd = findViewById(R.id.btn_end);
23.     btnPlay = findViewById(R.id.btn_play);
24.     btnPrev.setOnClickListener(this);
25.     btnPlay.setOnClickListener(this);
26.     btnEnd.setOnClickListener(this);
27.     // 进度条拖动事件监听
28.     seekBar = findViewById(R.id.seek_bar);
29.     seekBar.setOnSeekBarChangeListener(new SeekBar.
                                             OnSeekBarChangeListener() {
30.         @Override
31.         public void onProgressChanged(SeekBar seekBar,int progress,
                                             boolean fromUser){
32.             if (player == null) {
33.                 seekBar.setProgress(0);
34.             } else {
35.                 player.seekTo(progress);
36.             }
37.         }
38.         @Override
39.         public void onStartTrackingTouch(SeekBar seekBar) {
40.             if (player == null) {
41.                 Toast.makeText(MainActivity.this, "音乐播放器还未开始",
42.                     Toast.LENGTH_SHORT).show();
43.             }
44.         }
45.         @Override
46.         public void onStopTrackingTouch(SeekBar seekBar) {
47.         }
48.     });
49. }
```

步骤 3：创建 Handler 对象更新进度条

每隔 1 秒发送一条当前播放位置的消息给 Handler 对象，播放位置通过调用 MediaPlayer 的 getCurrentPosition()方法获取，调用 Handler 对象的 sendMessageDelayed()方法延时 1 秒发送，代码如下：

```
1.  private void updateProgress() {
2.      // 使用 Handler 每间隔 1s 发送一次消息，通知进度条更新
```

```
3.        Message msg = Message.obtain();
4.        // 使用 MediaPlayer 获取当前播放时间
5.        msg.arg1 = player.getCurrentPosition();
6.        handler.sendMessageDelayed(msg, 1000);
7.    }
```

创建 Handler 对象接收消息，更新 SeekBar 的进度及当前播放的时间，时间格式为"mm:ss"，然后调用 updateProgerss()方法发送消息，代码如下：

```
1.  private final Handler handler = new Handler(Looper.getMainLooper()) {
2.      @Override
3.      public void handleMessage(@NonNull Message msg) {
4.          // 更新进度条和当前时间
5.          int progress = player.getCurrentPosition();
6.          seekBar.setProgress(progress);
7.          tvCurrent.setText(formatTime(progress));
8.          // 继续定时发送数据
9.          updateProgress();
10.     }
11. };
```

时间格式的设置通过创建 formatTime()方法实现，代码如下：

```
1.  // 格式化时间
2.  public static String formatTime(int time) {
3.      SimpleDateFormat sdf = new SimpleDateFormat("mm:ss", Locale.CHINA);
4.      return sdf.format(new Date(time));
5.  }
6.
```

步骤 4：创建 MediaPlayer 播放音频

了解 MediaPlayer 的使用方法之后，在 MainActivity 类中创建 start()方法播放功能，主要包括三个部分的功能实现：

- 设置播放源、播放音频文件，设置播放结束的事件监听器。
- 设置 SeekBar 的进度及更新播放时间。
- 更改播放按钮的图片，转变为暂停按钮的图片。

代码如下：

```
1.  private void start() {
2.      if (player == null) {
3.          // 设置播放源，并开始播放
4.          player = MediaPlayer.create(getApplicationContext(), R.raw.music);
5.      }
6.      player.start();
7.      // 设置播放结束的监听器
8.      player.setOnCompletionListener(new MediaPlayer.OnCompletionListener() {
9.          @Override
10.         public void onCompletion(MediaPlayer mp) {
11.             seekBar.setProgress(0); // 还原 seekbar
12.             Log.i(TAG, "播放完成");
13.             btnPlay.setBackground(ResourcesCompat.getDrawable(
14.                 getResources(),R.drawable.btn_play,MainActivity.
                    this.getTheme()));
15.         }
16.     });
17.     // 设置进度条，并更新
18.     int totalTime = player.getDuration();
19.     seekBar.setProgress(0);
```

```
20.        seekBar.setMax(totalTime);
21.        tvTotal.setText(formatTime(totalTime));
22.        updateProgress();
23.        // 更新按钮图片
24.        if (player.isPlaying()) {
25.            btnPlay.setBackground(ResourcesCompat.getDrawable(
26.                    getResources(), R.drawable.btn_pause, this.getTheme()));
27.        } else {
28.            btnPlay.setBackground(ResourcesCompat.getDrawable(
29.                    getResources(), R.drawable.btn_play, this.getTheme()));
30.        }
31. }
```

步骤 5：处理按钮事件

按钮的事件处理主要是播放按钮的事件处理，如果 player 对象为 null，则调用 start()方法创建 player 对象，如果 player 对象处于播放状态则改为暂停状态，并修改按钮图片为暂停图片，反之，则修改 player 对象的状态为播放状态，按钮图片改为播放图片，代码如下：

```
1.  @Override
2.  public void onClick(View view) {
3.      switch (view.getId()) {
4.          case R.id.btn_start:
5.              Toast.makeText(this, "第一首歌曲", Toast.LENGTH_SHORT).
                 show();
6.              break;
7.          case R.id.btn_play:
8.              if (player == null) {
9.                  start();
10.             } else if (player.isPlaying()) {
11.                 player.pause();
12.                 btnPlay.setBackground(ResourcesCompat.getDrawable(
13.                         getResources(), R.drawable.btn_play, this.
                            getTheme()));
14.             } else {
15.                 player.start();
16.                 btnPlay.setBackground(ResourcesCompat.getDrawable(
17.                         getResources(), R.drawable.btn_pause, this.
                            getTheme()));
18.             }
19.             break;
20.         case R.id.btn_end:
21.             Toast.makeText(this, "最后一首", Toast.LENGTH_SHORT).show();
22.             break;
23.     }
24. }
```

步骤 6：释放资源

重写 Activity 的 onDestroy()方法，释放 MediaPlayer 资源，代码如下：

```
1.  @Override
2.  protected void onDestroy() {
3.      super.onDestroy();
4.      // 在activity销毁时回收资源
5.      if (player != null && player.isPlaying()) {
6.          player.stop();
7.          player.release();
8.          player = null;
9.      }
10. }
```

通过以上案例，完成了 MediaPlayer 播放音频文件的基本流程，需要注意的是 MediaPlayer 是基于状态的，包括 Idle、Initialized、Prepared、Started、Paused 和 PlaybackCompleted 等状态，如图 8.3 所示。

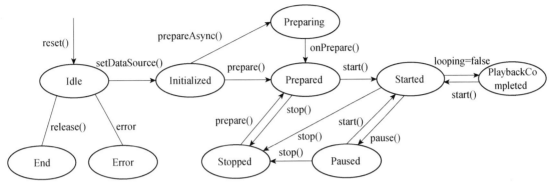

图 8.3 MediaPlayer 的状态转换

编码过程必须遵守 MediaPlayer 的状态转换调用相应的方法，状态切换时需要注意以下几点：

① Started->Paused->Stopped 的转换是单向的，无法再从 Stopped 直接转换到 Started，需要进入 Prepared 状态重新装载，才可以重新播放。

② Initialized 状态需要装载数据，才可以调用 start()播放，如果使用 prepareAsync()方法异步准备，需要等待准备完成后再开始播放，需要使用回调方法 setOnPreparedListener()进行监听。

③ End 状态游离在其他状态之外，在任何状态皆可切换，一般在不需要继续使用 MediaPlayer 时，才会使用 release()回收资源。

④ Error 状态也游离在其他状态之外，只有在 MediaPlayer 发生错误时才会转换，通常使用回调方法 setOnErrorListener()监听错误的处理。

在实际项目开发中，一般使用 Service 播放音频，以便于在锁屏或后台播放音频，而不是直接在 Activity 中使用。

8.3 视频播放

Android 提供了常见的视频编解码机制，使用 MediaPlayer 类也可以方便地实现视频播放的功能，但 MediaPlayer 没有提供图像输出的界面，常见的视频输出可以使用 VideoView、SurfaceView、TextureView 等 View 组件。

8.3.1 VideoView

VideoView 是 Android 实现视频播放的最简单方式，借助它可以实现一个简单的视频播放器。VideoView 类可以从本地、网络等不同的来源读取视频文件，计算和维护视频的画面尺寸，以使其适用于任何布局的管理器，VideoView 支持 mp4、avi、3gp 等格式的视频。VideoView 提供的视频播放的方法如表 8.2 所示。

表 8.2 VideoView 控件的常用方法

方法名称	功能描述
setVideoPath()	设置要播放的视频文件的位置
setVideoPath(String path)	以文件路径方式设置视频源
setVideoURI(Uri uri)	以网络或本地 Uri 方式设置视频源
start()	开始或继续播放视频
pause()	暂停播放视频
resume()	将视频从头开始播放
seekTo()	从指定的位置开始播放视频
isPlaying()	判断当前是否正在播放视频
getCurrentPosition()	获取当前播放的位置
getDuration()	获取视频文件的时长
setMediaController(MediaController)	设置 MediaController 控制器

图 8.4 视频播放的界面

接下来通过一个案例讲解通过 VideoView 控件实现一个简单的视频播放器。

【按钮 8-2】使用 VideoView 控件播放视频，可以播放、暂停、继续，界面如图 8.3 所示，使用 SeekBar 控件显示播放进度，单击播放按钮，将当前播放的进度和视频时长使用 TextView 控件显示，播放按钮也变为暂停按钮，播放运行界面如图 8.4 所示。

步骤 1：创建项目

启动 Android Studio，创建名为 D0802_VideoPlayer 的项目，选择 Empty Activity，将包名改为 com.example.videoplayer，单击 Finish 按钮，等待项目构建完成。

步骤 2：设计界面布局，初始化界面控件对象

首先在 res 目录下创建 raw 目录，将一个 mp4 视频文件复制到此目录。

然后，打开 activity_main.xml 文件，在根布局下添加 VideoView 控件，设置相关属性，示例代码如下：

```
1.  <VideoView
2.      android:id="@+id/video_view"
3.      android:layout_width="match_parent"
4.      android:layout_height=" wrap_content " />
```

接着，在 MainActivity 类的 onCreate()中获取 VideoView 对象，调用 setVideoPath()方法将视频文件加载到 VideoView 对象，也可以调用 setVideoURI()方法加载 Uri 路径的本地或网络视频文件。

对于加载时间较长的文件可以监听 OnPrepared 事件，当视频文件都解析到内存之后，在 onPrepared()方法中调用 start()方法进行播放。图 8.4 中的进度条、播放/暂停和快进等功能通过调用 setMediaController()绑定 VideoView 的媒体控制器即可完成，无须在布局中添加控件，代码如下：

```
1.  private VideoView videoView;
2.  @Override
3.  protected void onCreate(Bundle savedInstanceState) {
4.      super.onCreate(savedInstanceState);
```

```
5.        setContentView(R.layout.activity_main);
6.        String path = "android.resource://" + getPackageName() + "/" +
                        R.raw.android;
7.        videoView = findViewById(R.id.video_view);// 初始化 VideoView 控件
8.        videoView.setVideoPath(path);       // 设置视频文件路径
9.        videoView.setMediaController(new MediaController(this));
          // 设置 VideoView 的控制器
10.       // 监听 Prepared 事件，准备好之后再播放
11.       videoView.setOnPreparedListener(new MediaPlayer.OnPreparedListener() {
12.           @Override
13.           public void onPrepared(MediaPlayer mp) {
14.               videoView.start();      // 开始播放
15.               videoView.requestFocus();   // 设置当前播放器窗口设置为焦点
16.           }
17.       });
18. }
```

以上代码实现视频文件的播放的编码非常简单，原因是 VideoView 封装了 MediaPlayer 和 SurfaceView，视频播放控制的生命周期都已封装好。但遇到比较复杂的、自定义程度较高的布局，VideoView 就有点力不从心了，需要使用 SurfaceView 和 MediaPlayer 组合，其中，MediaPlayer 用于播放视频，SurfaceView 控件用于显示视频图像。

8.3.2 SurfaceView

SurfaceView 继承自 View，但与普通 View 有较大区别，普通 View 与它的宿主窗口共享一个绘图 Surface，而 SurfaceView 提供了一个独立的绘图 Surface 嵌入到视图结构层次中，因此就可以在独立的线程中渲染，而不占用主线程资源，这一特性可以实现复杂而高效的 UI 界面，而且 SurfaceView 的双缓冲技术也能有效解决反复局部刷新带来的闪烁问题。双缓冲技术通俗地讲就是有两个缓冲区，后台缓冲区接收数据，当绘制完整后提交到前台缓冲区显示，提高了渲染效率和刷新速度。SurfaceView 控件广泛用于游戏开发中，如绘制游戏背景、动画等复杂 UI 界面。

SurfaceView 的工作原理如图 8.5 所示，SurfaceView 负责将 Surface 显示到屏幕上，通过 SurfaceHolder 访问和控制 Surface 的大小、格式等，以及监听 Surface 的变化。一个 SurfaceView 包含一个 SurfaceHolder。SurfaceHolder 的 Callback 接口提供了 surfaceCreated()、surfaceChanged() 和 surfaceDestroyed()三个方法监听 Surface 的生命周期，其中，surfaceCreated()方法在第一次创建 Surface 时回调，完成视频播放初始化等工作；surfaceChanged()方法在 Surface 变化时回调，如开启线程循环实现各种动画效果的绘制；surfaceDestroyed()方法则在 Surface 销毁时回调，完成资源回收等工作。示例代码如下：

```
1.  surfaceView.getHolder().addCallback(new SurfaceHolder.Callback() {
2.      @Override
3.      public void surfaceCreated(@NonNull SurfaceHolder holder) {
4.          // Surface 创建时触发，初始化 MediaPlayer 对象，并播放视频
5.      }
6.      @Override
7.      public void surfaceChanged(@NonNull SurfaceHolder holder, int format,
8.                                 int width, int height) {
9.          // Surface 的大小发生变化时触发
10.     }
11.     @Override
12.     public void surfaceDestroyed(@NonNull SurfaceHolder holder) {
13.         // Surface 销毁时触发，停止播放，释放资源
14.         if (player != null && player.isPlaying()) {
```

```
15.            player.stop();
16.            player.release();
17.            player = null;
18.        }
19. });
```

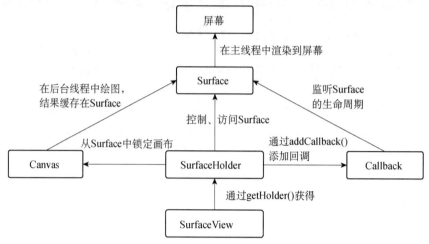

图 8.5 SurfaceView 的工作原理

接下来讲解如何使用 SurfaceView 配合 MediaPlayer 播放视频，整个过程大致包括以下步骤。
（1）创建 SurfaceView 组件
在布局中添加 SurfaceView 控件，代码如下：

```
1. <SurfaceView
2.     android:id="@+id/surface_view"
3.     android:layout_width="match_parent"
4.     android:layout_height="match_parent" />
```

然后在 MainActivity 类中调用 findViewById()方法获取 SurfaceView 对象，调用 setkeepScreenOn（true）设置屏幕高亮显示及实现 SurfaceHolder.Callback 接口，代码如下：

```
1. // 获取 SurfaceView 对象
2. surfaceView = findViewById(R.id.surface_view);
3. // 设置屏幕高亮
4. surfaceView.setKeepScreenOn(true);
5. // 设置 SurfaceHolder 的回调接口
6. surfaceView.getHolder().addCallback(new SurfaceHolder.Callback() {...});
```

SurfaceView 的 getHolder()方法返回 SurfaceHolder 对象，然后调用 addCallback()方法给 SurfaceView 提供一个回调对象监听 Surface 的创建、变化和销毁的生命周期变化。
（2）创建 MediaPlayer 组件
与使用 MediaPlayer 播放音频的方法相同，可以使用构造方法或静态方法 create()创建。

```
1. // 创建 MediaPlayer 对象
2. player = new MediaPlayer();
3. // 设置视频文件路径
4. String path = "android.resource://" + getPackageName() + "/" +
                R.raw.android;
5. player.setDataSource(SurfaceViewActivity.this, Uri.parse(path));
```

（3）设置视频输出
视频输出到 SurfaceView 是通过调用 MediaPlayer 的 setDisplay()方法实现的，该方法的参数是 SurfaceView 的 SurfaceHolder 对象，代码如下：

```
1. // SurfaceView 控件与 MediaPlayer 关联
2. player.setDisplay(surfaceView.getHolder());
```

（4）控制视频播放

调用 MediaPlayer 的 play()、pause()和 stop()等方法，控制视频的播放、暂停、停止等。

```
1. if (isPlaying) {    //视频的播放与暂停
2.     isPlaying = false;
3.     player.pause();
4. } else {
5.     isPlaying = true;
6.     player.start();
7. }
```

SurfaceView 提供了两个线程：UI 线程和渲染线程，在使用过程中需要注意以下两点。

● 所有 SurfaceView 和 SurfaceHolder.Callback 的方法都应在 UI 主线程中调用，渲染线程所要访问的各种变量应该做同步处理。

● 由于 Surface 只在 surfaceCreated()和 surfaceDestroyed()之间有效，所以要确保渲染线程访问的是合法有效的 Surface。

 ## 8.4　动画和过渡

目前的手机应用越来越重视用户体验，为了实现更好的界面展示和状态转换的效果，合理使用动画能够获得友好的用户体验和外观风格。Android 提供了多种类型动画，分为以下两类。

● 视图动画：视图动画作用于 View 控件，分为逐帧动画和补间动画。逐帧动画是顺序播放一系列图片产生动画效果，类似于动画片；补间动画通过对场景中的 View 不断做渐变、旋转、缩放或平移等变换产生动画效果。

● 属性动画：属性动画是补间动画的增强版，通过动态改变对象的属性达到动画效果。

动画实现包括两种方式：XML 格式的资源文件和编码实现，接下来详细讲解这两类动画的实现，所有的动画都在 D0803_Animation 项目中实现，项目的主界面、逐帧动画和补间动画的界面如图 8.6～图 8.8 所示。

图 8.6　主界面布局界面

图 8.7　逐帧动画的界面

图 8.8　补间动画的界面

8.4.1 逐帧动画

逐帧动画的英文为 Frame Animation，采用类似胶片电影的实现原理，定义每一帧为一张图片，按照事先准备好的顺序进行播放，利用人眼的视觉偏差产生动画效果。

使用逐帧动画，首先准备几张连续播放的图片放到 drawable 目录中，图标可以放到 mipmap 目录中，然后在 res/drawable 目录中创建根标签为 animation-list 的 XML 资源文件，代码如下：

```xml
1.  <?xml version="1.0" encoding="utf-8"?>
2.  <animation-list
3.      android:oneshot="false"
4.      xmlns:android="http://schemas.android.com/apk/res/android">
5.      <item android:drawable="@mipmap/pic_1" android:duration="100" />
6.      <item android:drawable="@mipmap/pic_2" android:duration="100" />
7.      <item android:drawable="@mipmap/pic_3" android:duration="100" />
8.      <item android:drawable="@mipmap/pic_4" android:duration="100" />
9.      <item android:drawable="@mipmap/pic_5" android:duration="100" />
10.     <item android:drawable="@mipmap/pic_6" android:duration="100" />
11.     <item android:drawable="@mipmap/pic_7" android:duration="100" />
12.     <item android:drawable="@mipmap/pic_8" android:duration="100" />
13. </animation-list>
```

此 XML 资源文件的标签和属性的含义如下。

- animation-list：根节点的标签名，表示逐帧动画，item 标签表示每一帧的内容。
- android:oneshot：控制动画是否循环播放，true 表示不循环播放，false 表示循环播放。
- Android:duration：表示每一帧持续播放的时间，毫秒为单位。

然后可以在 XML 布局文件中设置 ImageView 的 background 或 src 属性，示例代码如下：

```xml
1.  <ImageView
2.      android:id="@+id/image_view"
3.      android:layout_width="wrap_content"
4.      android:layout_height="wrap_content"
5.      android:background="@drawable/frame_animation" />
```

也可以调用 setImageResource()或 setBackgroundResource()方法设置创建的 XML 资源文件，示例代码如下：

```java
1.  // 设置动画
2.  imageView.setImageResource(R.drawable.frame_animation);
3.  // 获取动画对象
4.  frameDrawable = (AnimationDrawable) imageView.getDrawable();
5.  // 启动动画
6.  frameDrawable.start();
```

8.4.2 补间动画

补间动画只需要指定动画的开始帧、结束帧图片，动画的中间过程则通过系统计算进行填充，而无须像逐帧动画提供所有帧的图片。Android 系统提供了 4 种类型的补间动画，分别是位移动画、缩放动画、旋转动画和透明度动画 Alpha，应用于 Activity 的切换效果及视图的出场效果。

补间动画的实现方式有两种，一种是定义 XML 动画资源文件，另一种是使用 Java 代码生成，补间动画的 XML 文件中的所有属性都可以使用 setXxx()方法进行设置。补间动画的 XML 元素与 Java 类之间的关系如表 8.3 所示。

表 8.3 补间动画的 XML 元素与 Java 类之间的关系

XML 元素	Java 类	功能描述
<translate>	TranslateAnimation	位置移动动画
<scale>	ScaleAnimation	缩放动画
<rotate>	RotateAnimation	旋转动画
<alpha>	AlphaAnimation	渐变透明度动画
<set>	AnimationSet	多种动画的组合

补间动画的公共属性包括以下几个。
- android:duration：动画时间，单位为毫秒。
- android:startOffset：动画延迟执行时间。
- android:fillBefore：动画播放完成之后视图是否停留在动画开始的状态，默认为 true。
- android:fillAfter：动画播放完成之后视图是否停留在动画结束的状态，优先于 fillBefore 值，默认为 false。
- android:fillEnable：是否应用 fillBefore 值，对 fillAfter 值无影响，默认为 true。
- android:restart：重复播放动画模式，restart 代表正序重放，reverse 代表倒序回放，默认为 restart。
- android:repeatCount：重放次数，infinite 表示无限重复。

1. 位移动画

使用 XML 方式创建动画资源文件的方法是在 res 目录上右击，选择 New->Android Resource File，打开创建资源文件的对话框，资源类型选择 animation，创建根元素为 translate 的位移动画资源文件 anim_translate.xml，设置属性使得 x 坐标值增加 100，y 坐标值减少 100，代码如下：

```
1.  <?xml version="1.0" encoding="utf-8"?>
2.  <translate xmlns:android="http://schemas.android.com/apk/res/android"
3.      android:duration="1000"
4.      android:fromXDelta="0"
5.      android:fromYDelta="0"
6.      android:toXDelta="100"
7.      android:toYDelta="-100" />
```

XML 文件中的位移动画的特有属性包括：
- android:fromXDelta：水平方向 x 移动的起始值。
- android:toXDelta：水平方向 x 移动的结束值。
- android:fromYDelta：竖直方向 y 移动的起始值。
- android:toYdelta：竖直方向 y 移动的结束值。

以上属性的取值可以为数值、百分数、百分数 p 三种类型，其中，数值表示以当前 View 左上角坐标加上响应数值的 px 值为初始点；百分数表示以当前 View 的左上角加上当前 View 宽高的百分数作为初始点；百分数 p 表示以当前 View 的左上角加上父控件宽高的百分数作为初始点。以 50、50%和 50%p 为例，它们的含义及区别如图 8.9 所示。

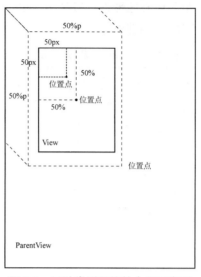

图 8.9 三种类型取值的含义示意图

然后，在布局文件中添加 ImageView 控件，Activity 类文件实现播放位移动画，示例代码如下：

```
1.  // 初始化 ImageView 控件
2.  ImageView imageView = findViewById(R.id.image_view);
3.  // 加载 anim 目录下的 xml 动画
4.  Animation animation = AnimationUtils.loadAnimation(this,
                          R.anim.anim_translate);
5.  // 播放动画
6.  imageView.startAnimation(animation);
```

也可以使用 Java 代码实现以上动画效果，通过调用 TranslateAnimation 类的构造方法创建，构造方法的 4 个参数分别用于设置 fromXDelta、toXDelta、fromYDelta 和 toYDelta 的值，这些值对应 XML 文件中的属性，它们的类型都是 float 类型，代码如下：

```
1.  // 创建位移动画，x 坐标增加 100，y 坐标减少 100
2.  Animation animation = new TranslateAnimation(
3.        0f,      // x 移动起始位置
4.        100f,    // x 移动结束位置
5.        0f,      // y 移动起始位置
6.        -100f);  // y 移动结束位置
7.  animation.setDuration(1000);    // 设置动画持续时间
8.  imageView.startAnimation(animation);   // 播放动画
```

2. 缩放动画

缩放动画可以实现 View 的缩放操作，具体的实现步骤与位移动画类似，缩放动画特有的属性如下。

- android:fromXScale：x 方向相当于自身的起始缩放比例。
- android:toXScale：x 方向相当于自身的结束缩放比例。
- android:fromYScale：y 方向相当于自身的起始缩放比例。
- android:toYScale：y 方向相当于自身的结束缩放比例。

- android:pivotX：缩放起点的 *x* 坐标。
- android:pivotY：缩放起点的 *y* 坐标。

缩放起点坐标的取值也有数值、百分数、百分数 p 三种类型，它们的含义与 translation 动画的含义相同，View 控件以控件中心点为缩放起点放大 1.5 倍，XML 文件示例代码如下：

```xml
1.  <?xml version="1.0" encoding="utf-8"?>
2.  <scale xmlns:android="http://schemas.android.com/apk/res/android"
3.      android:duration="1000"
4.      android:fromXScale="1"
5.      android:fromYScale="1"
6.      android:pivotX="50%"
7.      android:pivotY="50%"
8.      android:toXScale="1.5"
9.      android:toYScale="1.5" />
10.
```

使用 Java 实现以上动画效果的代码如下：

```java
1.  // 创建缩放动画，以中心点为缩放起点放大1.5倍
2.  Animation animation = new ScaleAnimation(
3.          1.0f,    // x 起始缩放比例
4.          1.5f,    // x 结束缩放比例
5.          1.0f,    // y 起始缩放比例
6.          1.5f,    // y 结束缩放比例
7.          Animation.RELATIVE_TO_SELF, 1f,   // pivotX 的类型和取值
8.          Animation.RELATIVE_TO_SELF, 1f);  // pivotY 的类型和取值
9.  animation.setDuration(1000);  // 设置动画持续时间
10. imageView.startAnimation(animation);
```

第 7、8 行代码用于设置缩放起点 x、y 的类型和取值，有三种取值类型 Animation.ABSOLUTE、Animation.RELATIVE_TO_SELF 和 Animation.RELATIVE_TO_PARENT，对应 XML 文件中的数值、百分比和百分比 p。

3. 旋转动画

旋转动画实现 View 的旋转动画效果，旋转动画特有的属性如下。
- android:fromDegrees：旋转起始角度。
- android:toDegrees：旋转结束角度。
- android:pivotX：旋转中心点的 *x* 轴坐标。
- android:pivotY：旋转中心点的 *y* 轴坐标。

pivotX 和 pivotY 的含义及取值与 translation 相同，控件旋转 90 度的示例代码如下：

```xml
1.  <rotate xmlns:android="http://schemas.android.com/apk/res/android"
2.      android:duration="1000"
3.      android:fromDegrees="0"
4.      android:pivotX="50%"
5.      android:pivotY="50%"
6.      android:toDegrees="90" />
```

使用 Java 实现以上动画效果的代码如下：

```java
1.  private float fromDegree = 0.0f;
2.  private float toDegree = 0.0f;
3.  private float rotateDegree = 90.0f;
4.  private void rotateAnimation() {
5.      // 创建旋转动画，围绕中心点顺时针旋转90度
```

```
6.      toDegree = fromDegree + rotateDegree;
7.      Animation animation = new RotateAnimation(
8.              fromDegree,   //起始角度
9.              toDegree,     //结束角度
10.             RotateAnimation.RELATIVE_TO_SELF, 0.5f,  // pivotX 的类型和
                                                              取值
11.             RotateAnimation.RELATIVE_TO_SELF, 0.5f); // pivotY 的类型和
                                                              取值
12.     animation.setFillAfter(true);  // 设置动画播放完停留在目前的位置
13.     animation.setDuration(1000);   // 设置动画持续时间
14.     fromDegree = toDegree;
15.     imageView.startAnimation(animation);
16. }
```

以上代码设置了起始、结束和旋转角度,能完成每执行一次就顺时针旋转 90 度,运行 4 次即可旋转一周。

4. 透明度动画

透明度动画可以实现 View 透明度的渐变过程,透明度为 0 表示完全透明,它特有的属性包括:

- android:fromAlpha:透明度起始值。
- android:toAlpha:透明度结束值。

透明度从完全不透明到透明度为 0.2 的示例代码如下:

```
1. <alpha xmlns:android="http://schemas.android.com/apk/res/android"
2.     android:fromAlpha="1.0"
3.     android:toAlpha="0.2"
4.     android:duration="1000" />
```

使用 Java 实现以上透明度动画的代码如下:

```
1. // 创建透明度动画,透明度从不透明到 0.2
2. Animation animation = new AlphaAnimation(1.0f, 0.2f);
3. animation.setDuration(1000); // 设置动画持续时间
4. imageView.startAnimation(animation);
```

5. 组合动画

这 4 补间动画也可以组合使用,在 XML 文件中使用<set>标签将这几种补间动画任意组合,实现想要的动画效果,比如实现点赞按钮模拟人竖大拇指的效果,可以设置在按下时放大 1.2 倍并逆时针旋转 20 度的组合动画,示例代码如下:

```
1.  <?xml version="1.0" encoding="utf-8"?>
2.  <set xmlns:android="http://schemas.android.com/apk/res/android">
3.      <scale
4.          android:duration="200"
5.          android:fromXScale="1"
6.          android:fromYScale="1"
7.          android:pivotX="1"
8.          android:pivotY="1"
9.          android:toXScale="1.2"
10.         android:toYScale="1.2" />
11.     <rotate
12.         android:duration="200"
13.         android:fromDegrees="0"
14.         android:pivotX="0.5"
```

```
15.            android:pivotY="0.5"
16.            android:toDegrees="-20" />
17. </set>
```

以上两个动画同时开始，如果想按先后顺序开始，可以设置 android:startOffset 属性延迟单个动画播放时间。另外，组合动画的 scale 及缩放动画的 repeatCount 属性和 fillBefore 均无效。

以上组合动画可以使用 AnimationSet 类实现，使用它的 addAnimation()方法添加单个动画，示例代码如下：

```
1.  // 创建一个动画集合
2.  AnimationSet animationSet = new AnimationSet(true);
3.  // 放大1.2倍的动画
4.  ScaleAnimation scaleAnimation = new ScaleAnimation(
5.          1.0f, // x 起始缩放比例
6.          1.2f, // x 结束缩放比例
7.          1.0f, // x 起始缩放比例
8.          1.2f, // y 结束缩放比例
9.          Animation.RELATIVE_TO_SELF, 1f,
10.         Animation.RELATIVE_TO_SELF, 1f);
11. scaleAnimation.setDuration(200); // 动画持续时间
12. // 逆时针旋转 20 度的动画
13. RotateAnimation rotateAnimation = new RotateAnimation(
14.         0.0f,  // 起始角度
15.         -20f,  // 结束角度
16.         RotateAnimation.RELATIVE_TO_SELF, 0.5f,  // 设定 x 旋转中心点
17.         RotateAnimation.RELATIVE_TO_SELF, 0.5f); // 设定 y 旋转中心点
18. rotateAnimation.setDuration(200); // 动画持续时间
19. // 将动画添加到动画集合
20. animationSet.addAnimation(scaleAnimation);
21. animationSet.addAnimation(rotateAnimation);
22. // 启动动画
23. imageView.startAnimation(animationSet);
```

8.4.3 属性动画

逐帧动画和补间动画使用简单且方便，但由于这两种动画只能满足简单的动画需求而无法扩展，另外也只能改变 View 的显示效果而无法真正改变 View 的属性。针对它们的局限性，Android 3.0 提供了一种全新的动画模式-属性动画，属性动画是对 View 动画的扩展，它的作用对象不仅限于 View 视图，还能作用于任何 Java 对象。属性动画是在一段指定时间内将对象的属性从一个值动态改变到另一个值实现动画效果。

属性动画的自定义属性包括：

- 时长（Duration）：指定动画总共完成所需要的时间，默认为 300 毫秒。
- 时间插值器（Time Interpolator）：基于当前动画已消耗时间的函数，用于计算属性值。
- 重复计数（Repeat Count）：指定动画是否重复执行及执行次数，也可以指定动画向反方向回退操作。
- 动画集（AnimatorSet）：定义一组动画，设置同时执行或者顺序执行。
- 延迟刷新时间（Frame refresh delay）：指定动画的刷新频率，默认为每 10ms 刷新一帧，但真实的刷新频率取决于整个系统的运行状态。

属性动画的运行机制是通过不断改变 View 的属性值实现的,具体步骤包括:
(1)通过 get()方法获取目标对象的属性值。
(2)调用时间插值器方法修改属性值。
(3)调用 set()方法更改目标对象的属性值,达到动画效果。

Animator 类是属性动画的基类,它提供了创建动画的基本结构,是一个抽象类,它有两个重要的具体实现类:ValueAnimator 和 ObjectAnimator 类。

1. ValueAnimator

ValueAnimator 类是整个属性动画框架的核心类,负责动画初始值和结束值之间的动画过渡的计算,它本身不提供对 View 的任何操作,只是一个值变化器,需要通过动画的 AnimatorUpdateListener 监听器指定动画的开始、结束、取消和重复等事件发生时的处理。

ValueAnimator 类有如下三个重要方法。

- ValueAnimator.ofInt(int…values):默认采用 IntEvaluator 整型估值器将初始值以整型数的形式过渡到结束值,可以传入多个整型值。
- ValueAnimator.ofFloat(float…values):默认采用 FloatEvaluator 整型估值器将初始值以浮点数的形式过渡到结束值。
- ValueAnimator.ofObject(int…values):采用自定义对象估值器 TypeEvaluator 将初始值以对象的形式过渡到结束值。

属性动画也有两种实现方式:XML 方式和代码方式,XML 方式通过 res 创建根元素为 <animator> 的 animation 类型的 XML 资源文件即可,属性动画与视图动画并不在同一个目录,使用线性插值器 LinearInterpolator 从 0 移动到 500 的代码如下,属性的含义都很明确,不再一一描述。

```
1.  <animator xmlns:android="http://schemas.android.com/apk/res/android"
2.      android:duration="2000"
3.      android:interpolator="@android:anim/linear_interpolator"
4.      android:valueFrom="0"
5.      android:valueTo="500"
6.      android:valueType="floatType" />
```

然后在 Java 类中使用 AnimatorInflate 类的 loadAnimator()方法加载 XML 文件,通过监听 ValueAnimator 的 AnimatorUpdateListener 事件设置 View 控件的位移方法,实现动画效果,示例代码如下:

```
1.  private TextView tvXmlValue;
2.  public void ofFloatByXML() {
3.      // 加载 XML 动画文件
4.      ValueAnimator animator = (ValueAnimator) AnimatorInflater.loadAnimator(
5.              this, R.animator.animator_value);
6.      // 监听 ValueAnimator 的更新事件
7.      animator.addUpdateListener(new ValueAnimator.AnimatorUpdateListener() {
8.          @Override
9.          public void onAnimationUpdate(ValueAnimator animation) {
10.             // 获取 ValueAnimator 移动的值
11.             float progress = (float) animation.getAnimatedValue();
12.             // 设置 TextView 的 Y 轴移动的值
13.             tvXmlValue.setTranslationY(progress);
14.         }
15.     });
16.     // 启动动画
```

```
17.         animator.start();
18.     }
```

VauleAnimator 的 start() 只是根据 LinearInterpolator 线性插值器计算在 2 秒之内从 0 到 500 的移动速率，并不能实现 TextView 的移动动画，而是通过第 7 行的 AnimatorUpdateListener 监听器监听移动数值的变化，在回调方法 onAnimationUpdate() 中更新 TextView 的 translationY 的值，实现 TextView 在 Y 轴的移动动画。

通过 Java 代码实现以上动画的代码如下所示，其中第 4~8 行完成了 XML 文件的动画属性设置，由此可以看出，简单的属性动画使用代码动态生成也非常简单方便。

```
1.  private TextView tvCodeValue;
2.  public void ofFloatByCode() {
3.      // 创建 ValueAnimator
4.      ValueAnimator valueAnimator = ValueAnimator.ofFloat(0f, 500f);
5.      // 设置线性变化插值器
6.      valueAnimator.setInterpolator(new LinearInterpolator());
7.      // 设置持续时长
8.      valueAnimator.setDuration(2000);
9.      // 监听 ValueAnimator 的更新事件
10.     valueAnimator.addUpdateListener(new ValueAnimator.
                                        AnimatorUpdateListener() {
11.         @Override
12.         public void onAnimationUpdate(ValueAnimator animation) {
13.             float progress = (float) animation.getAnimatedValue();
14.             // 更新 TextView 的 Y 轴移动的值
15.             tvCodeValue.setTranslationY(progress);
16.         }
17.     });
18.     // 启动动画
19.     valueAnimator.start();
20. }
```

插值器 Interpolator 的作用是在指定的时间内使用特定的方法计算变化值。Android 系统内置的插值器及其含义如表 8.4 所示。例如，LinearInterpolator 提供匀速变化的移动，AccelerateInterpolator 则提供加速度变化的移动。图 8.10 可以帮助理解这两种插值器的含义。另外还可以通过实现 TimeInterpolator 接口自定义插值器。

表 8.4 Android 系统提供的内置插值器

插值器类名称	含义描述
AccelerateDecelerateInterpolator	变化率开始和结束时缓慢，中间阶段加快
AccelerateInterpolator	变化率加速度变化
AnticipateInterpolator	先反向变化，然后再急速正向变化
AnticipateOvershootInterpolator	先反向变化，再急速正向变化，然后超过设定值，最后返回到最终值
BounceInterpolator	变化率在结束时回弹
CycleInterpolator	在指定数量的周期内循环
DecelerateInterpolator	变化率减速变化，与 AccelerateInterpolator 相反
LinearInterpolator	变化率恒定不变，匀速变化
OvershootInterpolator	急速正向变化，超出最终值后返回
TimeInterpolator	插值器的父接口，用于实现自定义的插值器

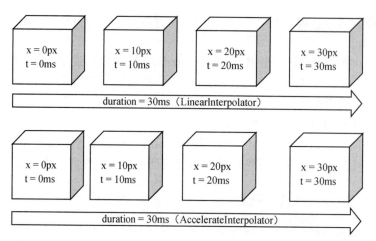

图 8.10 LinearInterpolator 和 AccelerateInterpolator 的运动示意图

2. ObjectAnimator

ObjectAnimator 是 ValueAnimator 的子类，它直接对对象的属性进行动画操作，极大地简化了为对象添加动画的过程，而且也无须实现 AnimatorUpdateListene。两者的区别为：ValueAnimator 了类先改变值，然后手动赋值给对象的属性实现动画，间接操作对象属性；ObjectAnimator 类也是先改变值，然后自动赋值给对象的属性实现动画，直接操作对象属性。但它们的用法非常类似，ValueAnimator 设置的动画使用 ObjectAnimator 的 Java 代码可以这样写：

```
1.  public void ofFloatObject() {
2.      // 创建 ObjectAnimator 对象
3.      ObjectAnimator animator=ObjectAnimator.ofFloat(tvCodeValue,
                            "translationY",0,500);
4.      // 设置插值器
5.      animator.setInterpolator(new LinearInterpolator());
6.      // 设置时长，并启动动画
7.      animator.setDuration(2000).start();
8.  }
```

上述代码比 ValueAnimator 简洁，其中 ofFloat()方法包含多个参数，第一个参数为动画作用对象，第二个参数是对象的属性，这些属性的含义如表 8.5 所示，第三个参数是可变参数，为这个属性将会达到的一系列值，如开始值和结束值等。

表 8.5 View 的动画相关的属性及含义

属性名称	对应的 set 方法	含义描述
alpha	setAlpha(float)	控制 View 的透明度
translationX	setTranslationX(float)	控制 X 方向的位移
translationY	setTranslationY(float)	控制 Y 方向的位移
scaleX	setScaleX(float)	控制 X 方向的缩放比例
scaleY	setScaleY(float)	控制 Y 方向的缩放比例
rotationX	setRotationX(float)	控制以 X 为轴的旋转度数
rotationY	setRotationY(float)	控制以 Y 为轴的旋转度数
rotation	setRotation(float)	控制以屏幕方向为轴的旋转度数

ObjectAnimator 使用 XML 动画资源文件时，根元素为 objectAnimator，示例代码如下：

```
1.  <?xml version="1.0" encoding="utf-8"?>
2.  <objectAnimator xmlns:android="http://schemas.android.com/apk/res/android"
3.      android:duration="2000"
4.      android:interpolator="@android:anim/linear_interpolator"
5.      android:valueFrom="0"
6.      android:valueTo="500"
7.      android:propertyName="y"
8.      android:valueType="floatType" />
```

XML 资源也是使用 AnimatorInflater 类的 loadAnimator()静态方法进行加载的，然后调用 setTarget()方法设置动画对象，启动即可，而无须添加动画更新的监听器。示例代码如下：

```
1.  private TextView tvXmlObject;
2.  public void ofFloatByObjectXML() {
3.      // 加载 XML 动画文件
4.      ObjectAnimator animator = (ObjectAnimator) AnimatorInflater.
                                          loadAnimator(
5.              this, R.animator.animator_object);
6.      // 设置动画的目标对象
7.      animator.setTarget(tvXmlObject);
8.      // 启动动画
9.      animator.start();
10. }
```

3. AnimatorSet

以上单一动画实现的效果有限，实际项目开发中更多的是同时使用多个动画产生炫酷的动画效果，Android 提供了 AnimatorSet 类实现复杂的组合动画，这个类既可以同时对多个对象进行动画组合，也可以控制单个对象的多个动画的执行顺序。AnimatorSet 可以调用 playTogether()或 playSequentially()方法将所有的动画一次性添加到 AnimatorSet 中同时播放或顺序播放；还可以调用 play()方法构建 AnimatorSet.Builde 对象，然后利用 Builder 提供的以下方法添加动画，并设置各个动画执行顺序的依赖关系。另外，AnimatorSet 对象还可以相互嵌套使用。

- Play（Animator）：播放当前动画。
- after（Animator）：将当前动画插入到传入的动画之后执行。
- after（long）：将当前动画延迟指定的毫秒后执行。
- before（Animator）：将当前动画插入到传入的动画之前执行。
- with（Animator）：将当前动画和传入的动画同时执行。

接下来使用 AnimatorSet.Builde 对象实现"在平移过程中伴随旋转动画，平移完成后变化透明度"的组合动画，示例代码如下：

```
1.  private TextView tvAnimatorSet;
2.  public void animatorSetByCode() {
3.      // 设置需要组合的动画效果
4.      // 平移动画：移动 300px，再回到原有的位置
5.      ObjectAnimator translation = ObjectAnimator.ofFloat(
6.              tvAnimatorSet, "translationX", 0, 300, 0);
7.      // 旋转动画：旋转 360 度
8.      ObjectAnimator rotate = ObjectAnimator.ofFloat(tvAnimatorSet,
                                      "rotation",0f,360f);
9.      // 透明度动画：从不透明到完全透明再回到不透明
10.     ObjectAnimator alpha = ObjectAnimator.ofFloat(tvAnimatorSet,
```

```
                              "alpha",1f,0f,1f);
11.     // 创建组合动画对象
12.     AnimatorSet animSet = new AnimatorSet();
13.     animSet.setDuration(5000);
14.     // 根据需求组合动画
15.     animSet.play(translation).with(rotate).before(alpha);
16.     // 启动动画
17.     animSet.start();
18. }
```

8.4.4 布局动画

Android 除了提供以上类型的动画用于 View 对象，还提供了 LayoutTransition 类用于 ViewGroup 布局改变时的动画。LayoutTransition 是在 Android 3.0 中引入的，用于当前 ViewGroup 的子 View 添加、删除、隐藏及显示时定义布局容器和子 View 的动画。

布局动画有添加、删除两种状态的改变而执行 4 种不同的动画，分别为：
- APPEARING：ViewGroup 添加子 View 时触发。
- DISAPPEARING：ViewGroup 移除子 View 时触发。
- CHANGE_APPEARING：ViewGroup 添加子 View 使得其他子 View 发生变化时触发。
- CHANGE_DISAPPEARING：ViewGroup 移除子 View 使得其他子 View 发生变化时触发。
- CHANGING：Android 4.1 版本新增的类型，子 View 在 ViewGroup 的位置变化时触发。

添加布局动画的方法非常简单，与之前所有的动画方法类似，它既可以使用 XML 进行设置，也可以使用 Java 代码直接控制。

使用 XML 方式设置布局动画非常简单，只需要给子 View 所在的 ViewGroup 的 XML 添加属性 android:layoutAnimation=true 即可加载系统提供的默认的过渡动画。

```
1. <LinearLayout xmlns:android="http://schemas.android.com/apk/res/android"
2.     ...
3.     android:animateLayoutChanges="true">
```

添加 View 的默认动画是先执行 CHANGE_APPEARING 动画，延迟一段时间后执行 APPEARING 动画，删除 View 的默认动画则是执行 DISAPPEARING 消失动画，延迟一段时间后受影响的其他 View 会执行 CHANGE_DISAPPEARING 动画补上相应的位置。

除了默认的动画效果，也可以通过 LayoutTransition 自定义布局动画，步骤非常简单，包括：
（1）创建一个 LayoutTransition 实例。
（2）使用上一节讲解的方法创建各种属性动画。
（3）调用 setAnimator()方法为 LayoutTransition 设置创建的动画。
（4）最后调用 ViewGroup 的 setLayoutTransition()方法添加动画。

示例代码如下：

```
1. private LinearLayout container;
2. private LayoutTransition layoutTransition;
3. @Override
4. protected void onCreate(Bundle savedInstanceState) {
5.     super.onCreate(savedInstanceState);
6.     setContentView(R.layout.activity_layout);
7.     // 初始化布局控件
```

```
8.      container = findViewById(R.id.layout_contrainer);
9.      // 创建自定义布局动画
10.     addCustomTransition();
11.     // 添加布局动画
12.     container.setLayoutTransition(layoutTransition);
13. }
```

addCustomTransition()方法自定义了 APPEARING、DISAPPEARING 两个状态的动画，代码如下：

```
1.  private void addCustomTransition() {
2.      // 创建 LayoutTransition 对象
3.      layoutTransition = new LayoutTransition();
4.      // 添加 View 时的 APPEARING 动画，X 方向从 0.5 到 1 的放大，时长为 2 秒
5.      ObjectAnimator animator = ObjectAnimator.ofFloat(null, "scaleX",
                                    0.5f, 1)
6.              .setDuration(2000);
7.      layoutTransition.setAnimator(LayoutTransition.APPEARING, animator);
8.      // 移除 View 时的 DISAPPEARING 动画，Y 方向的向下移动 300px 再回到原点，
           时长 2 秒
9.      animator = ObjectAnimator.ofFloat(null, "translationY", 0, 300, 0)
10.             .setDuration(2000);
11.     layoutTransition.setAnimator(LayoutTransition.DISAPPEARING, animator);
12. }
```

至此，讲解了常用动画的基本用法，属性动画和布局动画的运行效果如图 8.11、图 8.12 所示。动画的高级技巧将不在本书涉及。

图 8.11　属性动画的效果图

图 8.12　布局动画的效果图

 8.5　本章小结

本章主要讲解了 Android 的多媒体开发应具备的音视频播放相关知识，包括 MediaPlayer 类、VideoView 控件、SurfaceView 控件的使用；同时也讲解了常用的视图动画、属性动画和布局动画的知识，以及使用 XML 方式和 Java 方式创建动画的基本方法。通过本章的学习，希望读者能在项目中融合多媒体技术，开发具有良好用户体验的应用程序，为以后能够开发

更复杂的项目做好准备。

习 题

一、选择题

1. Android 中（　　）就是播放一系列的图片资源。
   A. 属性动画　　　　　B. 补间动画　　　　　C. 逐帧动画　　　　　D. 位移动画

2. 下列关于补间动画说法中正确的是（　　）。
   A. 补间动画和帧动画类似　　　　　　　B. frameAnimation 属于补间动画
   C. 补间动画不会改变控件的真实坐标　　D. 以上说法都不正确

3. 针对以下的动画设置代码进行判断，选项中哪一项解释是正确的（　　）。

```
<animation-list android:id = "selected" android:oneshot = "false" >
    <item android:drawable = "@drawable/wheel0" android:duration = "1000" />
    <item android:drawable = "@drawable/wheel1" android:duration = "1000" />
</animation-list>
```

   A. 这是一段补间动画设置，该动画由两幅图片组成，每帧画面显示 1 秒
   B. 这是一段透明度渐变动画设置，该动画由两幅图片组成，每帧画面显示 1 秒
   C. 这是一段帧动画设置，该动画由两幅图片组成，每帧画面显示 1 秒
   D. 这是一段帧动画设置，该动画由两幅图片组成，每帧画面显示 1000 秒

4. MeidiaPlayer 播放视频使用（　　）组件进行显示视频。
   A. SurfaceView　　　B. VideoView　　　C. View　　　D. ViewHolder

5. 下列选项中，属于获取 SurfaceHolder 类的方法的是（　　）。
   A. newInstance()　　　　　　　　　B. getHolder()
   C. getSurfaceHolder()　　　　　　D. new SurfaceHolder()

6. 下列选项中，属于设置 MediaPlayer 数据源的方法的是（　　）。
   A. create()　　　B. setDataSource()　　　C. load()　　　D. setDataPath()

7. 下列选项中，属于 MediaPlayer 支持的音频类型的是（　　）。
   A. AudioManager.STREAM_MUSIC　　　　B. AudioManager.STREAM_RING
   C. AudioManager.STREAM_ALARM　　　　D. AudioManager.STREAM_NOTIFICTION

8. 下列属于补间动画相关类的是（　　）。
   A. TranslateAnimation　　　　　　　B. FrameAnimation
   C. RotateAnimation　　　　　　　　D. AlphaAnimation

9. 以下关于 Frame 动画说法正确的是（　　）。
   A. Frame 动画可以顺序播放事先准备好的图片
   B. Frame 动画和补间动画原理一样
   C. Frame 动画在 values 目录下创建
   D. Frame 动画可以设置动画的执行时间

10. 在 Android 中，下列关于视频播放的实现的描述中错误的有（　　）。
    A. 使用 VideoView 播放视频时需要 MediaPlayer 配合
    B. 使用 SurfaceView 播放视频时需要 MediaPlayer 配合
    C. 使用 VideoView 播放视频可以改变播放的位置和大小

D. 使用 SurfaceView 播放视频可以改变播放的位置和大小

二、填空题

1. Android 提供了 4 种补间动画，分别是_____、_____、_____和_____。

2. 使用 VideoView 播放视频，设置播放视频路径的方法是_____。

3. Android 中_____动画就是播放一系列的图片资源。

4. 使用 SurfaceView 实现 SurfaceHolder.Callback 接口时_____方法是 Surfaceview 初始化的方法。

5. MeidiaPlayer 播放音乐设置数据源的方法是_____。

三、问答题

1. 简述使用 MediaPlayer 播放音频的步骤。

2. 简述使用 VideoView 播放视频的步骤。

3. 简述补间动画的工作原理以及特点。

四、编程题

1. 编写一个逐帧动画的应用代码。

2. 使用 Service 作为后台服务，重构音乐播放器的案例。

第 9 章 进阶技术

以上章节讲解的内容都是 Android 应用开发的基础,本章将讲解 Android 开发的一些高阶技术,包括手势处理、传感器及 Android Jetpack 的相关技术,特别是 Android Jetpack,能大大降低开发 Android 的难度。

本章学习目标:

- 掌握手势的实现方式
- 了解基础传感器的应用
- 了解 Android Jetpack 工具集

 9.1 手势处理

智能手机的应用离不开与屏幕的交互,而手势则是与屏幕交互的最自然方式。手势交互的普及,降低了人与设备之间的沟通门槛,丰富的手势操作不仅可以让用户更顺畅地使用 App,也会直接影响到 App 的使用体验。

9.1.1 手势简介

手势是指用户手指在屏幕上的连续触碰的行为,比如:手指在屏幕上右滑关闭界面、在屏幕上画一个圆等都是手势,应用程序中的手势是指多个持续的触摸事件在屏幕上形成特定的形状。当用户手指触碰屏幕时,会产生三种动作:按下 Down、移动 Move 和抬起 Up,这三种动作的组合可以产生各种各样的手势。因此,手势控制分为触碰动作和触碰行为,触碰动作是用户手指在屏幕上如何动作,而触碰行为是指触碰动作在不同情境下的不同结果,如轻触屏幕的动作可以是开启/关闭指示或取消等行为,放大屏幕图片的行为则可以使用捏放或双击等动作实现。

Android 支持多种触摸手势,View 组件提供了 OnTouchListener 接口用于监听 MotionEvent 触碰事件,它的 onTouch() 方法则用于处理屏幕的触碰事件。但这个方法对于处理复杂手势相对麻烦,Android 提供了 GestureDetector 类用于检测复杂的手势。

Android 的手势支持体现在两方面:手势检测与手势识别,接下来分别进行讲解。

9.1.2 手势检测

Android 的手势检测用于识别单击、长按、双击、缩放、滑动、拖曳、返回等多种手势。Android 提供了用于手势检测的 GestureDetector 类，它的内部定义了 3 个监听接口和 1 个类，分别是 OnGestureListener 接口、OnDoubleTapListener 接口、OnContextClickListener 接口及 SimpleOnGestureListener 类。GestureDetector 类与 onTouchEvent()方法结合可以识别一些简单的手势。

OnGestureListener 接口提供了用于监听单击、短按、抬起、长按、滚动和滑动等手势的方法，分别为：

- onDown(MotionEvent e)：按下屏幕时回调。
- onShowPress(MotionEvent e)：按下 100ms 后没有松开或者移动回调。
- onSingleTapUp(MotionEvent e)：手指松开时若没有回调 onLongPress()和 onScroll()，那么松开后则会回调此方法，但无法区分是单击还是双击的抬起。
- onLongPress(MotionEvent e)：长按后回调，不再触发其他回调，直到松开。不同的手机长按的时间有可能不同，默认时间为 100ms+500ms。
- onScroll(MotionEvent e1, MotionEvent e2, float distanceX, float distanceY)：手指滑动一段位移、接收到 MOVE 事件时回调，e1 和 e2 分别为之前 DOWN 事件和当前 MOVE 事件，distanceX 和 distanceY 是当前 MOVE 时间和上一个 MOVE 事件的位移量。
- onFling(MotionEvent e1, MotionEvent e2, float velocityX, float velocityY)：执行抛操作后回调，抛操作是 MOVE 事件后抬起瞬间的 X 或 Y 方向的数据达到 50px/s 的操作，velocityX 和 velocityY 分别是用户抬起瞬间的移动速度，一般情况下回调 onFling()必然会回调与 onScroll()方法。

OnDoubleTapListener 接口用于监听双击和单击手势，该接口中的方法包括：

- onSingleTapConfirmed(MotionEvent e)：单击屏幕时触发，不会触发双击事件。
- onDoubleTap(MotionEvent e)：双击屏幕时触发。
- onDoubleTapEvent(MotionEvent e)：触发了 onDoubleTap()方法之后输入事件时会触发，可以实现双击后的控制，如双击后按下、移动和抬起等。

Android 6.0 新增了 OnContextClickListener 接口，用于监听鼠标或触摸板的单击手势，这个接口只有一个方法 onContextClick(MotionEvent e)，它在右击鼠标或触摸板时触发。

SimpleOnGestureListener 类实现了 3 个接口的所有回调方法的简单实现，因此继承该类时，只需要实现需要回调的方法即可。

一个完整的手势从用户首次轻触屏幕开始，跟踪用户手指的位置移动，捕获到用户手指离开屏幕时结束，在整个交互过程中，Android 系统按以下逻辑进行处理。

（1）触屏瞬间会触发 MotionEvent 事件。
（2）OnTouchListener 监听该事件，它的 onTouch()方法获得 MotionEvent 对象。
（3）通过 GestureDetector 对象转发 MotionEvent 对象给 OnGestureListener 接口。
（4）OnGestureListener 根据 MotionEvent 对象封装的信息，实现需要的功能逻辑。

创建 GestureDetector 对象，重写 OnGestureListener 接口方法的示例代码如下：

```
1.  // 定义手势检测器实例
2.  private GestureDetector detector;
```

```
3.      @Override
4.      protected void onCreate(Bundle savedInstanceState) {
5.          super.onCreate(savedInstanceState);
6.          setContentView(R.layout.activity_main);
7.          // 创建 GestureDetector 对象
8.          detector = new GestureDetector(this, new GestureDetector.
                OnGestureListener() {
9.              @Override
10.             public boolean onDown(MotionEvent e) {
11.                 // 用户轻触屏幕
12.                 Log.i(TAG, "单击 - onDown");
13.                 return true;
14.             }
15.             @Override
16.             public void onShowPress(MotionEvent e) {
17.                 // 用户轻触屏幕,尚未松开或拖动
18.                 Log.i(TAG, "短按 - onShowPress");
19.             }
20.             @Override
21.             public boolean onSingleTapUp(MotionEvent e) {
22.                 // 用户轻击屏幕后抬起
23.                 Log.i(TAG, "抬起 - onSingleTapUp");
24.                 return true;
25.             }
26.             @Override
27.             public boolean onScroll(MotionEvent e1, MotionEvent e2,
28.                                 float distanceX, float distanceY) {
29.                 // 用户按下屏幕 & 拖动
30.                 Log.i(TAG, "滚动 - onScroll");
31.                 return true;
32.             }
33.             @Override
34.             public void onLongPress(MotionEvent e) {
35.                 // 用户长按屏幕
36.                 Log.i(TAG, "长按 - onLongPress");
37.             }
38.             @Override
39.             public boolean onFling(MotionEvent e1, MotionEvent e2,
40.                                 float velocityX, float velocityY) {
41.                 // 用户按下屏幕、快速移动后松开
42.                 Log.i(TAG, "滑动 - onFling " + velocityX + ", " + velocityY);
43.                 // 右滑关闭 acitivity
44.                 if (velocityX > 10 && velocityX > Math.abs(velocityY)) {
45.                     Log.i(TAG, "关闭 Activity");
46.                     finish();
47.                 }
48.                 return true;
49.             }
50.         });
51. }
```

重写 Activity 的 onTouchEvent()方法,调用 MotionEvent 的 getAction()方法获取事件类型,其中 ACTION_DOWN,ACTION_MOVE 和 ACTION_UP 分别对应按下、移动、抬起三种动作,示例代码如下所示:

```
1.  // 重写 onTouch()方法
2.  @Override
3.  public boolean onTouchEvent(MotionEvent event) {
```

```
4.     // 获取当前操作的类型码
5.     final int action = event.getActionMasked();
6.     switch(action) {
7.         case (MotionEvent.ACTION_DOWN):
8.             Log.d(TAG, "手指按下动作");
9.             break;
10.        case (MotionEvent.ACTION_MOVE):
11.            Log.d(TAG, "手指移动动作");
12.            break;
13.        case (MotionEvent.ACTION_UP):
14.            Log.d(TAG, "手指抬起动作");
15.            break;
16.        case (MotionEvent.ACTION_CANCEL):
17.            Log.d(TAG, "手指移动从当前控件转移到了外层控件");
18.            break;
19.        case (MotionEvent.ACTION_OUTSIDE):
20.            Log.d(TAG, "手指的移动操作超出了当前的屏幕边界");
21.            break;
22.     }
23.     return false;
24. }
```

在 onTouchEvent() 方法中将触屏事件交给 GestureDetector 处理的示例代码如下：

```
1. @Override
2. public boolean onTouchEvent(MotionEvent event) {
3.     // 将手势操作交给 GestureDetector 进行检测
4.     return detector.onTouchEvent(event);
5. }
```

onTouchEvent() 检测到按下、移动和抬起的运行结果如图 9.1 所示。

图 9.1　MotionEvent 事件的动作类型

GestureDetector 的触屏行为的运行结果如图 9.2 所示。

图 9.2　GestureDetector 的触屏行为

除了可以对 Activity 的触屏手势进行监听，还可以对 View 组件进行监听，代码如下：

```
1. View myView = findViewById(R.id.my_view);
2. myView.setOnTouchListener(new OnTouchListener() {
3.     public boolean onTouch(View v, MotionEvent event) {
```

```
4.               // 此处的处理与上文中的处理一致，不再进行描述
5.      ......
6.               return true;
7.      }
8.  });
```

需要注意的是，根据 Android 的事件分发机制，Android 的事件由 Activity 沿着 Activity->ViewGroup->View 由上往下调用 dispatchTouchEvent 进行分发，当传递到子 View 时，再沿着 View->ViewGroup->Activity 从下往上调用 onTouchEvent 进行处理。当事件的回调方法返回 true 则说明该事件被处理而停止传递，返回 false 则回传给父控件的 onTouchEvent 进行处理，如图 9.3 所示。因此，在 View 组件调用 onTouchEvent()方法处理 ACTION_DOWN 事件后不要返回 false，这样会使得系统将无法针对 ACTION_MOVE 和 ACTION_UP 事件做出响应处理。

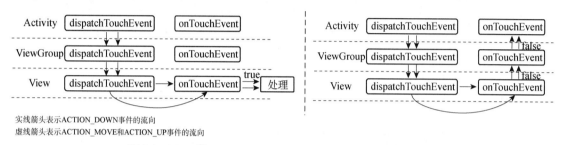

实线箭头表示ACTION_DOWN事件的流向
虚线箭头表示ACTION_MOVE和ACTION_UP事件的流向

图 9.3 View 的 onTouchEvent 返回 true 和 false 的事件流向

实现 onOnGestureListener 接口需要重写所有的六个事件方法，相对烦琐，Android 提供了 GestureDetector.SimpleOnGestureListener 类的 OnGestureListener 接口的所有方法的空实现，因此，继承 SimpleOnGestureListener 类只需重写关注的手势方法即可，以下示例代码创建 GestureDetector 对象时的手势检测监听器继承了 GestureDetector.SimpleOnGestureListener 类并重写了 onFling()。

```
1.  GestureDetector detector = new GestureDetector(this,
2.          new GestureDetector.SimpleOnGestureListener() {
3.      @Override
4.      public boolean onFling(MotionEvent e1, MotionEvent e2,
5.                          float velocityX, float velocityY) {
6.          if (e1.getX() - e2.getX() > 50) {
7.              Log.i(TAG, "从右往左滑动");
8.          } else if (e2.getX() - e1.getX() > 50) {
9.              Log.i(TAG, "从左往右滑动");
10.         }
11.         return super.onFling(e1, e2, velocityX, velocityY);
12.     }
13. });
```

手势检测需要注意以下几点。

● 有返回值的手势事件方法的结果都会返回到 View 的 onTouchEvent 方法中，会影响 View 的事件分发。

● onDown()方法应返回 true，否则就只能触发 onDown()和 onLongPress()事件，其他的手势时间都不能触发，因为影响了事件的分发过程。

● onSingleTapUp 事件和 onSingleTapConfirmed 事件的区别：onSingleTapUp 事件发生后可能会触发 onDoubleTap 事件；而 onSingleTapConfirmed 事件是严格的单击事件，后续不会

触发 onDoubleTap 事件。

9.1.3 手势识别

Android 还提供了手势识别功能，可以将多个持续的触屏事件形成的特定形状的手势添加到指定文件进行保存，当用户下次出现这个手势会被识别。Android 的手势识别包括以下几个基本的类。

- GestureLibrary：手势库的类，用以存储手势。
- GestureLibraries：创建手势库的工具类。
- GestureOverlayView：用于手势编辑的透明覆盖层组件，类似一个绘图组件。

GestureLibraries 类提供了 4 个静态方法用于加载手势，分别从公开的文件目录、私有数据的文件目录和项目的 raw 目录进行加载，返回 GestureLibrary 对象，方法定义如表 9.1 所示。

表 9.1 GestureLibraries 类的加载手势库的静态方法

方法名称	功能描述
fromFile(String path)	从公开目录加载，path 为包含文件名的目录名称字符串
fromFile(File path)	从公开目录加载，path 为包含文件名的 File 对象
fromPrivateFile(Context context, String name)	从应用的私有文件目录下的名为 name 的文件中加载
fromRawResource(Context context, int resId)	从 resId 代表的资源中加载

GestureLibrary 类用于管理手势库的添加、获取、识别、删除等功能，方法描述如表 9.2 所示。

表 9.2 GestureLibrary 类的常用方法

方法名称	功能描述
addGesture(String name, Gesture gesture)	将名为 name 的手势对象添加到手势库
save()	添加或删除手势之后，调用该方法保存手势库
getGestureEntries()	获取手势库所有的手势名称的集合
getGestures(String name)	获取名为 name 的所有手势的集合
recognize(Gesture gesture)	识别与 gesture 匹配的所有手势
removeEntry(String name)	删除名为 name 的所有手势
removeGesture(String name, Gesture gesture)	删除名为 name 的手势

GestureOverlayView 组件用于绘制手势，常用的 XML 属性及监听器的定义如表 9.3 所示。XML 属性也可以使用相应的 set 方法进行设置。

表 9.3 GestureOverlayView 的常用属性及监听器的定义

属性/监听器类的名称	功能描述
android:eventsInterceptionEnabled	定义当手势被识别后是否拦截该手势动作
android:fadeDuration	当用户画完手势效果淡出的毫秒数
android:fadeEnabled	用户画完之后手势是否自动淡出
android:gestureColor	设置手势的颜色

续表

属性/监听器类的名称	功能描述
android:gestureStrokeType	设置笔画的类型是单笔还是多笔绘制
android:gestureStrokeWidth	设置笔画的粗细
OnGestureListener	手势监听器
OnGesturePerformedListener	手势绘制完成监听器
OnGesturingListener	手势绘制过程中的监听器

Android 的增加和识别手势的基本步骤如下。

（1）设置 GestureOverlayView 界面控件

在布局文件中加入 GestureOverlaytView 控件，设置表 9.3 列出的一些属性，示例代码如下：

```
1.  <android.gesture.GestureOverlayView
2.      android:id="@+id/gesture_overlay"
3.      android:layout_width="match_parent"
4.      android:layout_height="match_parent"
5.      android:gestureColor=" @android:color/holo_green_light"
6.      android:fadeDuration="500"
7.      android:gestureStrokeWidth="8"
8.      android:gestureStrokeType="multiple"/>
```

（2）加载手势库

从应用的私有数据文件目录下加载文件名为 gestures 的手势库文件，如果不能加载则无法继续。

```
1.  private GestureLibrary library;
2.  @Override
3.  protected void onCreate(Bundle savedInstanceState) {
4.      super.onCreate(savedInstanceState);
5.      setContentView(R.layout.activity_overlay);
6.      // 获取应用的私有数据目录的路径
7.      String path = getExternalFilesDir("").getAbsolutePath();
8.      // 从gestures文件中获取GestureLibrary对象，若没有文件则创建
9.      library = GestureLibraries.fromFile(path + "/gestures");
10.     if (!library.load()) {
11.         Toast.makeText(this, "不能加载手势库", Toast.LENGTH_SHORT).show();
12.         finish();
13.     }
14. }
```

（3）绑定监听器

绑定 GestureOverlayView 的 OnGesturePerformedListener 监听器，当绘制完成时将手势保存到文件中：

```
1.  // 绑定监听器
2.  gestureView.addOnGesturePerformedListener(
3.      new GestureOverlayView.OnGesturePerformedListener() {
4.      // 手势绘制完成时调用
5.      @Override
6.      public void onGesturePerformed(GestureOverlayView overlay,
                                       Gesture gesture) {
7.          OverlayActivity.this.gesture = gesture;
8.          // 加载对话框的界面布局
9.          View dialog
```

```
          getLayoutInflater().inflate(R.layout.dialog_save, null,false);
10.                 gesturePic = dialog.findViewById(R.id.iv_show);
11.                 gestureName = dialog.findViewById(R.id.et_name);
12.                 // 根据Gesture的手势创建位图
13.                 Bitmap bitmap = gesture.toBitmap(128, 128, 10, 0xffff0000);
14.                 gesturePic.setImageBitmap(bitmap);
15.                 // 使用对话框显示界面
16.                 new AlertDialog.Builder(OverlayActivity.this).setView(dialog)
17.                         .setPositiveButton("保存", new DialogInterface.
                                                   OnClickListener() {
18.                             @Override
19.                             public void onClick(DialogInterface dialog,
                                                       int which) {
20.                                 saveFile();
21.                             }
22.                         })
23.                         .setNegativeButton("取消", null)
24.                         .show();
25.             }
26.         });
```

第 7 行代码将绘制完成的 Gesture 对象参数赋给定义的 Gesture 对象，然后使用对话框调用 saveFile()方法进行保存，第 20 行的 saveFile()方法调用了 GestureLibrary 的 addGesture()方法将绘制的手势添加到手势库，调用它的 save()方法进行保存。示例代码如下：

```
1.  private void saveFile() {
2.      if (gesture != null) {
3.          // 添加绘制的手势
4.          library.addGesture(gestureName.getText().toString(), gesture);
5.          // 保存手势库
6.          if (library.save()) {
7.              Toast.makeText(this, "手势库文件保存成功", Toast.LENGTH_
                              SHORT).show();
8.          } else {
9.              Toast.makeText(this, "手势库文件保存失败", Toast.LENGTH_
                              SHORT).show();
10.         }
11.     }
12. }
```

执行以上代码，保存手势库文件的目录如图 9.4 所示。

（4）识别手势

调用 GestureLibrary 对象的 recognize()方法获取手势库与绘制手势匹配的所有手势的集合，Prediction 对象代表匹配的手势对象，这个对象包含两个属性：name 表示手势名称，score 表示相似度，相似度的值越大表示手势图形越相似。返回值集合按照 score 的倒序排列，因此第 1 个 Prediction 就是最匹配的手势，根据此相似度值识别手势，示例代码如下：

图 9.4　手势库保存的目录

```
1.  if (gesture != null) {
2.      // 获取识别的手势集合，按score的值倒序排列
3.      ArrayList<Prediction> predictions = library.recognize(gesture);
4.      if(predictions.size() > 0) {
5.          Prediction prediction = predictions.get(0);
6.          // 判断相似度，值越大相似度越高
7.          if (prediction.score > 1.0) {
```

```
8.              Toast.makeText(OverlayActivity.this, "手势为" + prediction.
                    name,
9.              Toast.LENGTH_SHORT).show();
10.         }
11.     }
12. }
```

绘制手势、保存手势库及手势识别的运行结果如图 9.5 所示。

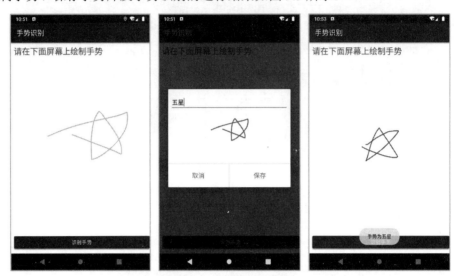

图 9.5　手势绘制、保存和识别的运行效果

 ## 9.2　传感器开发

Android 系统提供了对多种传感器的支持，如磁场、温度、压力、加速度等。Android 设备内置的传感器基本都能提供高精度的原始数据，非常适合用来监测设备的移动、定位或监测设备周围环境的变化。例如，游戏可以跟踪设备的重力传感器的读数，推断用户的各种手势和动作，如倾斜、摇晃或晃动等，指南针应用则可以利用磁场传感器和加速度传感器报告方位，微信的摇一摇功能则是利用加速度传感器实现的。

9.2.1　传感器简介

传感器是一种用来探测外界信号、物理条件或化学组成的物理设备，它能将探测的信息传输给其他设备或部件，Android 平台支持的传感器包括三大类。

- 运动传感器 Motion Sensor：用于沿三个轴方向测量加速力和旋转力，包括加速度计、重力传感器、陀螺仪和旋转矢量传感器。
- 环境传感器 Environmental Sensor：用于测量各种环境参数，如空气温度和压力、照明和湿度，包括气压计、光度计和温度计。
- 位置传感器 Position Sensor：用于测量设备的物理位置，包括方向传感器和磁力计。

Android 提供 android.hardware.Sensor 类表示传感器，该类定义了相应的整型常量代表不同的传感器类型，常见的传感器的整型常量值如表 9.4 所示。

表 9.4 Android 系统支持的常见传感器类型

传感器名称	内部整数值	Sensor 类定义的类型常量	常见用途
加速度传感器	1	Sensor.TYPE_ACCELEROMETER	动态检测摇晃、倾斜等
磁力传感器	2	Sensor.TYPE_MAGNETIC_FIELD	创建罗盘
方向传感器	3	Sensor.TYPE_ORIENTATION	确定设备位置（已废弃）
陀螺仪传感器	4	Sensor.TYPE_GYROSCOPE	旋转检测旋转、转动等
光线传感器	5	Sensor.TYPE_LIGHT	控制屏幕亮度
压力传感器	6	Sensor.TYPE_PRESSURE	监测气压变化
温度传感器	7	Sensor.TYPE_TEMPERATURE	监测温度（已废弃）
接近传感器	8	Sensor.TYPE_PROXIMITY	通话过程中手机的位置
重力传感器	9	Sensor.TYPE_GRAVITY	动态检测摇晃、倾斜等
线性加速度传感器	10	Sensor.TYPE_LINEAR_ACCELERATION	监测单个轴向上的加速度
旋转矢量传感器	11	Sensor.TYPE_ROTATION_VECTOR	动态检测和旋转检测
湿度传感器	12	Sensor.TYPE_RELATIVE_HUMIDITY	监测露点、绝对湿度和相对湿度
温度传感器	13	Sensor.TYPE_AMBIENT_TEMPERATURE	监测气温

9.2.2 使用传感器

Android 系统提供了传感器框架访问设备内置的传感器，传感器框架提供多个类和接口执行各种与传感器相关的任务，表 9.5 描述了传感器的常用的类和接口定义。

表 9.5 传感器常用的类及接口描述

类/接口名称	类描述
SensorManager 类	传感器管理类，提供访问传感器列表、注册或注销传感器的事件监听、获取方位信息等方法，通过 getSystemService()方法获取它的实例对象
Sensor 类	传感器类，提供访问传感器技术参数的方法，包括名称、类型、版本及供应商等
SensorEvent 类	传感器事件类，提供传感器事件相关的信息，包括传感器数据、传感器类型、数据精度及触发事件的时间等
SensorEventListener 接口	传感器事件监听器，包括传感器精度变化、数据变化的两个回调方法。

使用传感器主要完成两个任务：识别传感器对象及技术参数、监控传感器事件，具体分为以下步骤。

（1）获取 SensorManager 对象

调用 getSystemService（Context.SENSOR_SERVICE）方法获取，SensorManager 对象的 getSensorList()方法获取设备上的所有传感器的列表，示例代码如下：

```
// 获取手机的所有传感器
public void getAllSensors() {
    SensorManager manager = (SensorManager) getSystemService
                    (Context.SENSOR_SERVICE);
    List<Sensor> sensors = manager.getSensorList(Sensor.TYPE_ALL);
    for (Sensor sensor : sensors) {
        Log.i("MainActivity", sensor.toString());
    }
}
```

得到的传感器信息如图 9.6 所示。

```
I/MainActivity: {Sensor name="Goldfish 3-axis Accelerometer", vendor="The Android Open Source Project", version=1, type=1, maxRange=2.8, resolution=2.480159E-4, power=3.0, minDelay=10000}
I/MainActivity: {Sensor name="Goldfish 3-axis Gyroscope", vendor="The Android Open Source Project", version=1, type=4, maxRange=11.111111, resolution=0.001, power=3.0, minDelay=10000}
I/MainActivity: {Sensor name="Goldfish 3-axis Magnetic field sensor", vendor="The Android Open Source Project", version=1, type=2, maxRange=2000.0, resolution=1.0, power=6.7, minDelay=10000}
I/MainActivity: {Sensor name="Goldfish Orientation sensor", vendor="The Android Open Source Project", version=1, type=3, maxRange=360.0, resolution=1.0, power=9.7, minDelay=10000}
I/MainActivity: {Sensor name="Goldfish Ambient Temperature sensor", vendor="The Android Open Source Project", version=1, type=13, maxRange=80.0, resolution=1.0, power=0.001, minDelay=10000}
```

图 9.6 Sensor 对象的信息

（2）获取 Sensor 对象

SensorManager 的 getDefaultSensor（int type）方法用于获取指定类型的传感器对象，Sensor 对象提供了一些 getXxx()方法获取传感器的信息，如：获取名称的方法 getName()、获取版本号的方法 getVersion()、获取功率的方法 getPower()及获取最大量程的方法 getMaximumRange() 等，可以在实际使用过程中进行选择。

（3）注册 Sensor 对象的监听器

在 Activity 的 onResume()方法中调用 SensorManager 对象的 registerListener()方法为传感器注册监听器，获取传感器的数据。registerListener()方法有三个参数，分别为：实现 SensorEventListener 接口对象、传感器对象和获取传感器数据的频率，频率的取值包括：

- SensorManager.SENSOR_DELAY_FASTEST：延迟 10ms，速度最快，会造成手机电量的大量消耗，只有特别依赖于传感器数据的应用才推荐使用。
- SensorManager.SENSOR_DELAY_GAME：延迟 20ms，用于游戏的频率，一般有实时性要求的应用推荐使用。
- SensorManager.SENSOR_DELAY_UI：延迟 60ms，用于普通用户界面的频率，比较省电，系统开销也很小，但延迟较大。
- SensorManager.SENSOR_DELAY_NORMAL：延迟 200ms，正常频率，用于对实时性要求不高的应用，适用于大多数应用。

（4）重写 onAccuracyChanged()、onSensorChanged()方法

SensorEventListener 接口有两个方法，其中 onAccuracyChanged()在传感器的精度发生变化时调用，onSensorChanged()则是数据发生变化时调用，Sensor 注册监听器之后需要重写这两个方法。

onAccuracyChanged()的精度取值定义了 4 种常量：SENSOR_STATUS_ACCURACY_LOW 为低精度、SENSOR_STATUS_ACCURACY_MEDIUM 为中精度、SENSOR_STATUS_ACCURACY_HIGH 为高精度及 SENSOR_STATUS_UNRELIABLE 为不可靠精度。

onSensorChanged()方法传递的参数为 SensorEvent 对象，它包含传感器的数据信息，包括：数据的精度、生成数据的传感器对象、生成数据的时间戳及传感器记录的新数据等。

（5）注销 Sensor 对象

传感器管理属于底层系统服务，类似于 Service，使用结束后务必注销，避免传感器一直在后台运行造成电量消耗，影响手机性能。注销 Sensor 对象是在 Activity 类的 onPause()方法中调用 SensorManager 对象的 unregisterListener()方法来注销指定的传感器监听器的。

以上是传感器使用的步骤解析，使用光线传感器的示例代码如下所示：

```
1.  public class SensorActivity extends AppCompatActivity
2.          implements SensorEventListener {
3.      private SensorManager sensorManager;
4.      private Sensor sensor;
5.      private TextView tvLight;
6.      @Override
```

```
7.      protected void onCreate(Bundle savedInstanceState) {
8.          super.onCreate(savedInstanceState);
9.          setContentView(R.layout.activity_sensor);
10.         tvLight = findViewById(R.id.tv_light);
11.         // 获取感应器服务对象
12.         sensorManager = (SensorManager) getSystemService
                            (Context.SENSOR_SERVICE);
13.         // 获取光线传感器对象
14.         sensor = sensorManager.getDefaultSensor(Sensor.TYPE_LIGHT);
15.         if (sensor == null) {
16.             Toast.makeText(this, "此传感器不存在", Toast.LENGTH_SHORT).
                            show();
17.             // 进行响应的处理
18.         }
19.     }
20.     @Override
21.     protected void onResume() {
22.       super.onResume();
23.       // 注册监听器
24.       sensorManager.registerListener(this, sensor,SensorManager.
                            SENSOR_DELAY_NORMAL);
25.     }
26.     @Override
27.     protected void onDestroy() {
28.         super.onDestroy();
29.         sensorManager.unregisterListener(this); // 注销监听器
30.     }
31.     @Override
32.     public void onSensorChanged(SensorEvent event) {
33.         // 当传感器数据发生变化时，values 数组中第一个值是当前的光照强度
34.         float value = event.values[0];
35.         tvLight.setText("当前亮度 " + value + " lx(勒克斯)");
36.     }
37.     @Override
38.     public void onAccuracyChanged(Sensor sensor, int accuracy) {
39.         // 当传感器精度发生变化时
40.     }
41. }
```

光线传感器在真机与模拟器上运行结果不同，如图 9.7 所示。

图 9.7　真机和模拟器的光线传感器的光照数据

至此，讲解了 Android 传感器的概念及基本用法，由于各个品牌的手机内置的传感器不尽相同，在获取传感器对象时应进行是否存在的判断，避免出错；另外，建议在 onResume() 和 onPause() 方法中注册和注销传感器监听器，避免手机资源的消耗和电量的浪费。

9.3 Android Jetpack

9.3.1 Jetpack 简介

Android 从 2008 年推出 1.0 版本至今，十几年的发展已经使其成为相对成熟和稳定的手机操作系统，基于 Android 的开放性，优秀的第三方框架也数量庞大，如网络访问框架 Retrofit、事件总线框架 EventBus、数据库框架 GreenDAO、数据刷新框架 SmartRefreshLayout 等，以及 MVC、MVP 和 MVVM 等各种开发模式，可谓百家争鸣，它们在带来开发便利性的同时，也大大增加了选型和学习的成本，导致 Android 开发的技术良莠不齐，碎片化严重。为了帮助 Android 更快、更好地开发 App，Google 在 2018 I/O 大会发布了一系列辅助 Android 开发者的组件库，统称为 Android Jetpack，旨在帮助开发者快速构建出稳定、高性能、测试友好且向后兼容的 Android 应用。

Android Jetpack 是 Android SDK 之外的组件，包含架构、UI 界面、基础和行为等四大类的组件，如图 9.8 所示。

- 基础组件 Foundation：提供向后兼容性、测试以及 Kotlin 语言的支持；
- 架构组件 Architecture：提供设计稳健、可测试且易维护的应用；
- 行为组件 Behavior：提供应用与标准 Android 服务的集成；
- 界面 UI 组件：提供控件和辅助程序，使得应用简单易用，提升用户体验。

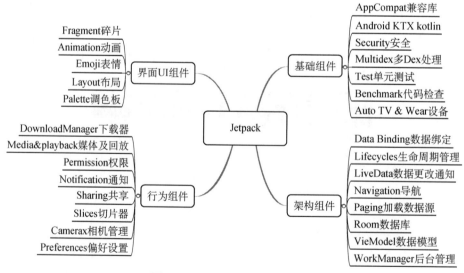

图 9.8　Android Jetpack 组件图

使用 Jetpack 组件具有以下优势：
- 轻松管理应用程序的生命周期。
- 构建可观察的数据对象，以便在基础数据库更改时通知视图。

- 存储在应用程序转换中未销毁的 UI 数据，界面重建后恢复数据。
- 轻松实现 SQLite 数据库。
- 系统自动调度后台任务的执行，优化性能。

尽管 Jetpack 极大地方便了 Android 开发，但 Jetpack 还远未成熟，在使用过程中也要关注新版本的发布及存在的一些缺点，如：数据双向绑定会导致 View 不可重用、DataBinding 实现数据绑定的同时也增加了 Bug 调试的难度、复杂业务使得 View 页面也复杂从而导致 Model 层的代码量激增等，因此，应深入理解 Jetpack 的组件，合理使用 Jetpack。

9.3.2 Jetpack 架构组件

Jetpack 不属于 Android Framework，它只是 Android 开发的一种辅助手段，帮助开发者解决一些常见问题，如：版本兼容、API 易用性和生命周期管理等，其中 Architecture 部分的组件（Android Architecture Components，AAC）组合成了一套完整的架构解决方案，如图 9.9 所示。该架构中的每个组件仅依赖其下一级的组件，整体架构可划分为三个部分。

① UI 层：包含 Activity 和 Fragment。
② ViewModel 层：既可以是 MVVM 的 VM、MVP 的 P，也可以是 UI 的数据适配，实现数据驱动 UI。
③ Repository 层：数据层，对上层屏蔽数据的来源，从本地、远程、数据库等数据源获取数据。

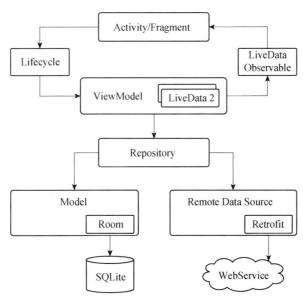

图 9.9 Android 应用的推荐架构的示意图

Android 系统为了提供向下兼容，推出了 Android Support Library 库，如大家都熟知的 support-v4、support-v7 库，但由于 Support 库的版本号与 SDK 高度耦合，所有的 Support 库的版本又必须保持一致，无法单独升级。为了解决 Support 库的版本管理问题，Google 在 2018 年的 I/O 大会上发布了 AndroidX 替换 Android Support Library 库，原有 Support 库的 API 都映射到 androidx 命名空间，AndroidX 的版本管理将不再依赖 SDK 的版本，使其能够应对更加频繁的更新。除了现有的支持库之外，AndroidX 还包含了最新的 Jetpack 组件，Android 之前的 AAC 在 Support 库中发布的组件，如 LiveData、ViewModel 等也被并入了 AndroidX。

综上所述，Jetpack 和 AndroidX 可以等同理解，都是对 Andriod 的 SDK 之外的功能组件集合的描述，Jetpack 确立了 Andriod 的标准化开发模式，代表了 Android 原生开发的未来方向。

由于 Jetpack 的组件众多，限于篇幅，仅讲解图 9.9 涉及的 Lifecycle、LiveData、ViewModel、等组件，首先讲解它们的基本概念，然后通过一个完整的综合案例使用这些组件。

在使用 Jetpack 的组件之前，首先需要在项目的 build.gradle 文件中添加 google()代码库，然后在 app/build.gradle 文件中声明依赖项，如：

```
1.  dependencies {
2.      def lifecycle_version = "2.3.2"
3.      // ViewModel
4.      implementation "androidx.lifecycle:lifecycle-viewmodel:
                        $lifecycle_version"
5.      // LiveData
6.      implementation "androidx.lifecycle:lifecycle-livedata:$lifecycle_
                        version"
7.      // Lifecycles only
8.      implementation "androidx.lifecycle:lifecycle-runtime:$lifecycle_
                        version"
9.      // Annotation processor
10.     annotationProcessor "androidx.lifecycle:lifecycle-compiler:
                            $lifecycle_version"
11. }
```

1. Lifecycle

Lifecycle 是 Jetpack 提供的、具有生命周期感知能力的组件，Lifecycle 组件采用观察者模式管理生命周期。当 Activity 或 Fragment 的生命周期发生变化时，通过向 Activity 或 Fragment 注册生命周期的回调监听，感应生命周期的变化，无须重写 Activity 的生命周期回调方法，主要涉及 Lifecycle、LifecycleOwner 和 LifecycleObserver 等类或接口，详细描述如表 9.6 所示。

表 9.6 Lifecycle 的主要类与接口描述

类/接口名称	描述说明
LifecycleOwner 接口	Lifecycle 拥有者，实现该接口就表明该类具有生命周期，生命周期改变的事件会被 LifecycleObserver 接收到。该接口只有一个方法 getLifecycle()用于返回 Lifecycle 对象
Lifecycle 类	生命周期抽象类，具有添加、移除监听的方法
LifecycleRegistry 类	Lifecycle 的实现类，协助组件处理生命周期，可处理多个观察者
LifecycleObserver 接口	Lifecycle 观察者，是一个空接口，通过注解方式观察生命周期方法
Lifecycle.State 类	生命周期枚举类，取值：DESTROYED、INITIALIZED、CREATED、STARTED 和 RESUMED
LifeCycle.Event 类	生命周期事件枚举类，与 Activity 的生命周期回调一一对应，取值：ON_CREATE、ON_START、ON_RESUME、ON_PAUSE、ON_STOP、ON_DESTROY、ON_ANY，一般与 OnLifecycleEvent 注解一起使用，用于在 LifecycleObserver 中标注方法属于哪个生命周期阶段

State 状态与 Event 事件并非一一对应，它们之间的对应关系如图 9.10 所示，其中，ON_CREATE 和 ON_STOP 事件对应 CREATED 状态，ON_START 和 ON_PAUSE 事件对应 STARTED 状态，ON_RESUME、ON_DESTROY 分别对应 RESUMED、DESTROYED 状态，ON_ANY（图中未标注）则没有对应的状态。例如，当 Activity 在 RESUME 状态时按下 Home 键，Activity 的状态则会变到 STARTED 状态，对应的 Event 为 ON_PAUSE。

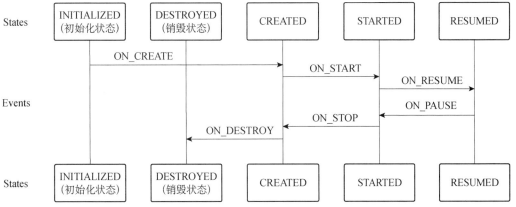

图 9.10 State 状态与 Event 事件之间的关系

Lifecycle 类的核心方法主要用于管理观察者。

● addObserver(LifecycleObserver)：用于添加一个 Observer 实例接收导致当前状态的所有事件。如果 Lifecycle 处于 RESUMED 状态，那观察者将收到 ON_CREATE、ON_START、ON_RESUME 三个事件。

● removeObserver()：从 Lifecycle 的观察者列表中删除给定的观察者，删除后将不再接收任何触发事件。

● getCurrentState()：返回生命周期所在的当前状态。

Lifecycle、Observer 和 LifecycleOwner 之间的关系如图 9.11 所示。

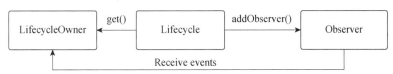

图 9.11 Lifecycle 与 Observer、LifecycleOwner 之间的关系

Lifecycle 的使用也非常简单，首先创建一个自定义类实现 LifecycleObserver 接口，通过 @OnLifecycleEvent 注解标识不同 Event 类型定义相应的处理方法，示例代码如下：

```
1.   public class MyLifecycleObserver implements LifecycleObserver {
2.       private static final String TAG = "MyLifecycleObserver";
3.       // 观察 onCreate()方法
4.       @OnLifecycleEvent(Lifecycle.Event.ON_CREATE)
5.       private void create() {
6.           Log.i(TAG, "MainActivity的onCreate()被调用");
7.       }
8.       // 观察 onPause()方法
9.       @OnLifecycleEvent(Lifecycle.Event.ON_PAUSE)
10.      private void pause() {
11.          Log.i(TAG, "MainActivity的onPause()被调用");
12.      }
13.      // 观察 onResume()方法
14.      @OnLifecycleEvent(Lifecycle.Event.ON_RESUME)
15.      private void resume() {
16.          Log.i(TAG, "MainActivity的onResume()被调用");
17.      }
18.      // 观察 onDestroy()方法
19.      @OnLifecycleEvent(Lifecycle.Event.ON_DESTROY)
20.      private void destroy() {
```

```
21.         Log.i(TAG, "MainActivity的onDestroy()被调用");
22.     }
23. }
```

然后在 MainActivity 类中调用 getLifecycle()获取 Lifecycle 对象，该对象的 addObserver() 方法用于添加该观察者对象，代码如下：

```
1. @Override
2. protected void onCreate(Bundle savedInstanceState) {
3.     super.onCreate(savedInstanceState);
4.     setContentView(R.layout.activity_main);
5.     // 添加观察者
6.     getLifecycle().addObserver(new MyLifecycleObserver());
7. }
```

这里要求 Activity 类必须继承自实现了 LifecycleOwner 接口的类，如 Androidx 包下的 AppCompatActivity、Fragment 类，只有实现了 LifecycleOwner 接口才能使得 Activity 或 Fragment 当前所处的生命周期状态 State 的变化会触发 Event 事件，该事件才能被注册的观察者对象接收处理，因此对于未实现 LifecycleOwner 接口的类则必须实现该接口，设置 State 状态，生命周期变化才能被感知，示例代码如下：

```
1. public class DemoActivity extends Activity implements LifecycleOwner {
2.     // 定义Lifecycle对象
3.     private LifecycleRegistry registry;
4.
5.     @Override
6.     protected void onCreate(@Nullable Bundle savedInstanceState) {
7.         super.onCreate(savedInstanceState);
8.         setContentView(R.layout.activity_main);
9.         // 创建Lifecycle对象，添加观察者
10.        registry = new LifecycleRegistry(this);
11.        registry.addObserver(new MyLifecycleObserver());
12.        registry.setCurrentState(Lifecycle.State.CREATED);
13.    }
14.    @Override
15.    protected void onResume() {
16.        super.onResume();
17.        // 设置State状态
18.        registry.setCurrentState(Lifecycle.State.RESUMED);
19.    }
20.    @Override
21.    protected void onPause() {
22.        super.onPause();
23.        registry.setCurrentState(Lifecycle.State.DESTROYED);
24.    }
25.    // 获取Lifecycle对象
26.    @NonNull
27.    @Override
28.    public Lifecycle getLifecycle() {
29.        return registry;
30.    }
31. }
```

修改 AndroidManifest.xml 文件，设置 MainActivity 为启动界面后运行，切换 MainActivity 的状态经历"启动->隐藏->恢复->销毁"的过程的日志输出如图 9.12 所示。

修改 AndroidManifest.xml 文件，设置 DemoActivity 为启动界面后运行，切换 DemoActivity 的状态经历"启动->隐藏->恢复->销毁"的过程的日志输出如图 9.13 所示。

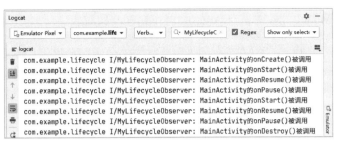

图9.12　Lifecycle 感知的 Activity 生命周期变化的日志

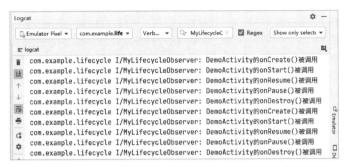

图9.13　Lifecycle 感知的 DemoActivity 生命周期变化的日志

比较两个 Activity 的日志输出发现，DemoActivity 按下 Home 键时会销毁界面，其原因是：它的 onPause()方法将 State 状态设成了 DESTROYED 触发 ON_DESTROY 事件，导致 onDestroy()方法被调用而销毁。若要使得 DemoActivity 的状态切换与 MainActivity 一致，应根据 State 与 Event 的对应关系设置 State 状态，即将 DemoActivity 类的 onPause()方法的 Lifecycle 状态设为 STARTED。

2. ViewModel

ViewModel 是 Jetpack 的核心组件，Google 官方的定义是：以注重生命周期的方式存储和管理界面相关的数据，用于解决 Android 应用因屏幕旋转等配置更改导致 Activity 或 Fragment 状态数据的丢失问题，实现数据与界面视图解耦，以及规范 MVVM 架构的实现，它具备两层含义。

● 注重生命周期的方式：ViewModel 的生命周期作用于整个 Activity，并在生命周期内保持局部单例，使得多个 Activity、Fragment 之间的数据可以共享，通信和维护都变得简单。

● 存储和管理界面相关的数据：ViewModel 最根本的职责就是维护 UI 界面的状态，也就是维护对应的数据。

ViewModel 是 MVVM 架构中的关联层，担当数据驱动的职责，提供和管理 UI 界面的数据。在 Activity、Fragment 等组件的整个生命周期过程中，有且只有一个 ViewModel 实例，因此它可以实现多个 Fragment 之间的数据共享。

由图 9.14 所示的 ViewModel 生命周期表明，当屏幕发生旋转而导致 Activity 被销毁重建时，ViewModel 并不会被销毁，它能帮助 Activity 在重建后维护数据状体，比 onSaveInstanceState()方法更具优势，onSaveInstanceState()方法只能恢复少量序列化或反序列化的数据，而 ViewModel 不仅支持大量数据，还不需要序列化、反序列化操作。

需要注意的是，ViewModel 应只负责管理界面 UI 的数据，不负责获取数据，而且要避免在 ViewModel 中引用 View、Activity 或者 Fragment 上下文，否则会造成内存泄露。如果必须引用上下文，ViewModel 可以继承自 AndroidViewModel 实现。

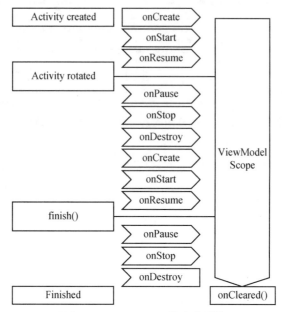

图 9.14　ViewModel 的生命周期

ViewModel 的使用非常简单，只需创建继承自 ViewModel 的类，该类负责为界面准备数据，可以结合接下来讲解的 LiveData 使用，最简单的示例代码如下：

```
1.  public class MyViewModel extends ViewModel {
2.      private int number = 0;
3.      public MyViewModel() {
4.      }
5.      public MyViewModel(int number) {
6.          this.number = number;
7.      }
8.      // 省略 getter/setter 方法
9.  }
```

获取 ViewModel 对象可以使用 ViewModelProvider 类的 get()方法，ViewModelProvider 类是 ViewModel 的提供者，提供获取 ViewModel 的入口，它依赖 ViewModelStore 类存储 ViewModel、Factory 接口生成和恢复 ViewModel，提供了三个构造方法，分别是：

● ViewModelProvider(ViewModelStoreOwner)：该构造函数创建不传参的 ViewModel 对象。

● ViewModelProvider(ViewModelStoreOwner，Factory)：该构造函数使用 Factory 对象传递参数。

● ViewModelProvider(ViewModelStore，Factory)：直接使用 ViewModelStore 对象，Factory 对象传递参数用于传递参数。

ViewModelStoreOwner 关联绑定生命周期，Androidx 包的 Fragment 和 AppCompatActivity 都默认实现了 ViewModelStoreOwner 接口，因此在 MainActivity 类中获取 ViewModel 对象可以传递 this，示例代码如下：

```
1.  public class MainActivity extends AppCompatActivity {
2.      private TextView tvData;
3.      private MyViewModel viewModel;
4.      @Override
5.      protected void onCreate(Bundle savedInstanceState) {
```

```
6.         super.onCreate(savedInstanceState);
7.         setContentView(R.layout.activity_main);
8.         // 创建 ViewModel 对象
9.         viewModel = new ViewModelProvider(this).get(MyViewModel.
               class);    // 无参构造
10.        Log.i("MainActivity", viewModel.getNumber() + "");
11.        // 设置 TextView 的值
12.        tvData = findViewById(R.id.textView);
13.        tvData.setText(String.valueOf(viewModel.getNumber()));
14.        // +1 按钮的事件处理
15.        findViewById(R.id.btn_1).setOnClickListener(view -> {
16.            viewModel.setNumber(viewModel.getNumber() + 1);
17.            tvData.setText(String.valueOf(viewModel.getNumber()));
18.        });
19.        // -1 按钮的事件处理
20.        findViewById(R.id.btn_2).setOnClickListener(view -> {
21.            viewModel.setNumber(viewModel.getNumber() - 1);
22.            tvData.setText(String.valueOf(viewModel.getNumber()));
23.        });
24.    }
25. }
```

ViewModelProvider.Factory 接口负责实例化 ViewModel 对象的传递参数，它的 create()方法用于创建 ViewModel 实例对象。创建自定义 Factory 类传递 number 值的示例代码如下：

```
1.  public class MyViewModelFactory implements ViewModelProvider.Factory {
2.      private int number;
3.      public MyViewModelFactory(int number) {
4.          this.number = number;
5.      }
6.      @NonNull
7.      @Override
8.      public <T extends ViewModel> T create(@NonNull Class<T> modelClass) {
9.          return (T) new MyViewModel(number);
10.     }
11. }
```

在 MainActivity 类中使用上述的第 2 个构造方法创建 ViewModel 对象并传递参数，代码如下：

```
1.  @Override
2.  protected void onCreate(Bundle savedInstanceState) {
3.      super.onCreate(savedInstanceState);
4.      setContentView(R.layout.activity_main);
5.      // 创建 ViewModel 对象
6.      viewModel = new ViewModelProvider(this, new MyViewModelFactory(11))
7.              .get(MyViewModel.class);   // 有参构造
8.      Log.i("MainActivity", viewModel.getNumber() + "");
9.  }
```

3. LiveData

LiveData 是 Jetpack 架构组件 Lifecycle 库的一部分，是一个可感知生命周期可观察容器类。与常规的可观察类不同，LiveData 具有生命周期的感知能力，也就是它遵循 Activity、Fragment 或 Service 等组件的生命周期，并且它仅更新处于活跃生命周期状态的应用组件。LiveData 的作用是持有一份给定的数据，并在生命周期变化中观察它，只要数据被观察者订阅，当数据发生变化时就会通知观察者做出响应。使用 LiveData 具有以下优势。

- 确保 UI 界面的数据状态始终保持一致。
- 没有内存泄露：观察者绑定到 Lifecycle 对象，在其生命周期 DESTROYED 后自动解除绑定。
- 不会因为 Activity 停止而崩溃：如 Activity 执行 finish()方法后，就不会收到任何 LiveData 事件。
- 无须手动处理生命周期：UI 组件只需观察相关数据，不会停止或恢复观察，LiveData 自动管理这些操作，因为 LiveData 可以时刻感知生命周期状态的改变。
- 始终保持最新数据：在生命周期从非激活状态变为激活状态，会始终保持最新数据，如后台 Activity 返回前台后，能立即收到最新状态的数据。
- 适当的配置更改：当 Activity、Fragment 发生屏幕旋转等更改而重建时，会立即收到最新的可用数据。
- 资源共享：LiveData 适合用于组件 Activity、Fragment 之间的通信及共享数据资源。

LiveData 类的常用方法的描述如表 9.7 所示。

表 9.7　LiveData 类的主要方法描述

方法名称	功能描述
setValue(T)	在主线程中设置 LiveData 持有的数据
postValue(T)	在子线程中设置 LiveData 持有的数据
getValue()	获取 LiveData 持有的数据
observer(LifecycleOwner, Observer)	根据 Lifecycle 的状态进行不同处理。 • DESTROYED 状态：忽略该调用； • 非 DESTROYED 状态：Observer 被添加到 LiveData 的观察者列表中，并与 LifecycleOwner 绑定，当它的生命周期状态变成 DESTROYED 时 Observer 被自动移除
removeObserver(Observer)	移除观察者 Observer
removeObservers(LifecycleOwner)	移除与该 LifecycleOwner 所绑定的所有观察者
hasActiveObservers()	检查 LiveData 中是否有活跃的观察者
observeForever(Observer)	将 Observer 添加到活跃的观察者列表中，该列表始终处于 ACTIVE 状态，不会自动从观察者列表移除，必须调用 removeObserver()移除
onActive()	当有活跃状态的订阅者订阅 LiveData 时回调
onInactive()	当没有活跃状态的订阅者订阅 LiveData 时回调

官方建议 LiveData 与 ViewModel 配合使用，ViewModel 用于存放 UI 界面数据，界面只需关心数据的展示，并希望数据变化时能及时得到通知并做出更新，业务逻辑及数据变化交给 ViewModel 处理，而 ViewModel 中需要被 UI 感知的数据则交给 LiveData，具体实现步骤如下。

（1）创建继承自 ViewModel 类的子类，创建 LiveData 对象

ViewModel 子类创建后，添加 LiveData 类型的实例对象，由于 LiveData 是抽象类，一般使用子类 MutableLiveData，调用 getValue()、setValue()等方法获取或更新数据，示例代码如下：

```
1.    public class MyVMLiveData extends ViewModel {
2.        private MutableLiveData<Integer> number;
3.        // 获取数据
4.        public MutableLiveData<Integer> getNumber() {
```

```
5.         if (number == null) {
6.             number = new MutableLiveData<>();
7.             number.setValue(0);
8.         }
9.         return number;
10.    }
11.    // 设置数据
12.    public void addNumber(int number) {
13.        this.number.setValue(this.number.getValue() + number);
14.    }
15. }
```

（2）创建和注册 Observer 对象，观察 LiveData 对象

在 Activity 或 Fragment 等 UI 控制器类中创建 Observer 对象，重写 onChanged()方法，当 LiveData 的数据发生变化时回调该方法。

LiveData 的 observe()方法将 Observer 对象注册到 LiveData 对象，使得 Observer 与 LiveData 建立订阅关系，当 LiveData 数据发生变化时通知 Observer 更新数据，示例代码如下：

```
1.  public class LiveDataActivity extends AppCompatActivity {
2.      private TextView tvData;
3.      private MyVMLiveData viewModel;
4.      @Override
5.      protected void onCreate(Bundle savedInstanceState) {
6.          super.onCreate(savedInstanceState);
7.          setContentView(R.layout.activity_live_data);
8.          //初始化控件对象
9.          tvData = findViewById(R.id.textView);
10.         // 创建 ViewModel 对象
11.         viewModel = new ViewModelProvider(this).get(MyVMLiveData.class);
12.         viewModel.getNumber().observe(this, new Observer<Integer>() {
13.             @Override
14.             public void onChanged(Integer integer) {
15.                 // 设置 TextView 的值
16.                 tvData.setText(String.valueOf(integer));
17.             }
18.         });
19.     }
20. }
```

（3）更新 LiveData 对象

按钮事件更新 TextView 控件数据的事件处理就可以直接调用 ViewModel 类中的 addNumber()方法来更新 LiveData 的数据，代码重构为以下形式：

```
1.  @Override
2.  protected void onCreate(Bundle savedInstanceState) {
3.      ...
4.      // +1 按钮的事件处理
5.      findViewById(R.id.btn_1).setOnClickListener(view -> {
6.          viewModel.addNumber(1);
7.      });
8.      // -1 按钮的事件处理
9.      findViewById(R.id.btn_2).setOnClickListener(view -> {
10.         viewModel.addNumber(-1);
11.     });
12. }
```

LiveData 与 ViewModel 的配合规范了 Android 的 MVVM 实现模式，在实际项目开发中推荐使用。

① 职责分离：Activity、Fragment 只需负责展示数据，无须承担配置更改时保留数据的责任，这个职责交给 ViewModel 完成，使得视图与数据完全解耦。

② 简化清理无用数据的步骤：当 Activity、Fragment 负责清理数据时，需要编写大量代码；而 ViewModel 只需将无用数据放到 onCleared()方法中，这些资源在 Activity 结束时就会自动清除。

③ 减少类的膨胀：由于职责的转移，Activity、Fragment 无须实现处理请求、状态持久性和注销数据的代码，使用 ViewModel 可以缓解这些类的代码膨胀，使各个类的职责尽可能单一。

④ 容易测试：职责分离使测试变得容易，并且可以编写更细粒度的测试用例。

4. Paging

Android 中很多的应用都会使用列表方式展示数据，为了避免一次性加载大量数据，对数据进行分页加载是极其常见的需求。分页加载可以对数据进行按需加载，在不影响用户体验的前提下，大大提升应用程序的性能。Jetpack 提供了 Paging 组件用于分页处理，借助于它可以轻松在 RecyclerView 中分页加载和展示大型数据集，进行快速、无限滚动。相比于其他分页解决方案，Paging 组件具备与 LiveData、Room 和 RecyclerView 结合的优势。

Paging 支持三种数据架构类型。
- 网络：对网络数据进行分页加载，配合使用 Retrofit 库。
- 数据库：对设备存储的数据库进行分页加载，配合使用 Room 库。
- 网络+数据库：先网络数据缓存到数据库，然后再对数据库的数据进行分页加载。

使用 Paging 库，需要先了解它的相关库定义，具体描述如表 9.8 所示。

表 9.8 Paging 库的核组件描述

类/接口名称	作用描述
PagedList	以分页方式异步加载数据的集合，并传递给 PagedListAdapter
PagedList.Config	PagedList 的配置类，设置初始化数量和界面预取数量等参数
PagedListAdapter	RecyclerView 的适配器，同时负责通知 PagedList 何时加载更多数据
DataSource	在子线程中执行数据载入的逻辑，它本身不存储数据，获取的数据交给 PagedList 存储
DataSource.Factory	DataSource 工厂类，提供 DataSource 的实例
LivePagedListBuilder	用于生成 LiveData<PagedList>，参数为 DataSource.Factory
BoundaryCallback	数据到达边界时的回调接口

当 RecyclerView 滑动时会触发 PagedListAdapter 类的 onBindViewHolder()方法，它会调用 getItem()方法通知 PagedList 载入更多数据，PagedList 根据 PagedList.Config 的配置通知 DataSource 执行具体的数据载入工作。当一条新的数据插入到数据库，DataSource 完成初始化后，LiveData 后台线程就会创建一个新的 PagedList，这个新的 PagedList 会被发送到 UI 线程的 PagedListAdapter 中，PagedListAdapter 使用 DiffUtil 对比现有的 item 和新的 item，对比之后，PagedListAdapter 通过调用 PagedListAdapter 的 notifyItemInserted()方法将新的 item 插入到适当的位置，具体流程如图 9.15 所示。

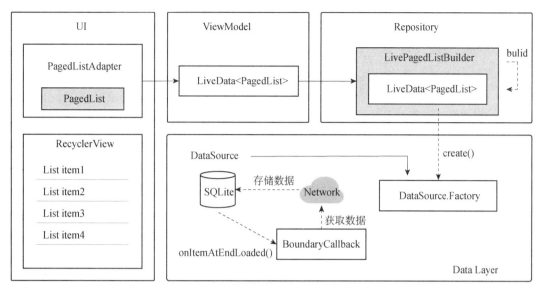

图 9.15 Paging 的工作流程

下面以查询数据库显示在 RecyclerView 为例讲解 Paging 的使用步骤,并结合 Room 数据库框架。

(1) 加载依赖

在 app/build.gradle 文件中添加 paging 的依赖库。

```
1.  dependencies {
2.      def paging_version = "2.1.2"
3.      implementation "androidx.paging:paging-runtime:$paging_version"
4.  }
```

(2) 创建数据源

数据源 DataSource<Key,Value>是一个抽象类,其中 Key 为加载数据的条件,Value 为应加载数据的实体类。DataSource 针对不同的数据分页的加载策略提供了三个抽象子类用于不同的场景。

● PositionalDataSource<T>:最简单的 DataSource 类型,基于 index 加载特定范围的数据,T 为数据类型,适用于目标数据总数固定的场景,如本地数据库的数据。

● PageKeyedDataSource<key,T>:基于页码加载数据,数据容量可动态自增,适用于后端 API 提供了分页数据的场景。

● ItemKeyedDataSource<key,T>:根据前一页的某个 item 的信息加载下一页的数据,适用于目标数据的加载依赖特定条目的信息。

如果没有使用 Room 数据库框架,则需要自定义实现 DataSource,Room 框架可以直接获取 DataSource.Factory 对象,大大简化了获取 DataSource 对象的方式,示例代码如下:

```
1.  @Dao
2.  public interface StudentDao {
3.      @Query("SELECT * FROM Student ")
4.      DataSource.Factory<Int, Student> getAllStudent();
5.      @Insert(onConflict = OnConflictStrategy.IGNORE)
6.      void insert(Student student);
7.  }
```

另外,还需要使用 Room 框架创建数据库,参考第 5 章的 Room 框架的讲解进行创建,其中涉及的实体类 Student 自行创建，此处不再赘述，只给出示例代码：

```
1.  @Database(entities = {Student.class}, version = 1, exportSchema
                    = false)
2.  public abstract class StudentDatabase extends RoomDatabase {
3.      public static String DB_NAME = "student.db";
4.      private static final int NUMBER_OF_THREADS = 4;
5.      private static volatile StudentDatabase INSTANCE;
6.      // 数据库写操作的线程池
7.      public static final ExecutorService writeExecutor =
8.              Executors.newFixedThreadPool(NUMBER_OF_THREADS);
9.      // 单例模式
10.     public static StudentDatabase getInstance(final Context context) {
11.         if (INSTANCE == null) {
12.             synchronized (StudentDatabase.class) {
13.                 if (INSTANCE == null) {
14.                     INSTANCE = Room.databaseBuilder(context,
15.                             StudentDatabase.class, StudentDatabase.
                            DB_NAME)
16.                         .allowMainThreadQueries()
17.                         .build();
18.                 }
19.             }
20.         }
21.         return INSTANCE;
22.     }
23.     // 获取 Dao 的抽象方法
24.     public abstract StudentDao getNoteDao();
25.     // 清除 Database 实例
26.     public void cleanUp() {
27.         INSTANCE = null;
28.     }
29. }
```

（3）构建 LiveData<PagedList>

PagedList 是一个集合类，它负责从数据源取出数据，控制加载数据的方式及将数据的变更反映到 UI 上。Room 库可以直接获取 DataSource.Factory 类型的数据；使用 PageList.Config 类控制加载数据的配置参数，它提供了 4 个可选的配置，分别是：初始加载的数量 InitialLoadSizeHint；预加载数量 PrefetchDistance；分页数量 PageSize；是否启用占位符 PlaceholderEnabled。

变更的数据反映到 UI 上的含义就是将 DataSource.Factory 数据源转变为 LiveData<PagedList>类型的可观察者对象，Paging 提供了 LivePagedListBuilder 类实现这个转换。示例代码如下：

```
1.  public class StudentViewModel extends AndroidViewModel {
2.      private final StudentDao dao;
3.      public StudentViewModel(@NonNull Application application) {
4.          super(application);
5.          dao = StudentDatabase.getInstance(application).getNoteDao();
6.      }
7.      public LiveData<PagedList<Student>> getAllStudents() {
8.          return new LivePagedListBuilder<>(dao.queryAll(),
9.                  new PagedList.Config.Builder()
10.                     .setInitialLoadSizeHint(30)// 初次加载的数量，
                                                   默认为 pageSize*3
11.                     .setPageSize(10)           // 分页加载的数量
12.                     .setPrefetchDistance(10)   // 预取数据的距离，
                                                   默认为 pageSize
```

```
13.                     .setEnablePlaceholders(false) // 是否启用占位符
14.                     .build())
15.             .build();
16.     }
17. }
```

（4）创建 PagedListAdapter

PagedListAdapter 是一个特殊的 RecyclerView 的适配器类，它继承自 RecyclerView.Adapter，使用方法也类似，PagedListAdapter 的 getItem()会触发 DiffUtil 对新旧数据之间进行差量计算，使得 Paging 库具备差量更新的能力，PagedListAdapter.ItemCallback 接口中的两个方法用于定义数据的比较规则，所以需要在构造方法中实现该接口，具体代码如下：

```
1.  public class StudentAdapter extends
2.          PagedListAdapter<Student, StudentAdapter.ViewHolder> {
3.      protected StudentAdapter() {
4.          super(new DiffUtil.ItemCallback<Student>() {
5.              // 实现新旧数据的比较规则
6.              @Override
7.              public boolean areItemsTheSame(@NonNull Student oldItem,
8.                                             @NonNull Student newItem) {
9.                  return oldItem.getId() == newItem.getId();
10.             }
11.             @Override
12.             public boolean areContentsTheSame(@NonNull Student oldItem,
13.                                               @NonNull Student
                                                    newItem) {
14.                 return oldItem.getName().equals(newItem.getName());
15.             }
16.         });
17.     }
18.     @NonNull
19.     @Override
20.     public ViewHolder onCreateViewHolder(@NonNull ViewGroup parent,
                                             int viewType) {
21.         View view = LayoutInflater.from(parent.getContext())
22.                 .inflate(R.layout.item_student, parent, false);
23.         return new ViewHolder(view);
24.     }
25.     @Override
26.     public void onBindViewHolder(@NonNull ViewHolder holder,
                                     int position) {
27.         final Student student = getItem(position);
28.         if (student == null) {
29.             holder.tvName.setText("loading");
30.         } else {
31.             holder.tvName.setText(student.getName());
32.         }
33.         holder.itemView.setOnClickListener(v -> Log.i("Adapter",
                                                student.toString()));
34.     }
35.     public static class ViewHolder extends RecyclerView.ViewHolder {
36.         TextView tvName;
37.         public ViewHolder(@NonNull View itemView) {
38.             super(itemView);
39.             tvName = itemView.findViewById(R.id.tv_name);
40.         }
41.     }
42. }
```

（5）监听数据

接下来在 MainActivity 中初始化 RecyclerView 控件并设置它的适配器，使用 LiveData 监听加载的数据，调用 sumbitList()方法将数据提交给 PagedListAdapter，PagedListAdapter 在后台线程中对比新旧数据的差异，最后更新 RecyclerView，代码如下：

```java
public class MainActivity extends AppCompatActivity {
    @Override
    protected void onCreate(Bundle savedInstanceState) {
        super.onCreate(savedInstanceState);
        setContentView(R.layout.activity_main);

        initView();
        mockData();
        getData();
    }
    // 创建模拟数据
    private void mockData() {
        // 创建100个模拟数据写入数据库
        final Student[] students = new Student[100];
        for (int i = 0; i < 100; i++) {
            Student student = new Student();
            student.setName("student" + i);
            students[i] = student;
        }
        StudentDao studentDao = StudentDatabase.getInstance(this).
                    getNoteDao();
        studentDao.insert(students);
    }
    // 数据绑定，监测加载
    private void getData() {
        // 获取ViewModel对象
        ViewModelProvider viewModelProvider = new ViewModelProvider(this,
                new ViewModelProvider.AndroidViewModelFactory
                (getApplication()));
        StudentViewModel studentViewModel=viewModelProvider.get
                    (StudentViewModel.class);
        // 监视数据变化
        studentViewModel.getAllStudents()
                .observe(this, new Observer<PagedList<Student>>() {
            @Override
            public void onChanged(PagedList<Student> students) {
                adapter.submitList(students);
            }
        });
    }
    // 初始化Recyclver、设置适配器
    private StudentAdapter adapter;
    private void initView() {
        adapter = new StudentAdapter();
        RecyclerView rvStudent = findViewById(R.id.rv_student);
        rvStudent.setLayoutManager(new LinearLayoutManager(this));
        rvStudent.addItemDecoration(new DividerItemDecoration(this,
                DividerItemDecoration.VERTICAL));
        rvStudent.setAdapter(adapter);
    }
}
```

最后总结一下 Paging 的基本流程为：

① 使用 DataSource 从数据库或网络中获取数据。

② 将数据保存到 PagedList 中。
③ 将 PagedList 中的数据交给 PagedListAdapter。
④ PagedListAdapter 在后台线程中通过 Diff 对比新旧数据，并反馈给 RecyclerView。
⑤ RecyclerView 刷新数据。

本小节只是对 Paging 进行了简单的概述，Paging 的高级应用较为复杂，限于篇幅，不再深入。

5. DataBinding

Android 项目中的布局文件一般只负责 UI 组件的布局，布局及数据加载都是通过代码提供的方法完成的，Activity 或 Fragment 类需要编写大量的 View 定义与初始化、控件操作的代码，如 findViewById()、setText()、setOnClickListener()等。为了减少这些不必要的代码，Google 在 2015 年 I/O 大会上发布了 DataBinding，它是一个实现数据与 UI 绑定的框架，现在是 Jetpack 的一部分。借助于它，可以通过声明式布局绑定应用程序的布局和逻辑，使得数据变化的同时布局的 UI 组件也同步更新。使用 DataBinding 具备以下优势。

- 使用简单，主要以声明的方式实现。
- 减少了大量不必要的代码。
- 不再使用 findViewById()等方法。
- 防止内存泄露。
- 自动进行空检测以避免空指针异常。
- 代码更简洁、可读性更强，结构层次也更清晰。

相比于普通布局，使用 DataBinding 的布局文件在最外层增加了 layout 标签，使用 data 标签导入类和声明变量等数据描述，包含的标签及含义如表 9.9 所示。

表 9.9　DataBinding 涉及的标签及含义

标签名称	含义描述
layout	布局的根节点，只能包裹一个 View，且不能包裹 merge 标签
data	DataBinding 的数据，只能有一个 data 标签，支持简单的表达式，表达式的格式为@{...}
variable	data 的子标签，用于声明变量，type 属性指定变量的类，name 属性指定布局中使用的变量名称
import	data 的子标签，用于导入外部类，type 属性指定类的路径
include	在 View 标签中使用，与普通布局中的 include 一样使用，用于引入其他布局文件

示例代码如下：

```xml
1.  <?xml version="1.0" encoding="utf-8"?>
2.  <layout xmlns:android="http://schemas.android.com/apk/res/android"
3.      xmlns:app="http://schemas.android.com/apk/res-auto"
4.      xmlns:tools="http://schemas.android.com/tools"
5.      <!-- 数据描述 -->
6.      <data>
7.          <import type="com.example.databinding.MyViewModel" />
8.          <variable name="data" type="MyViewModel" />
9.          <variable name="myHandler" type="com.example.databinding.
                      MyHandler" />
10.     </data>
11.     <androidx.constraintlayout.widget.ConstraintLayout
12.         android:layout_width="match_parent"
13.         android:layout_height="match_parent"
```

```
14.            tools:context=".MainActivity">
15.        <TextView
16.            android:id="@+id/textView"
17.            android:layout_width="wrap_content"
18.            android:layout_height="wrap_content"
19.            android:text="@{String.valueOf(data.number)}"
20.            android:onClick="@{myHandler::onClick}"
21.            android:textSize="30sp"
22.            app:layout_constraintBottom_toBottomOf="parent"
23.            app:layout_constraintLeft_toLeftOf="parent"
24.            app:layout_constraintRight_toRightOf="parent"
25.            app:layout_constraintTop_toTopOf="parent"
26.            app:layout_constraintVertical_bias="0.4" />
27.        <Button
28.            android:id="@+id/button"
29.            android:layout_width="wrap_content"
30.            android:layout_height="wrap_content"
31.            android:onClick="@{()->data.add()}"
32.            android:text="Button"
33.            app:layout_constraintBottom_toBottomOf="parent"
34.            app:layout_constraintEnd_toEndOf="parent"
35.            app:layout_constraintStart_toStartOf="parent"
36.            app:layout_constraintTop_toTopOf="parent" />
37.    </androidx.constraintlayout.widget.ConstraintLayout>
38.</layout>
```

DataBinding 给控件属性赋值使用@{}绑定表达式，如上述代码中的第 19 行中的 @{data.number}，其中的 data 就是<variable>的 name 属性值，data.number 调用 type 属性值 MyViewModel 类的 getNumber()方法。

DataBinding 表达式支持的运算符如表 9.10 所示。

表 9.10 绑定表达式的常用运算符

运算类型	运算符
算术	加、减、乘、除、求余（+、-、*、/、%）
逻辑	与、或(&&、||)
一元	+、-、!、~
移位	>>、>>>、<<
关系	==、>、<、>=、<=（使用符号<时，要换成<）
其他	字符拼接+、instanceof、?:、()、[]

对于数组、List、Map 等集合数据采用下标[]访问它们的元素，如：

```
1. <data>
2.     <import type="java.util.Map"/>
3.     <import type="java.util.List"/>
4.     <variable name="users" type="List&lt;String>"/>
5.     <variable name="classmates" type="Map&lt;String, String>"/>
6.     <variable name="index" type="int"/>
7.     <variable name="key" type="String"/>
8. </data>
```

在布局文件中使用这些元素：

```
1. <!-- 绑定 List 元素 -->
2. <TextView
3.     android:id="@+id/textView"
```

```
4.         android:text="@{list[index]}" />
5. <!-- 绑定 Map 元素 -->
6. <TextView
7.         android:id="@+id/textView"
8.         android:text="@{map[key]}" />
9. <!-- 绑定 Map 元素 -->
10. <TextView
11.        android:id="@+id/textView"
12.        android:text="@{map.key}" />
```

以上代码使用单向的 DataBinding 绑定给 View 组件设置值，除此之外，还可以使用"@={}"实现双向绑定，并同时监听用户的更新事件。如：

```
1. <CheckBox
2.     android:id="@+id/ck_remember"
3.     android:checked="@={viewmodel.rememberMe}"/>
```

除了属性绑定之外，DataBinding 还能为 View 绑定事件处理函数，有以下两种机制。

● 方法引用：在表达式中使用方法，DataBinding 会将方法与对象包装成一个 Listener 并设置给 View。

● 监听绑定：Lambda 表达式，也会生成一个 Listener，供事件触发时调用。

方法引用与监听绑定的主要区别为：

● 方法引用的监听器是在数据绑定时创建的，监听绑定是在触发事件时创建的。

● 方法引用不能是表达式，监听绑定可以是表达式。

● 方法引用限制了绑定方法的参数列表，返回值必须和监听器中的方法一致，而监听绑定只限制 Lambda 表达式中的语句返回值与监听器中的方法一致。

如果将事件处理直接分配给事件处理方法，则使用方法绑定表达式，该表达式的值为调用的方法名称，如数据对象定义了以下方法：

```
1. public class MyHandlers {
2.     public void onClick(View view) {
3.         // 事件处理
4.     }
5. }
```

绑定表达式就可以为 View 分配一个单击监听器，如代码中的第 9 行的数据变量的定义及第 20 行的方法引用。需要特别注意的是，@{myHandler::onClick}表达式中的 onClick()方法签名必须与 android:onClick 监听器对象中的方法签名完全匹配。如果想自定义事件处理方法，则可以使用监听绑定，如上述代码中的第 31 行就调用了 MyViewModel 的 add()方法，由于监听绑定只允许 Lambda 表达式，所以无法监听复杂事件。

了解以上 DataBinding 的基本概念之后，接下来总结一下使用它的具体步骤。

（1）启用 DataBinding 库

从 Android Studio 3.6 版本开始，DataBinding 库就内置在 Gradle 插件中，因此无须在 app/build.gradle 文件中添加任何依赖项，但 Android Studio 3.6 和 Android 4.0 启动它的方式不太一样，Android Studio 4.0 及以上版本的配置如下：

```
1. android {
2.     buildFeatures {
3.         dataBinding = true
4.     }
5. }
```

Android Studio 4.0 以下版本的配置如下：

```
1.  android {
2.      dataBinding {
3.          enable = true
4.      }
5.  }
```

（2）定义数据类

在使用 DataBinding 之前，首先创建一个数据对象类用于提供数据，可以使用 ViewModel 结合 LiveData 提供生命周期感知的数据对象，示例代码如下：

```
1.  public class MyViewModel extends ViewModel {
2.      private MutableLiveData<Integer> number;
3.      // 获取数据
4.      public MutableLiveData<Integer> getNumber() {
5.          if (number == null) {
6.              number = new MutableLiveData<>();
7.              number.setValue(0);
8.          }
9.          return number;
10.     }
11.     // 设置数据
12.     public void add() {
13.         number.setValue(number.getValue() + 1);
14.     }
15. }
```

（3）布局绑定数据源

启用 DataBinding 之后，打开布局文件，选中根布局后，按住 Alt + 回车键，单击 Convert to data binding layout，生成 DataBinding 的布局结构，添加绑定的数据类 MyViewModel 描述，参见以上定义的布局实例代码。

（4）数据绑定

布局文件创建后，单击菜单 Build->Make Project，系统根据 XML 布局文件名生成 Binding 类名，默认情况下，Binding 类名是 XML 文件名以大写字母开头、移除下划线，然后追加 Binding 生成，如 activity_main.xml 生成 ActivityMainBinding.java，如果 module 的包名为 com.example.databinding，Binding 类则被放在 com.example.databinding.databinding 包中，如图 9.16 所示。

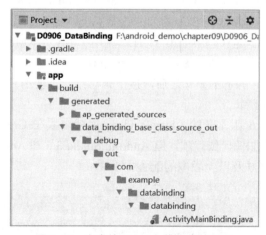

图 9.16　生成的 Binding 绑定类的位置

这个类中包含布局文件的所有绑定关系，根据布局文件的绑定表达式给 View 控件赋值。编译时产生的 Binding 类主要完成了两件事：解析 layout 文件，根据 data 标签定义成员变量；解析 layout 文件，根据绑定表达式生成绑定代码。Binding 创建后还调用 DataBindingUtil 类的静态方法 setContentView()来创建 Binding 类的对象，并与 layout 文件绑定，示例代码如下：

```
1.  ActivityMainBinding binding;
2.  @Override
3.  protected void onCreate(Bundle savedInstanceState) {
4.      super.onCreate(savedInstanceState);
5.      binding = DataBindingUtil.setContentView(this, R.layout.activity_
                                                main);
6.      MyViewModel myViewModel = new ViewModelProvider(this).
                                  get(MyViewModel.class);
7.      binding.setData(myViewModel);
8.      binding.setLifecycleOwner(this);
9.  }
```

上述代码的第 5 行的 DataBindingUtil 是创建 DataBinding 对象的工具类，调用它的 setContentView()方法完成设置 Activity 的布局文件，并返回 DataBinding 对象。对于 Fragment、ListView 或 RecyclerView 等，也可以调用它的 inflate()方法加载布局文件。

获取 DataBinding 对象之后，就可以通过该对象直接访问 View 组件了，如上述代码中的第 7 行。第 8 行的 setLifecycleOwner()方法用于设置 LiveData 的生命周期所有者。

（5）事件处理

使用方法引用或监听绑定设置 View 组件的事件处理方法，此处不再赘述。

在第 2 章中曾经讲解过 ViewBinding，从名称上看它与 DataBinding 很相似，那它们之间有什么区别呢？总结如表 9.11 所示。

表 9.11　DataBinding 与 ViewBinding 的区别

区别点	DataBinding	ViewBinding
支持程度	包含了 ViewBinding 所有的功能	仅支持绑定 View
布局文件配置	需要在布局文件中添加 layout 标签	不需要在布局文件中添加 layout 标签
Build.gradle 配置	dataBinding = true	viewBinding = true
是否支持据绑定	支持双向绑定	否
执行效率	数据绑定开销较大	高于 DataBinding

6. Navigation

Navigation 是 Jetpack 中的新组件，它采用类似 layout 的可视化方法，通过在 XML 中添加元素并指定导航的起始和目的地，建立界面之间的路由关系，在 Activity 中调用导航 action 跳转到目的地，简化了应用程序界面之间导航的实现过程。它提供了可视化的界面导航图，大大简化界面导航的实现过程。

Navigation 导航是在 res 目录中创建 Navigation 类型的 XML 资源文件，打开如图 9.17 所示的可视化编辑界面，整个界面分为三个区域，左侧为 Navigation 的结构展示区域，中部为 Navigation 的编辑区域，右侧为属性编辑区域。

图中清晰地表达了两个 Fragment 之间的跳转关系，从图中结构可以看出，Navigation 主要包括以下三个关键的元素。

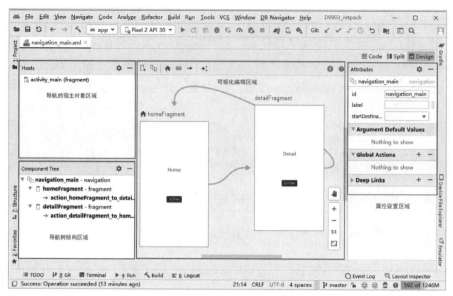

图 9.17 Navigation 的可视化编辑界面

（1）导航图 Navigation Gragh

Navigation Graph 包含所有导航的 Activity 或 Fragment 的 XML 资源，图中的界面节点表示界面。界面之间的连线是 action，表示跳转关系，它可以配置动画、出栈行为和启动选项等，action 箭头所指的方向表示目标界面入栈，箭头的反方向表示目标界面出栈。图 9.10 所示的导航图的代码如下：

```xml
1.  <?xml version="1.0" encoding="utf-8"?>
2.  <navigation xmlns:android="http://schemas.android.com/apk/res/android"
3.      xmlns:app="http://schemas.android.com/apk/res-auto"
4.      xmlns:tools="http://schemas.android.com/tools"
5.      android:id="@+id/navigation_main"
6.      app:startDestination="@id/home_fragment">
7.      <fragment
8.          android:id="@+id/home_fragment"
9.          android:name="com.example.jetpack.HomeFragment"
10.         android:label="Home"
11.         tools:layout="@layout/fragment_home">
12.         <action
13.             android:id="@+id/action_homeFragment_to_detailFragment"
14.             app:destination="@id/detail_fragment"
15.             app:enterAnim="@anim/fragment_fade_enter"
16.             app:exitAnim="@anim/fragment_fade_exit" />
17.     </fragment>
18.     <fragment
19.         android:id="@+id/detail_fragment"
20.         android:name="com.example.jetpack.DetailFragment"
21.         android:label="Detail"
22.         tools:layout="@layout/fragment_detail">
23.         <action
24.             android:id="@+id/action_detailFragment_to_homeFragment"
25.             app:destination="@id/home_fragment" />
26.     </fragment>
27. </navigation>
```

上述代码中，为根标签，它的 app:startDestination 属性用于指定启动页面；<fragment>标签代表一个 fragment，<action>标签定义界面跳转的行为，它的 app:destination

属性用于指定跳转的目标界面,app:enterAnim、app:exitAnim 属性指定进入、退出动画。

(2)宿主对象 NavHostFragment

NavHostFragment 宿主对象是用于显示导航图中的 Fragment 的空白容器,属性 app:name、app.defaultNavHost 和 app:navGraph 分别设置宿主对象 NavHostFragment、默认导航及导航图的 id,MainActivity 作为宿主对象的示例代码如下:

```xml
1.  <?xml version="1.0" encoding="utf-8"?>
2.  <androidx.constraintlayout.widget.ConstraintLayout
3.      xmlns:android="http://schemas.android.com/apk/res/android"
4.      xmlns:app="http://schemas.android.com/apk/res-auto"
5.      xmlns:tools="http://schemas.android.com/tools"
6.      android:layout_width="match_parent"
7.      android:layout_height="match_parent"
8.      tools:context=".MainActivity">
9.      <fragment
10.         android:id="@+id/fragment_host"
11.         android:layout_width="match_parent"
12.         android:layout_height="match_parent"
13.         app:layout_constraintBottom_toBottomOf="parent"
14.         app:layout_constraintEnd_toEndOf="parent"
15.         app:layout_constraintStart_toStartOf="parent"
16.         app:layout_constraintTop_toTopOf="parent"
17.         app:defaultNavHost="true"
18.         android:name="androidx.navigation.fragment.NavHostFragment"
19.         app:navGraph="@navigation/navigation_main" />
20. </androidx.constraintlayout.widget.ConstraintLayout>
```

调用 NavigationUI 类的 setupActionBarWithNavController()方法设置回退导航,重写 MainActivity 类的 onSupportNavigateUp()方法的示例代码如下:

```java
1.  private NavController controller;
2.  @Override
3.  protected void onCreate(Bundle savedInstanceState) {
4.      super.onCreate(savedInstanceState);
5.      setContentView(R.layout.activity_main);
6.      // 设置 ActionBar
7.      controller = Navigation.findNavController(this, R.id.fragment_host);
8.      NavigationUI.setupActionBarWithNavController(this, controller);
9.  }
10. @Override
11. public boolean onSupportNavigateUp() {
12.     return controller.navigateUp();
13. }
```

(3)管理器 NavController

NavController 类的字面含义已经清楚地表达了这个类的作用,它是导航的控制者,用于管理 NavHost 中的所有界面对象的跳转。Android 提供了以下方法用于获取 NavController 对象。

- NavHostFragment.findNavController(Fragment):通过 Fragment 对象获取
- Navigation.findNavController(Activity, int viewId):通过 id 获取
- Navigation.findNavController(View):通过 View 对象获取

NavController 的 navigate()方法传入 XML 文件中的 action 即可导航到目标界面。HomeFragment 导航到 DetailFragment 的示例代码如下:

```
1.  @Override
2.  public void onActivityCreated(@Nullable Bundle savedInstanceState) {
3.      super.onActivityCreated(savedInstanceState);
4.      getView().findViewById(R.id.btn_home)
5.              .setOnClickListener(view -> {
6.                  NavController controller = Navigation.
                                            findNavController(view);
7.                  controller.navigate(R.id.action_homeFragment_to_
                                            detailFragment);
8.              });
9.  }
```

HomeFragment 和 DetailFragment 的导航跳转如图 9.18 所示。

图 9.18　Navigation 界面导航

两个 Fragment 之间的数据传递依旧使用 Bundle 对象实现，调用 NavController 的 navigate() 方法将 Bundle 对象传递给目标界面，代码如下：

```
1.  @Override
2.  public void onActivityCreated(@Nullable Bundle savedInstanceState) {
3.      super.onActivityCreated(savedInstanceState);
4.      getView().findViewById(R.id.btn_home)
5.              .setOnClickListener(view -> {
6.                  //使用 Bundle 传递数据
7.                  Bundle bundle = new Bundle();
8.                  bundle.putString("data", "HomeFragment");
9.                  NavController controller = Navigation.findNavController
                                            (view);
10.                 controller.navigate(R.id.action_homeFragment_to_
                                            detailFragment,bundle);
11.             });
12. }
```

目标界面 DetailFragment 调用 getArguments() 方法获取 Bundle 传递的数据，代码如下：

```
1.  @Override
2.  public void onActivityCreated(@Nullable Bundle savedInstanceState) {
3.      super.onActivityCreated(savedInstanceState);
4.      // 获取数据并写入 TextView
5.      final TextView textView = getView().findViewById(R.id.textView);
6.      textView.setText("接收到的数据: " + getArguments().getString("data"));
7.  }
```

9.3.3 综合应用

本小节通过"记事本"项目综合使用 Jetpack 的架构组件,采用 Google 推荐的 MVVM 架构搭建项目框架,如图 9.19 所示,具体包括:
- RecyclerView:UI 展示数据列表。
- Navigation:构建 Fragment 之间的导航。
- Paging:RecyclerView 数据的分页处理。
- ViewModel:以注重生命周期的方式管理 UI 界面数据。
- LiveData:驱动与监听 Model 数据的变化。
- DataBinding:以声明方式将可观察数据绑定在界面上。
- Room 框架:构建存储数据的 SQLite 数据库。

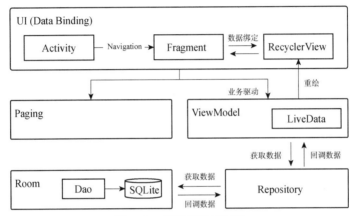

图 9.19 项目架构示意图

"记事本"的功能较为简单,主要包括笔记的增删改查及清空功能,展示笔记列表、添加和修改笔记和搜索的界面如图 9.20 所示,接下来讲解搭建的具体步骤。

图 9.20 展示列表、添加、搜索和删除的界面

步骤1：创建项目

启动 Android Studio，创建名为 D0907_MyNote 的项目，选择 Empty Activity，将包名改为 com.example.note，单击 Finish 按钮，等待项目构建完成。搭建项目的包结构，如图 9.21 所示。

图 9.21　项目的包结构

步骤2：添加依赖

在 app/build.gradle 文件中添加以下依赖及 Java 1.8 和数据绑定的配置：

```
1.  android {
2.      ...
3.      compileOptions {
4.          sourceCompatibility JavaVersion.VERSION_1_8
5.          targetCompatibility JavaVersion.VERSION_1_8
6.      }
7.      // 数据绑定
8.      buildFeatures {
9.          dataBinding true
10.     }
11. }
12.
13. dependencies {
14.     ...
15.     implementation 'androidx.lifecycle:lifecycle-extensions:2.2.0'
16.     implementation 'androidx.navigation:navigation-fragment:2.3.2'
17.     implementation 'androidx.navigation:navigation-ui:2.3.2'
18.     implementation 'androidx.paging:paging-runtime:2.1.2'
19.     implementation 'androidx.room:room-runtime:2.2.6'
20.     annotationProcessor 'androidx.room:room-compiler:2.2.6'
21. }
```

步骤3：创建 Room 数据库

使用 Room 数据库需要创建 JavaBean 实体类、DAO 接口、RoomDataBase 数据库类、Repository 类等。本案例的数据库只有一张表 note，只需要一个实体类 Note，记录笔记的标题、内容和记录时间等属性，Note 实体类的设计如下：

```
1.  @Entity(tableName = "note")
2.  public class Note implements Serializable {
3.      @PrimaryKey(autoGenerate = true)
```

```
4.     private int id;
5.     @ColumnInfo(name = "title")
6.     private String title;
7.     @ColumnInfo
8.     private String content;
9.     @ColumnInfo(name = "update_time")
10.    private Date updateTime;
11.    // 格式化Date
12.    public String getUpdateTimeFormatted() {
13.        SimpleDateFormat sdf=new SimpleDateFormat("yyyy-MM-dd hh:
                           mm:ss", Locale.CHINA);
14.        return sdf.format((updateTime));
15.    }
16.    // 省略getter/setter方法
17. }
```

Note 实体类使用的 Date 类型的属性需要创建 TypeConverter 类型转换器类，实现 Date 类型与 Room 的已知类型之间的相互转换，示例代码如下：

```
1. public class DateTypeConverter {
2.     @TypeConverter
3.     public static Date toDate(Long timestamp) {
4.         return timestamp == null ? null : new Date(timestamp);
5.     }
6.     @TypeConverter
7.     public static Long fromDate(Date date) {
8.         return date.getTime();
9.     }
10. }
```

然后创建实现笔记的增删改查的 DAO 接口，其中的分页查询使用 Paging 库，返回类型为 DataSource.Factory，示例代码如下：

```
1.  @Dao
2.  public interface NoteDao {
3.     @Insert
4.     void insert(Note... notes);
5.     @Update
6.     void update(Note... notes);
7.     @Delete
8.     void delete(Note... notes);
9.     @Query("delete from note")
10.    void deleteAllNotes();
11.    @Query("select * from note order by update_time desc")
12.    DataSource.Factory<Integer, Note> queryAll();
13.    @Query("select * from note where content like :pattern order by
                update_time desc")
14.    DataSource.Factory<Integer, Note> queryWithPattern(String pattern);
15. }
```

接着，创建 RoomDatabase 类提供获取 NoteDao 的抽象方法。为避免 UI 性能下降，Room 不允许在主线程进行查询，但若查询返回的是 LiveData，查询将自动在后台线程上异步运行，则无须遵守此规则，示例代码如下：

```
1.  @Database(entities = {Student.class}, version = 1, exportSchema = false)
2.  @TypeConverters({DateTypeConverter.class})
3.  public abstract class NoteDatabase extends RoomDatabase {
4.     public static String DB_NAME = "note.db";
5.     private static final int NUMBER_OF_THREADS = 4;
6.     private static volatile NoteDatabase INSTANCE;
```

```
7.      // 数据库写操作的线程池
8.      public static final ExecutorService writeExecutor =
9.          Executors.newFixedThreadPool(NUMBER_OF_THREADS);
10.     // 单例模式
11.     public static NoteDatabase getInstance(final Context context) {
12.         if (INSTANCE == null) {
13.             synchronized (NoteDatabase.class) {
14.                 if (INSTANCE == null) {
15.                     INSTANCE = Room.databaseBuilder(context,
16.                             NoteDatabase.class, NoteDatabase.DB_NAME)
17.                             .allowMainThreadQueries()
18.                             .build();
19.                 }
20.             }
21.         }
22.         return INSTANCE;
23.     }
24.     // 获取 Dao 的抽象方法
25.     public abstract NoteDao getNoteDao();
26.     // 清除 Database 实例
27.     public void cleanUp() {
28.         INSTANCE = null;
29.     }
30. }
```

接下来创建 NoteRepository 类，调用 NoteDao 接口方法完成增删改查功能，并将 DataSource 类型的数据转为 LiveData 类型的数据，示例代码如下：

```
1.  public class NoteRepository {
2.      private final NoteDao noteDao;
3.      private final LiveData<PagedList<Note>> notesLiveData;
4.      public NoteRepository(Context context) {
5.          noteDao = NoteDatabase.getInstance(context).getNoteDao();
6.          notesLiveData = new LivePagedListBuilder<>(noteDao.queryAll(),
7.                  new PagedList.Config.Builder()
8.                          .setPageSize(10)
9.                          .setPrefetchDistance(30)
10.                         .build())
11.                 .build();
12.     }
13.     public LiveData<PagedList<Note>> getAllNotes() {
14.         return notesLiveData;
15.     }
16.     public LiveData<PagedList<Note>> getNotesWithPattern(String pattern) {
17.         return new LivePagedListBuilder<>(noteDao.queryWithPattern
                                            ("%" +pattern+ "%"),
18.                 new PagedList.Config.Builder()
19.                         .setPageSize(10)
20.                         .setPrefetchDistance(30)
21.                         .build())
22.                 .build();
23.     }
24.     public void insertNotes(Note... notes) {
25.         NoteDatabase.writeExecutor.execute(() -> noteDao.insert(notes));
26.     }
27.     public void updateNotes(Note... notes) {
28.         NoteDatabase.writeExecutor.execute(() -> noteDao.update(notes));
29.     }
30.     public void deleteNotes(Note... notes) {
31.         NoteDatabase.writeExecutor.execute(() -> noteDao.delete(notes));
32.     }
```

```
33.    public void deleteAllNotes() {
34.        NoteDatabase.writeExecutor.execute(noteDao::deleteAllNotes);
35.    }
36. }
```

至此，Room 数据库相关的类都创建完成。

步骤 3：创建 ViewModel

创建 NoteViewModel 类负责保存和处理 UI 所需的 Note 数据。NoteViewModel 继承自 AndroidViewModel 类，便于获取上下文参数 Context 对象，示例代码如下：

```
1.  public class NotesViewModel extends AndroidViewModel {
2.      private final NoteRepository repository;
3.      public NotesViewModel(@NonNull Application application) {
4.          super(application);
5.          this.repository = new NoteRepository(application);
6.      }
7.      public LiveData<PagedList<Note>> getAllNotes() {
8.          return repository.getAllNotes();
9.      }
10.     public LiveData<PagedList<Note>> getWithPattern(String pattern) {
11.         return repository.getNotesWithPattern(pattern);
12.     }
13.     public void insertNotes(Note... notes) {
14.         repository.insertNotes(notes);
15.     }
16.     public void updateNotes(Note... notes) {
17.         repository.updateNotes(notes);
18.     }
19.     public void deleteNotes(Note... notes) {
20.         repository.deleteNotes(notes);
21.     }
22.     public void deleteAllNotes() {
23.         repository.deleteAllNotes();
24.     }
25. }
```

步骤 4：创建 Fragment

本案例使用 NotesFragment 和 AddFragment 分别展示笔记列表、添加/更新笔记内容，并使用 Navigation 库进行导航。首先进行这两个 Fragment 的界面布局设计，AddFragment 类对应的布局为 fragment_add.xml，NotesFragment 类对应的布局为 fragment_notes.xml 和列表项布局 item_notes.xml，并在 XML 中配置数据绑定的配置，item_notes.xml 的主要代码如下，其他布局的代码不在此列出。

```
1.  <layout xmlns:android="http://schemas.android.com/apk/res/android"
2.      xmlns:app="http://schemas.android.com/apk/res-auto">
3.      <data>
4.          <variable name="note" type="com.example.note.data.local.Note" />
5.          <variable name="onClickListener"
6.              type="android.view.View.OnClickListener"/>
7.      </data>
8.      <LinearLayout
9.          android:layout_width="match_parent"
10.         android:layout_height="wrap_content"
11.         android:onClick="@{onClickListener}" >
12.         <androidx.cardview.widget.CardView
13.             android:layout_width="match_parent"
14.             android:layout_height="match_parent"
15.             android:foreground="?selectableItemBackground">
```

```
16.            <androidx.constraintlayout.widget.ConstraintLayout
17.                android:layout_width="match_parent"
18.                android:layout_height="match_parent">
19.                <androidx.constraintlayout.widget.Guideline
20.                    android:id="@+id/guideline"
21.                    ...
22.                    app:layout_constraintGuide_percent="0.85" />
23.                <TextView
24.                    android:id="@+id/tv_title"
25.                    android:text="@{note.title}"
26.                    ... />
27.                <TextView
28.                    android:id="@+id/tv_time"
29.                    android:text="@{note.updateTimeFormatted}"
30.                    ... />
31.                <ImageView
32.                    android:id="@+id/imageView"
33.                    app:srcCompat="@drawable/ic_arrow_right"
34.                    ... />
35.            </androidx.constraintlayout.widget.ConstraintLayout>
36.        </androidx.cardview.widget.CardView>
37.    </LinearLayout>
38. </layout>
```

第 3~7 行代码用于设置数据绑定的两个数据变量 Note 对象和 OnClickListener 监听器对象，第 11、25 和 29 行分别是监听器和 Note 数据的绑定。

（1）导航图

activity_main.xml 是 fragment_add.xml 和 fragment_notes.xml 两个 Fragment 的导航宿主对象，创建 nav_main.xml 导航图，fragment_notes 将 Note 对象数据使用 argument 标签传递给 fragment_add，导航图的代码如下：

```
1.  <?xml version="1.0" encoding="utf-8"?>
2.  <navigation xmlns:android="http://schemas.android.com/apk/res/android"
3.      xmlns:app="http://schemas.android.com/apk/res-auto"
4.      xmlns:tools="http://schemas.android.com/tools"
5.      android:id="@+id/nav_main"
6.      app:startDestination="@id/notesFragment">
7.      <fragment
8.          android:id="@+id/notesFragment"
9.          android:name="com.example.note.ui.home.NotesFragment"
10.         android:label="笔记列表"
11.         tools:layout="@layout/fragment_notes">
12.         <action
13.             android:id="@+id/action_notesFragment_to_addFragment"
14.             app:destination="@id/addFragment"
15.             app:enterAnim="@anim/fragment_close_enter"
16.             app:exitAnim="@anim/fragment_close_exit" />
17.     </fragment>
18.     <fragment
19.         android:id="@+id/addFragment"
20.         android:name="com.example.note.ui.detail.AddFragment"
21.         android:label="添加笔记"
22.         tools:layout="@layout/fragment_add">
23.         <action
24.             android:id="@+id/action_addFragment_to_notesFragment"
25.             app:destination="@id/notesFragment" />
26.         <argument
27.             android:name="note"
28.             android:defaultValue="@null"
```

```
29.            app:argType="com.example.note.data.local.Note"
30.            app:nullable="true" />
31.
32.     </fragment>
33. </navigation>
```

（2）笔记列表 NotesFragment

NotesFragment 类展示笔记列表，它的适配器类 NotesAdapter 需要继承自 PagedListAdapter 类，并实现 DiffUtil.ItemCallback 接口；使用 item_note.xml 的生成绑定类 ItemNoteBinding 加载 XML 布局文件，并通过 binding 对象加载数据和设置监听器，示例代码如下：

```
1.  public class NotesAdapter extends PagedListAdapter<Note,
    NotesAdapter.ViewHolder> {
2.      public NotesAdapter() {
3.          // 设置数据比较规则的回调
4.          super(new DiffUtil.ItemCallback<Note>() {
5.              @Override
6.              public boolean areItemsTheSame(@NonNull Note oldItem,
7.                                             @NonNull Note newItem) {
8.                  return oldItem.getId() == newItem.getId();
9.              }
10.             @Override
11.             public boolean areContentsTheSame(@NonNull Note oldItem,
12.                                                @NonNull Note newItem) {
13.                 return oldItem.getContent().equals(newItem.
                        getContent()) &&
14.                         oldItem.getUpdateTime().equals(newItem.
                            getUpdateTime());
15.             }
16.         });
17.     }
18.     @NonNull
19.     @Override
20.     public ViewHolder onCreateViewHolder(@NonNull ViewGroup parent,
                                              int viewType) {
21.         ItemNoteBinding binding = ItemNoteBinding.inflate(LayoutInflater
22.                 .from(parent.getContext()), parent, false);
23.         return new ViewHolder(binding.getRoot());
24.     }
25.     @Override
26.     public void onBindViewHolder(@NonNull ViewHolder holder, int position) {
27.         final Note note = getItem(position);
28.         // 加载数据
29.         if (note == null) {
30.             holder.binding.tvTitle.setText("loading");
31.             holder.binding.tvTime.setText("loading");
32.         } else {
33.             holder.binding.setNote(note);
34.         }
35.         // 刷新界面
36.         holder.binding.executePendingBindings();
37.     }
38.     public static class ViewHolder extends RecyclerView.ViewHolder {
39.         ItemNoteBinding binding;    // 数据绑定对象
40.         public ViewHolder(@NonNull View itemView) {
41.             super(itemView);
42.             binding = DataBindingUtil.bind(itemView);
43.             // 设置监听器和事件处理
44.             binding.setOnClickListener(v -> {
45.                 final Bundle bundle = new Bundle();
```

```
46.                    bundle.putSerializable("note", binding.getNote());
47.                    final NavController controller = Navigation.
                           findNavController(v);
48.                    controller.navigate(R.id.action_notesFragment_to_
                                   addFragment, bundle);
49.             });
50.         }
51.     }
52. }
```

NotesFragment 类使用 LiveData<PagedList<Note>>对象观察数据的变化，具体代码如下：

```
1.  public class NotesFragment extends Fragment {
2.      // 视图层
3.      private NotesViewModel notesViewModel;
4.      private NotesAdapter adapter;
5.      // 数据层
6.      private LiveData<PagedList<Note>> notesLiveData;
7.      // 实时数据
8.      private List<Note> allNotes;
9.      // 生成绑定类
10.     private FragmentNotesBinding binding;
11.     // 操作标志，更新时上移，删除则保持不动
12.     private boolean undoFlag;
13.     @Override
14.     public View onCreateView(LayoutInflater inflater, ViewGroup container,
15.                     Bundle savedInstanceState) {
16.         // DataBinding
17.         binding = FragmentNotesBinding.inflate(inflater, container, false);
18.         return binding.getRoot();
19.     }
20.     @Override
21.     public void onActivityCreated(@Nullable Bundle savedInstanceState) {
22.         super.onActivityCreated(savedInstanceState);
23.         // 初始化 RecycleView
24.         adapter = new NotesAdapter();
25.         binding.rvNote.setAdapter(adapter);
26.         binding.rvNote.setLayoutManager(new LinearLayoutManager
                                   (this.getContext()));
27.         // 初始化当前页面的 ViewModel
28.         notesViewModel =new ViewModelProvider(getActivity()).
                           get(NotesViewModel.class);
29.         // 观察数据列表
30.         notesLiveData = notesViewModel.getAllNotes();
31.         notesLiveData.observe(getViewLifecycleOwner(), notes -> {
32.             // 获取当前显示列表的长度
33.             final int count = adapter.getItemCount();
34.             // 备份数据
35.             allNotes = notes;
36.             // 如果数据变化后的元素长度 > 变化前的个数，则为添加操作
37.             if (notes.size() > count && !undoFlag) {
38.                 new Timer().schedule(new TimerTask() {
39.                     @Override
40.                     public void run() {
41.                         binding.rvNote.smoothScrollToPosition(0);
42.                     }
43.                 }, 300);
44.             }
45.             if (undoFlag) {
46.                 undoFlag = false;
```

```
47.                    }
48.                    // 将观察数据注入 RecycleAdapter 中
49.                    adapter.submitList(notes);
50.                });
51.        }
52. }
```

(3) NotesFragment 的菜单处理

笔记列表 Toolbar 上的菜单包括查找和清空数据两个菜单项，在 res 目录中创建名为 main_menu 的菜单资源，代码如下：

```
1.  <?xml version="1.0" encoding="utf-8"?>
2.  <menu xmlns:android="http://schemas.android.com/apk/res/android"
3.      xmlns:app="http://schemas.android.com/apk/res-auto">
4.      <item
5.          android:id="@+id/app_bar_search"
6.          android:icon="@drawable/ic_search_black_24dp"
7.          android:title="@string/item_search"
8.          app:actionViewClass="android.widget.SearchView"
9.          app:showAsAction="always" />
10.     <item
11.         android:id="@+id/item_clear"
12.         android:icon="@drawable/ic_delete"
13.         android:title="@string/item_clear" />
14. </menu>
```

然后在 NotesFragment 类中重写 onCreateOptionsMenu()和 onOptionsItemSelected()方法加载菜单和处理菜单项事件，代码如下：

```
1.  public class NotesFragment extends Fragment {
2.      ...
3.      public NotesFragment() {
4.          // 显示菜单栏
5.          setHasOptionsMenu(true);
6.      }
7.      @Override
8.      public void onCreateOptionsMenu(@NonNull Menu menu,@NonNull
                                        MenuInflater inflater){
9.          super.onCreateOptionsMenu(menu, inflater);
10.         inflater.inflate(R.menu.main_menu, menu);
11.         // search 菜单项的处理
12.         final SearchView searchView = (SearchView) menu.findItem
                                          (R.id.app_bar_search)
13.                 .getActionView();
14.         // 设置 SearchView 的宽度
15.         final int maxWidth = searchView.getMaxWidth();
16.         searchView.setMaxWidth(((int) (maxWidth * 0.5)));
17.         // 设置 SearchView 的监听
18.         searchView.setOnQueryTextListener(new SearchView.
                                              OnQueryTextListener() {
19.             @Override
20.             public boolean onQueryTextSubmit(String query) {
21.                 return false;
22.             }
23.             @Override
24.             public boolean onQueryTextChange(String newText) {
25.                 // 查询数据
26.                 final String pattern = newText.trim();
27.                 notesLiveData = notesViewModel.getWithPattern(pattern);
28.                 // 移除所有的观察者
```

```
29.                    notesLiveData.removeObservers(getViewLifecycleOwner());
30.                    // 对 LiveData 进行观察
31.                    notesLiveData.observe(getViewLifecycleOwner(), notes -> {
32.                        allNotes = notes;
33.                        // 将观察的数据注入 RecycleAdapter 中
34.                        adapter.submitList(notes);
35.                    });
36.                    return true;
37.                }
38.            });
39.        }
40.        @Override
41.        public boolean onOptionsItemSelected(@NonNull MenuItem item) {
42.            if (item.getItemId() == R.id.item_clear) {
43.                new AlertDialog.Builder(getActivity())
44.                        .setTitle("清空数据")
45.                        .setPositiveButton("确定", (dialog, which) -> {
46.                            notesViewModel.deleteAllNotes();
47.                        })
48.                        .setNegativeButton("取消", null)
49.                        .show();
50.            }
51.            return super.onOptionsItemSelected(item);
52.        }
53.    }
```

（4）NotesFragment 的添加和删除功能

添加功能通过 FloatActionButton 按钮的单击时间跳转到导航图的 AddFragment，代码如下：

```
1.  public class NotesFragment extends Fragment {
2.      ...
3.      @Override
4.      public void onActivityCreated(@Nullable Bundle savedInstanceState) {
5.          ...
6.          // 初始化 FloatingActioinButton 控件，单击后跳转
7.          binding.fabAdd.setOnClickListener(v -> {
8.              final NavController controller = Navigation.
                                                 findNavController(v);
9.              controller.navigate(R.id.action_notesFragment_to_addFragment);
10.         });
11.     }
12. }
```

ItemTouchHelper 类是支持 RecyclerView 的侧滑删除和长按拖曳的工具类，它的 Callback 内部抽象类定义了实现这些方法的回调方法，本案例使用 Callback 的子类 SimpleCallback 简化操作，它只需要重写 onMove()、onSwiped()两个方法实现拖曳和滑动功能，通过重写 onChildDraw()方法绘制删除图标，具体代码如下：

```
1.  @Override
2.  public void onActivityCreated(@Nullable Bundle savedInstanceState) {
3.      ...
4.      // 滑动删除
5.      new ItemTouchHelper(new ItemTouchHelper.SimpleCallback(0,
6.              ItemTouchHelper.START | ItemTouchHelper.END) {
7.          @Override
8.          public boolean onMove(@NonNull RecyclerView recyclerView,
9.                                @NonNull RecyclerView.ViewHolder viewHolder,
10.                               @NonNull RecyclerView.ViewHolder target) {
```

```java
11.            // 上下拖曳
12.            return false;
13.        }
14.        @Override
15.        public void onSwiped(@NonNull RecyclerView.ViewHolder
                                viewHolder,int direction) {
16.            // 处理侧滑
17.            // 获取删除的 item
18.            final Note note = allNotes.get(viewHolder.
                                    getAdapterPosition());
19.            // 删除记录
20.            notesViewModel.deleteNotes(note);
21.            Snackbar.make(getActivity().findViewById(R.id.main_
                            fragment), "删除笔记",
22.                    Snackbar.LENGTH_SHORT)
23.                    .setAction("撤销", v -> {
24.                        undoFlag = true;
25.                        notesViewModel.insertNotes(note);
26.                    }).show();
27.        }
28.        // 显示删除图标
29.        Drawable icon = ContextCompat.getDrawable(getActivity(),
                            R.drawable.ic_delete);
30.        ColorDrawable background = new ColorDrawable(Color.LTGRAY);
31.        @Override
32.        public void onChildDraw(@NonNull Canvas c, @NonNull
                                    RecyclerView recyclerView,
33.                                @NonNull RecyclerView.ViewHolder
                                    viewHolder,
34.                                float dX, float dY, int actionState,
35.                                boolean isCurrentlyActive) {
36.            super.onChildDraw(c, recyclerView, viewHolder, dX, dY,
37.                    actionState, isCurrentlyActive);
38.            // 获取 ViewHolder 的 ItemView
39.            View itemView = viewHolder.itemView;
40.            // 计算删除图标的外边距
41.            int iconMargin = (itemView.getHeight() - icon.
                                getIntrinsicHeight()) / 2;
42.            // 计算删除图标的边距
43.            float translationX = 0f;
44.            int iconLeft, iconRight, iconTop, iconBottom;
45.            int backTop, backBottom, backLeft, backRight;
46.            backTop = itemView.getTop();
47.            backBottom = itemView.getBottom();
48.            iconTop = itemView.getTop() + (itemView.getHeight() -
49.                    icon.getIntrinsicHeight()) / 2;
50.            iconBottom = iconTop + icon.getIntrinsicHeight();
51.            if (dX > 0) {
52.                // 右滑
53.                backLeft = itemView.getLeft();
54.                backRight = itemView.getLeft() + (int) dX;
55.                background.setBounds(backLeft, backTop, backRight,
                                        backBottom);
56.                iconLeft = itemView.getLeft() + iconMargin;
57.                iconRight = iconLeft + icon.getIntrinsicWidth();
58.                icon.setBounds(iconLeft, iconTop, iconRight,
                                iconBottom);
59.            } else if (dX < 0) {
60.                // 左滑
61.                backRight = itemView.getRight();
```

```
62.                backLeft = itemView.getRight() + (int) dX;
63.                background.setBounds(backLeft, backTop, backRight,
                                        backBottom);
64.                iconRight = itemView.getRight() - iconMargin;
65.                iconLeft = iconRight - icon.getIntrinsicWidth();
66.                icon.setBounds(iconLeft, iconTop, iconRight, iconBottom);
67.            } else {
68.                background.setBounds(0, 0, 0, 0);
69.                icon.setBounds(0, 0, 0, 0);
70.            }
71.            // 在 Canvas 上绘制背景和图标
72.            background.draw(c);
73.            icon.draw(c);
74.        }
75.    }).attachToRecyclerView(binding.rvNote);
76. }
```

至此完成笔记列表界面的展示、查找、删除和清空等功能，添加和更新功能由 AddFragment 实现，功能相对简单，自行实现完成。

尽管本案例相对简单，但它采用了 MVVM 架构，并综合应用 Jetpack 的架构组件，大家在练习之后可以进一步拓展 Jetpack 组件的复杂应用。

9.4 本章小结

本章讲解了 Android 的一些高阶技术，包括手势处理、传感器开发和 Android Jetpack，对手势的基本概念、手势检测和手势识别等也进行了讲解，使读者对于手势开发有一个简单认识；然后讲解了 Android 传感器的概念及常用传感器的基本用法；最后讲解了最新的 Android Jetpack 组件库套件，对其中与开发密切相关的架构组件做了简单介绍，由于篇幅所限，只简单讲解了其中的 Lifecycle、ViewModel、LiveData、DataBinding、Paging 和 Navigation 组件的应用，需要注意的是，Jetpack 推出不久还不成熟，还在不断更新迭代、升级与完善中，预计后续将成为 Android 开发的主流框架，希望大家在了解这些高阶技术的基本概念之后，还应多读官方文档，及时更新知识体系。

习 题

一、思考题

1. 简述 Android 手势处理的基本原理。
2. 简述常用传感器的基本概念和应用场景。
3. 选择 Jetpack 的某一类的组件库，简述其中包含组件的基本原理。

二、编程题

1. 使用传感器开发一款模拟微信摇一摇的功能。
2. 使用 Jetpack 组件重构一个自己做过的 Android 项目。

附录 A　Android 项目开发规范

每个公司都有不同的项目开发规范,目的是保持代码的一致性,减少开发及沟通成本,提升代码的可读性和可维护性,提高团队的开发效率和质量。此处给出一般性的开发规范,仅供参考。

1. Android 的命名规范

好的命名能让开发者根据名称的含义快速理解程序的含义、厘清程序的结构。常用的命名形式有三种:首字母大写的帕斯卡命名、首字母小写的小驼峰方式和所有字母大写的下划线分割的方式,如表 A.1 所示。

表 A.1　常用的命名方式

类型	规范	举例
项目	小写字母,多个单词之间用中划线分隔	note-main
包	全部小写	com.example.note
类	帕斯卡命名方式	MainActivity
变量、方法	小驼峰命名方式	noteData
常量	全部大写,单词间用下划线分隔	TAG_NAME

Android 的命名不能使用中文,也不能使用拼音与英文混排的方式,正确的英文命名可以让代码更易于理解,避免歧义。即便使用纯拼音也应避免使用,国际通用的拼音名称除外。

Package 包名全部采用小写,不适用下划线,采用反域名命名规则,一般使用反域名加上项目名称作为前缀,然后根据 PBF(Package By Feature,按功能分包)方式命名子模块名称。

Android 类名的命名规则除了满足一般的要求之外,还应根据类的类型命名,类名通常是名词或名词短语,接口则可以用形容词或形容词短语,详情如表 A.2 所示。

表 A.2　Android 类的命名规则

分类	类型	描述	示例
Android 组件	Application	模块名 + Application	NoteApplication
	Activity	模块名 + Activity	SplashActivity
	Fragment	模块名 + Fragment	HomeFragment
	Service	模块名 + Service	MusicService
	BroadcastReceiver	功能名 + Receiver	NetworkReceiver
	ContentProvider	功能名 + Provider	ShareProvider

分类	类型	描述	示例
Android 组件	Dialog	功能名 + Dialog	LoginDialog
	Widget 小组件	功能名 + Widget	CircleWidget
	自定义 View	Custom+功能名+View/ViewGroup（组件名称）	CustomShapeButton
工具类	解析类	功能名 + Parser	JsonParser
	适配器类	功能名 + Adapter	NoteAdapter
	工具方法类	功能名 + Utils 或 Manager	LogUtils
	数据库类	功能名 + DBHelper	NoteDBHelper
	共享基础类	Base + 功能名	BaseActivity
接口	普通接口	Able/ible 结尾或 I 开头	Runnable

Android 的方法名采用小驼峰命名方式，通常为动词或动词短语，详情如表 A.3 所示。

表 A.3　Android 方法的命名规则

方法定义	功能描述	示例
initXxx()	初始化相关方法	initView()
isXxx()	返回值为 boolean 类型的方法	isVisible()
getXxx()	返回值的方法	getUserName()
handleXxx()	处理数据的方法	handleUserData()
showXxx()、displayXxx()	弹框或提示信息	showErrorInfo()
updateXxx()	更新数据	updateNote()
saveXxx()、insertXxx()	保存或插入数据	insertNote()
removeXxx()、deleteXxx()	移除数据或视图等	removeView()
drawXxx()	绘制相关的方法	drawCircle()

Android 资源的命名采用下划线的命名方式，所有字母必须为小写，常见的资源文件名或 id 的命名如表 A.4 所示。

表 A.4　Android 资源的命名规则

分类	类型	命名规则	示例
Layout 布局	Activity	activity_功能名	activity_main
	Fragment	fragment_功能名	fragment_note
	Dialog	dialog_功能名	dialog_login
	PopupWindow	ppw_模块_功能名	ppw_filter
	列表项的 Item	item_模块_功能名	item_user_list
菜单资源	Menu	menu_	menu_
控件 Id	TextView	tv_模块_功能名	tv_user_login
	EditText	et_功能名	et_user_name
	Button	btn_功能名	btn_login
	CheckBox	cb_功能名	cb_accept
	ImageView	iv_功能名	iv_header
	ViewPager	vp_功能名	vp_ads

续表

分类	类型	命名规则	示例
Drawable 资源	普通图标	ic_模块_功能	ic_note_delete
	背景样式	bg_模块_功能	bg_home_header
	展示图片	img_功能	img_loading
	样式选择器	selector_模块_功能	selector_note_button
	样式形状	shape_功能状态	shape_login_pressed
动画资源	补间动画	动画类型_方向_功能	淡入：fade_in_show

2. Android 的分包规范

包名的划分推荐采用按功能分包（PBF）的方式，而非按层分包（Package By Layer，PBL），PBF 与 PBL 相比，具备以下优势：

- package 内高内聚，package 间低耦合，PBL 降低了代码耦合，但带来了 package 耦合。
- package 有私有作用域。
- 更容易删除功能。
- 高度抽象，PBF 包名是对功能模块的抽象。
- 只需要通过 class 分离逻辑代码。

PBL 分包的代码结构示例如下：

```
com
└── domain
    └── app
        ├── App.java 定义 Application 类
        ├── Config.java 定义配置数据（常量）
        ├── base 基础组件
        ├── custom_view 自定义视图
        ├── data 数据处理
        │   ├── DataManager.java 数据管理器
        │   ├── local 本地的数据，比如 SharedPrefernces、Database、File
        │   ├── model 定义 model（数据结构及 getter/setter、equals 等，不含复杂操作）
        │   └── remote 远端数据
        ├── feature 功能
        │   ├── feature1 功能 1
        │   │   ├── Feature1Activity.java
        │   │   ├── Feature1Fragment.java
        │   │   ├── XxAdapter.java
        │   │   └── ... 其他 class
        │   └── ...其他功能
        ├── injection 依赖注入
        ├── util 工具类
        └── widget 小部件
```

3. Android 的代码样式规范

- 使用最新的稳定版的 IDE 进行开发。
- 采用统一的编码格式 UTF-8。
- 按照团队要求或默认模板进行代码格式化。
- 删除多余的导入包。
- 使用标准的大括号样式。
- 类成员定义的顺序推荐按照常量、字段、构造方法和回调、公有函数、私有函数和内部类或接口的顺序。
- 每行代码的长度应该不超过 160 个字符。
- 除赋值操作符之外，将换行符放在操作符之前，而函数链的换行符放在.之前，方法参数的换行符放在逗号后面。